Advances in
Aquatic Microbiology

Volume 2

Advances in Aquatic Microbiology

edited by
M. R. DROOP
Scottish Marine Biological Association
Dunstaffnage Marine Research Laboratory
Oban, Scotland
and
H. W. JANNASCH
Woods Hole Oceanographic Institution
Woods Hole, Massachusetts, USA

Volume 2

1980

ACADEMIC PRESS
A Subsidiary of Harcourt Brace Jovanovich, Publishers
LONDON NEW YORK TORONTO SYDNEY SAN FRANCISCO

Academic Press Inc. (London) Ltd
24–28 Oval Road
London NW1

US edition published by
Academic Press Inc.
111 Fifth Avenue,
New York, New York 10003

British Library Cataloguing in Publication Data

Advances in aquatic microbiology. Vol. 2

1. Aquatic microbiology
I. Droop, Michael Richmond
II. Jannasch, H. W.
576'.19'2 QR105 76-5988

ISBN 0-12-003002-0
ISSN 0140-2625

Printed in Great Britain by
W. & G. Baird Ltd.,
The Greystone Press, Antrim, Northern Ireland

Contributors

J. V. Landau
Department of Biology, Rensselar Polytechnic Institute, Troy, NY 12181, USA

J. J. Lee
Department of Biology, City College of Cuny, New York, NY 10031, USA

M. Legner
Hydrobiologická Laboratoř Československá akademie věd, Vltavska 17, Praha 5, Czeckoslovakia

D. H. McCarthy
Tavolek Inc., 2779 152nd Avenue, N.E., Redmond, Washington 98052, USA

D. H. Pope
Department of Biology, Rensselaer Polytechnic Institute, Troy, NY 12181, USA

G-Yull Rhee
Environmental Health Center, Division of Laboratories and Research, New York State Department of Health, Albany, NY 12201, USA

R. J. Roberts
Unit of Aquatic Pathobiology, University of Stirling, FK9 4LA, Scotland

J. W. M. Rudd
Department of Fisheries and the Environment, Freshwater Institute, 501 University Crescent, Winnipeg, Manitoba R3T 2N6, Canada

M. Shilo
Department of Microbiological Chemistry, The Hebrew University – Hadassah Medical School, Jerusalem, Israel

C. D. Taylor
Woods Hole Oceanographic Institution, Woods Hole, Massachusetts 02543, USA

M. Varon
Department of Microbiological Chemistry, The Hebrew University – Hadassah Medical School, Jerusalem, Israel

Preface

Having gone through a tedious "lag phase" microbial ecology has emerged in recent years as a bristling area of laboratory and field research. The pendulum is now in full swing and one wonders how to prevent it from reaching the opposite extreme too soon. One way to avert this coasting route is to take time off from data gathering and contemplate the actual progress made.

This is a situation where the published series of science advances become of value. They provide for the communication of thoughts that have no home in the common periodicals. There is, for example, the need for critical evaluation of novel techniques, many of which spring up, flourish and fade away. At other times it may be profitable to bridge areas of research that, for some reason, have been isolated from each other, or to combine the results of separate studies to arrive at new and surprising outlooks. A freedom of style is essential. Depending on the circumstance, the need may be for a brief essay, a longer monograph, a review-type discussion or a discourse on technical details. The common factor is that such articles are unsuited to most of the professional journals, which are designed for quick and efficient dissemination of information on everyday research progress.

This volume is an example *par excellence*. The bdellovibrio research which started with Stolp's discovery in 1962 has reached the stage at which an ecological assessment becomes possible. None appear better qualified than M. Shilo and his co-workers to handle this task. M. Shilo was recently honoured with the Fisher Award (of the American Society for Microbiology) for Environmental Microbiology, to a great extent for his work on *Bdellovibrio*. Under his leadership a "Dahlem Konferenz" was held in November 1978, focusing on strategies of life in extreme environments with a contribution on the survival of *Bdellovibrio* in low-nutrient habitats. The opening chapter in the present book, by

M. Varon and M. Shilo, combines a general review with the first ecological assessment of the bdellovibrio's parasitic role in microbial communities.

Microbiology of the deep sea has received new emphasis during the past decade. The fact that about 75 per cent of the biosphere, by volume, is exposed to hydrostatic pressures of 100 atm or more may have marked implications for the microbial turnover of organic and inorganic matter in the oceans. For the study of these processes, the knowledge of the physiological–biochemical basis of pressure effects in microbial metabolism is of obvious importance. J. V. Landau and D. H. Pope represent the leading team in this area of research and summarize in their chapter the most recent work.

Another topic that has received renewed attention during the last decade is the microbial production and consumption of methane, the ultimate product of carbon reduction in anoxic sediments. C. D. Taylor was a member of the original group studying methanogenesis at Urbana (University of Illinois). J. W. M. Rudd has pioneered modern limnological approaches measuring the microbial oxidation of methane in various types of lakes. Their chapter represents a cooperative product of two scientists with their roots in entirely different areas of research but with a joint interest in the role and fate of methane in natural environments.

Continuous culture studies of algae were preceded by chemostat research on exponentially multiplying prokaryotes. Little of the well worked out principles of population dynamics in bacterial continuous culture were recognized by phytoplanktologists when they discovered the advantages of open system kinetics for experimental studies. G-Yull Rhee in his chapter attempts to bridge this gap by combining the knowledge obtained from steady-state systems of bacteria and yeast with the particularities of photosynthetic growth in continuous culture, i.e. by organisms which receive their energy supply independently of the flow of culture medium. The large number of recent studies in this area supports the optimistic view that continuous culture experiments may help to elucidate the behaviour of natural populations.

Knowledge of the growth dynamics of protozoan populations has been advanced considerably by continuous culture experiments. M. Legner, drawing on the oldest tradition in this area, namely that of the Prague school, has extended his review-type chapter toward growth of protozoan populations in general. Our modern concept of the energy flow in freshwater and marine environments suffers from a dearth of new information on the microbe-grazing link in the food chain.

To avoid affecting the complexity of whole communities by experimental manipulation, J. J. Lee presents a qualitative model for the function of natural populations in the decomposition of marine plant litter. Knowing the levels of intermediate products, the relative rates of transformation, and the effect of certain substrates on the synthesis and activity of enzymes converting others, allows for a number of intelligent speculations on possible regulatory processes. Similar considerations can be made at the level of prey–predator systems in the micro- and meiofaunal regimes. On the basis of laboratory and field observations made over many years, the author is willing to speculate in the presence of the ever-ready criticism demanding numerical and reproducible data.

D. McCarthy and R. Roberts deal with a prominent area of research which most aquatic microbiologists tend to shirk. The ecology of microorganisms causing diseases of freshwater and marine fish or invertebrates is of more immediate practical consequence and certainly much more in the public eye than most other work in aquatic microbiology. Since pathogenic microorganisms, at least during part of their existence, are members of the free-living microbial population and depend on processes that have nothing to do with their pathogenicity, the traditional—in Europe more so than in the United States—and artificial segregation of veterinary microbiology from general and environmental microbiology is unfortunate, and every attempt to cross this line is highly welcome.

Some years ago the scope of this series of scientific advances was expanded, as indicated by a name change from "Advances in Microbiology of the Sea" to "Advances in Aquatic Microbiology". The present volume is the second under the new name. May it be useful to many colleagues.

Woods Hole, Massachusettes
January 1979

HOLGER W. JANNASCH

Contents

Ecology of aquatic bdellovibrios

M. VARON and M. SHILO

*Department of Microbiological Chemistry,
The Hebrew University—Hadassah Medical School,
Jerusalem, Israel*

1 Introduction

The small bacteria of the genus *Bdellovibrio*, which attack and lyse a wide array of Gram-negative bacteria, have intrigued many investigators by their predaceous way of life and their possible role in determining the dynamics of microbial populations in nature. The physiology, morphology and developmental cycle of the bdellovibrios have been extensively studied (for reviews see Shilo, 1969, 1974; Starr and Seidler, 1971; Starr and Huang, 1972; Stolp, 1973; Starr, 1975). However, our knowledge of their ecology is scant and often controversial. This review will, therefore, impinge only briefly on the former, and focus mainly on the ecological aspects of this group of microorganisms.

The bdellovibrios undergo a bi-phasic life-cycle in which a free-living, motile, and non-multiplying form alternates with an aflagellate form which grows and multiplies within the periplasmic space of specific host bacteria. The main stages of the life-cycle are schematically depicted in Fig. 1. A bdellovibrio attaches to the surface of a potential host bacterium, and penetrates its periplasmic space via a pore formed in the cell wall, losing its flagellum in the process. Usually, at the time of penetration, the infected host cell swells and changes into a spherical, though osmotically stable body, the "bdelloplast". After a lag period, the small and vibroid bdellovibrio begins to grow and elongates into a coiled filament, which, upon attaining full length, divides by multiple fission into a number of progeny cells. These acquire flagella and swarm out of the now empty bdelloplast, probably with the aid of a "late" enzyme capable of lysing the "ghost" from within.

All bdellovibrios, including soil types, are absolutely dependent on continuous water pathways of adequate dimension in order to reach their potential hosts. Certain aspects involving soil bdellovibrios will, therefore, also be included in this review of the aquatic bdellovibrios.

The uniqueness of *Bdellovibrio* lies in its obligate dependence on other bacteria for growth and multiplication, and in its nature of cell division, different from the binary fission familiar in most other procaryotes. These two properties distinguish bdellovibrios from other lytic bacteria that can utilize bacteria for growth. The last section (6) will be devoted to these lytic bacteria.

Bdellovibrios were originally described as bacterial parasites. However, unlike obligate bacterial endoparasites, a bdellovibrio appears to render its host a non-functional biological entity prior to developing within it. Early in the process of bdellovibrio attack, the host is immobilized (Stolp and Starr, 1963), its respiration shut off and its permeability control impaired (Rittenberg and Shilo, 1970); RNA and protein synthesizing capacities are damaged (Varon *et al.*, 1969) and rapid degradation of DNA (Matin and Rittenberg, 1972) and ribosomes (Hespell *et al.*, 1975) takes place. All this damage occurs before growth *per se* of the bdellovibrio (DNA synthesis or elongation) has been initiated. Furthermore, a normal developmental cycle proceeds just as readily upon infection of non-viable host bacteria (killed by uv irradiation or mild heating), the burst size being similar to that obtained in viable cells (Varon and Shilo, 1969b). Hespell (personal communication) found that the cellular composition of the bdellovibrios as well as their Y_{ATP} (gram cell material formed per mole adenosine-5′-triphosphate produced; Bauchop and Elsden, 1960) were identical whether grown on viable or dead *E. coli* cells (killed by heating at 55°C for 40 min). The parasite–host definition was, therefore, challenged by Rittenberg and his co-workers (Hespell *et al.*, 1973) who preferred the epithet predator–prey or predator–substrate organism. However, it has still not been confirmed whether irreversible damage ("killing") of the prey is a prerequisite for the development of the bdellovibrio cells: Milke and Stolp (cited by Starr, 1975) indicated that it was possible to cure bdellovibrio-infected hosts by the use of deoxycholate. The probability of "late" killing should be especially high when a bdellovibrio infects a large filamentous multinucleate host. Moreover, the common predator–prey model is insufficient to include some of the prominent characteristcs of the bdellovibrio system; no other system manifests such closely regulated prey breakdown coupled to and in synchrony with predator biosynthesis.

In an attempt to overcome the seemingly controversial semantics, Starr (1975) suggested that the interrelationship be considered within the framework of symbiosis in the broad biological context of de Bary's definition (1879). However, this definition, as those mentioned above, does not do justice to the unique features of the interaction between *Bdellovibrio* and the bacteria that are susceptible to it and allow its development.

Mutants capable of growth in complex media in the absence of host have been isolated by now from many different bdellovibrio strains. These mutants have been variously labelled as saprophytic (Stolp and

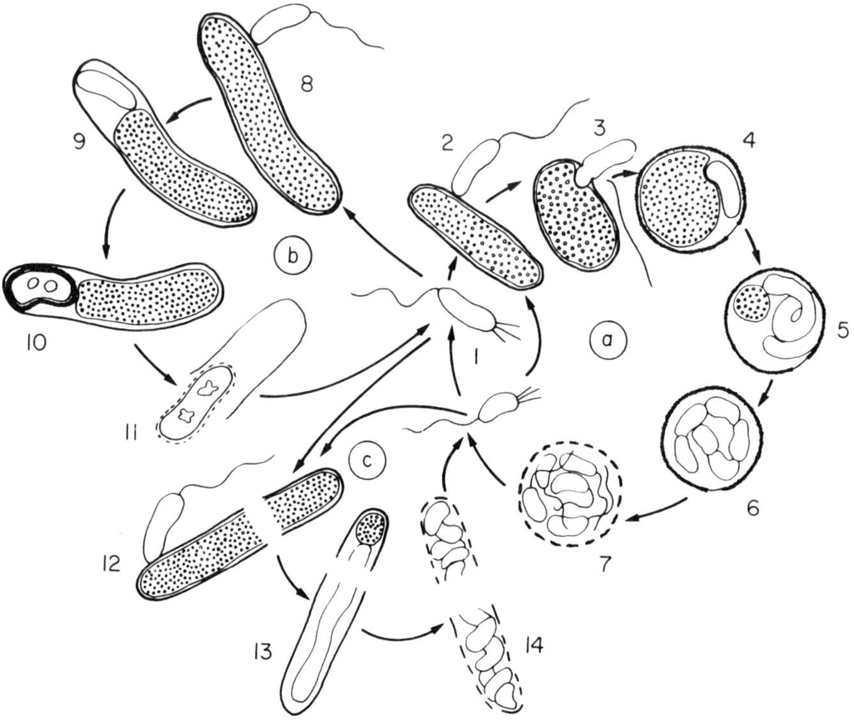

Fig. 1. The life-cycle of *Bdellovibrio*.

(a) The growth cycle of *Bdellovibrio* in a normal-sized host bacterium such as *E. coli*. The free bdellovibrio is a small, curved rod bearing a single sheathed flagellum at one pole of the cell and a number of fibres at the opposite pole (1). It may elongate to a limited extent in its free form before attaching to the host (2). Following attachment, the bdellovibrio penetrates through a pore formed in the host cell wall losing its flagellum in the process (3). The pore is smaller than the diameter of the bdellovibrio, and the latter becomes constricted as penetration proceeds (Burnham *et al.*, 1968). The infected host cell is converted into a swollen spherical body, the "bdelloplast" (4) (Scherff *et al.*, 1966; Starr and Baigent, 1966). The envelope of the host cell at this stage differs from that of the uninfected host in its morphology (Snellen and Starr, 1974, 1976), and loss of phage receptor sites (Varon, 1968), and composition (Rittenberg, personal communication). According to Abram and Davis (1970), the penetrating bdellovibrio is attached to the protoplast of the infected host. This may be supported by scar-like structures on the protoplast observed by scanning electron microscopy (Snellen and Starr, 1974). The penetration pore appears to be sealed off and a scar remains at its site (Shilo, 1969). Intraperiplasmic growth and elongation of the bdellovibrio into a C- or helical-shaped filament (Scherff *et al.*, 1966; Starr and Baigent, 1966) follows (5). During the growth phase, there is progressive shrinkage of the host protoplast. Finally, the long bdellovibrio filament divides by multiple fission (6); each resulting progeny cell acquires a flagellum during or just after division (7). Shortly before lysis of the ghost has been completed, the progeny bdellovibrio can microscopically be seen

Starr, 1963), host-independent (Shilo and Bruff, 1965), facultative (Diedrich *et al.*, 1970), host-independent-dependent (Shilo, 1973), non-parasitic (Varon *et al.*, 1974), non-symbiotic and facultative-symbiotic (Starr, 1975). None of these terms adequately describes the gamut of genetic properties possessed by the various mutants since they do not distinguish between the dependence of the bdellovibrios on other bacteria and their competence to utilize them. This has led to the commonly used genetic terminology: S^{d}*comp*$^{+}$ (for "symbiosis-dependent, symbiosis-competent") to describe the wild type; S^{in}*comp*$^{+}$ (for "symbiosis-independent, symbiosis-competent") to describe facultative mutants, and S^{in}*comp*$^{-}$ (for "symbiosis-independent, symbiosis-incompetent") to describe mutants capable of axenic growth but incapable of host attack (Varon and Seijffers, 1975). In the present review, we shall use the different terms at our convenience in accordance with the subject discussed.

The various bdellovibrio isolates, on the basis of common morphological and physiological properties, were grouped as a single genus (Bergey, 1974). They also share common antigens as recently described by Kramer and Westergaard (1977) and Schelling *et al.* (1977). Based on differences in cytochrome spectra (Seidler and Starr, 1969b), sensitivity to the antibiotic vibriostat, reassociations of DNA/DNA and DNA/RNA, genome size, DNA base composition, and enzyme electrophoretic migration rates, Seidler *et al.* (1972) classified the various isolates into three species: *B. bacteriovorus*, *B. starrii* and *B. stolpii*. The three species could also be differentiated on the basis of their susceptiblity to species-specific bacteriophages (Serban and Levisohn,

---

swimming around inside the now empty bdelloplast. The newly released progeny (1) can immediately attach to new host cells.

(b) Cyst-formation and germination of *Bdellovibrio* W in *Rhodospirillum rubum.* Under special conditions (starvation), *Bdellovibrio* W enters a cycle of cyst development (Tudor and Conti, 1977b). After attachment (8) and preparation (9), the encysting bdellovibrio increases in size, accumulates inclusion bodies, and becomes kidney-shaped and surrounded by a heavy outer layer (10). During germination (11), the outer wall is broken down, the inclusion material changes shape, and the germinant elongates, acquires a flagellum and is transformed into the infectious, free bdellovibrio cell (1).

(c) Development of *Bdellovibrio* in filamentous multinucleate *E. coli.* When a bdellovibrio infects a filamentous, mutlinucleate bacterium (12), the host does not become spherical (Kessel and Shilo, 1976), though a local "balooning" may be seen (Starr and Baigent, 1966). The bdellovibrio elongates into a very long filament (13), and divides upon exhaustion of the host protoplast. The number of progeny is in this case much higher (14) than that obtained in a normal-sized host.

1973). Serotyping by use of agglutination and indirect immunofluorescence techniques indicates that although all bdellovibrio strains tested share a common antigen, *B. bacteriovorus* is antigenically heterogenous and may be further subdivided into nine different types (Schelling *et al.*, 1977).

The marine bdellovibrios examined until now differ from the three species mentioned above, in a number of properties such as cation requirements and temperature optima (Taylor *et al.*, 1974; Marbach *et al.*, 1976). The most conspicuous difference lies in their DNA base composition: the GC content of their DNA is 33·4–38·6 mol per cent as compared to 50·4 ± 0·9 mol per cent in *B. bacteriovorus* and 41·8–43·5 mol per cent in *B. starrii* and *B. stolpii.* It is, therefore, doubtful whether marine bdellovibrios should be included within the same genus as the three other species, and whether the use of the predation complex as exclusive probe for defining the genus *Bdellovibrio* is justified. The marked variability observed among various bdellovibrios isolates in the frequency of mutation to host-independence ($10^5 - 10^8$, Marbach *et al.*, 1976) as well as the rare occurrence of cyst formation found only in two (Burger *et al.*, 1968; Mishustin and Nikitina, 1973) of the many bdellovibrio strains screened (Tudor and Conti, 1977a) further points to the heterogeneity of the group. The bdellovibrios appear, therefore, to consist of an assembly of different organisms with common morphological and physiological characteristics, small size, rapid motility, and regulatory mechanisms for host utilization, which may all be properties imposed by the predaceous way of life, and the smallness of the prey organism. They could have developed by a divergent evolution from one very ancient prototype or by convergent evolution of different ancestral organisms towards a common way of life.

2 The ecological advantages of the *Bdellovibrio* way of life

Penetration of the bdellovibrio into the periplasmic space of other bacteria ensures its nutritional requirements and makes it independent of external sources, thereby allowing growth even in ecosystems extremely poor in dissolved nutrients, such as oceans, lakes and rivers. The intraperiplasmic environment is also a competition-free niche, which may be of great value for a slowly growing bacteria such as *Bdellovibrio*.

Another advantage of growth inside the periplasmic space lies in the fact that the latter, being surrounded by a stable and rigid wall, impermeable to large and small molecules (Drucker, 1969; Crothers and Robinson, 1971), provides a protective environment for the bdellovibrio.

2.1 UTILIZATION OF THE HOST ORGANISM BY *BDELLOVIBRIO* FOR GROWTH

The potential host of a bdellovibrio is a bacterium often only 5 to 10 times larger than itself. Completion of a bdellovibrio's life-cycle and the production of a number of progeny sufficient for survival of the species in a given ecosystem necessitates both prevention of loss of host components and their efficient recycling.

Studies of the fate of host components following bdellovibrio infection showed that a high proportion of the *E. coli* carbon is incorporated into the bdellovibrio progeny, the rest used for bdellovibrio respiration (Matin and Rittenberg, 1972; Hespell *et al.*, 1973, 1975). Drucker (1969) found that only small amounts of ninhydrin-positive material, and even smaller amounts of β-galactosidase or uv-absorbing material, leaked out of the infected cells. No leakage of β-galactosidase was detected by Crothers and Robinson (1971) although the infected cells were found to be more permeable to *o*-nitrophenyl-β-D-galactoside and to the fluorescent dye, 8-anilino-1-napthalenesulphonic acid. The very low leakage level might indicate that the cytoplasmic membrane of the host remains an effective permeability barrier throughout the entire life cycle of the bdellovibrio, host components possibly being taken up only at contact areas (Snellen and Starr, 1974; Abram *et al.*, 1974). A second possibility, not excluding the first, is that the penetration pore becomes closed and the host cell wall sealed off in a manner preventing loss of internal components into the surrounding medium. The finding that a periplasmic enzyme (alkaline phosphatase) does not leak out of the infected bdelloplast (Drucker, 1969; Castro e Melo, 1975) supports the latter.

Detailed analysis of the energy efficiency involved in bdellovibrio intraperiplasmic growth showed the Y_{ATP} to be about 26 (Rittenberg and Hespell, 1975). This value is more than twice as high as the maximal values reported for bacteria growing in rich medium and close to the theoretical value calculated (Gunsalus and Shuster, 1961; Stouthamer, 1973). The high efficiency has been explained in part by the elegant findings of Rittenberg and his co-workers who showed that the macromolecules of the prey are degraded by the bdellovibrio, in a way well regulated in time and rate, to relatively large monomers directly utilized for its own biosynthetic needs. Thus, mononucleotides are derived from the prey's nucleic acids and used intact (Matin and Rittenberg, 1972; Hespell *et al.*, 1975; Rittenberg and Langley, 1975; Pritchard *et al.*, 1975), and fatty acids are derived from the lipids of the prey and used as such or after small alterations (Kuenen and Rittenberg, 1975). An important characteristic of the bdellovibrios enabling this phenomenon is their capacity to take up phosphorylated intermediates such as nucleoside monophosphates. This ability is rare among the bacteria (Yagil and Beacham, 1975) and found only in endosymbionts such as *Rickettsia* (Hatch, 1976) and *Chlamydia* (Winkler, 1976). Consequently, energy-rich bonds pre-existing in prey macromolecules are conserved, minimizing the energy requirements for predator growth. However, all these mechanisms do not suffice fully to explain the bdellovibrio's high Y_{ATP} value. Therefore, Rittenberg and Hespell (1975) have suggested that, in addition to the above, a coupling of energy production and utilization, more efficient than known in other biological systems, may be operative in the bdellovibrios.

2.2 PROTECTION OF *BDELLOVIBRIO* WITHIN THE PERIPLASMIC SPACE OF THE HOST

The intraperiplasmic bdellovibrio is considerably less susceptible than the free form to environmental stress. A comparison of the sensitivity of the free form versus that of the intraperiplasmic one was made using bdellovibrio mutants capable of both axenic and intraperiplasmic growth. Such mutants, when grown axenically, showed pronounced sensitivity to light (Friedberg, 1977): an exposure to an intensity of $1 \cdot 10^4$ erg cm^{-2} s^{-1} for 4 h reduced the viability by 99·9 per cent. The same mutant was resistant to 7-day exposure to the same irradiation when grown in a 2-membered culture. Irradiation of the free bdellovibrios, while greatly reducing their ability to form colonies, did not abolish their plaque-forming capacity. These experiments suggest that the bdelloplast not only protects the bdellovibrio against photodynamic damage, but also in some ways allows repair of damage inflicted prior to penetration.

The existence of bdellovibrios with DNA of low guanine plus cytosine (GC) content in the marine environment raises an interesting question. Such bacteria would be expected to be relatively sensitive to uv light due to their high thymine content and the increased probability of thymine dimer formation. Indeed, the bacteria normally found in the oceanic water–air interface, exposed to strong uv radiation, are of high GC content (Marshall, 1976). Survival of marine bdellovibrios with low GC content in the ocean surface layer may be attributable to their intraperiplasmic existence.

Bdellovibrios growing in axenic culture are sensitive to the osmotic pressure of the medium: the optimal growth rate is achieved at an osmosity of 0·03. Above and below this value, growth can be considerably slower (Varon and Seijffers, 1977). Intraperiplasmically growing bdellovibrios, on the other hand, grow well at much lower osmosities (Varon and Shilo, 1969b).

Another example of the protective value of the bdelloplast is seen in the case of bacteriophage (bdellophage) attack. Only the free bdellovibrio is vulnerable to attack; bdellophages cannot penetrate the host cell and, therefore, the intraperiplasmic bdellovibrio escapes infection (Varon and Levisohn, 1972).

It has been shown that bdellovibrios are susceptible more than other bacteria to various physical and chemical factors (Abram and Davis, 1970; Wehr and Klein, 1971; Corberi and Solaro, 1972; Castro e Melo, 1975). Their unique predatory way of life, which renders specific protective mechanisms superfluous, might account for this phenomenon.

2.3 CONTROL OF GROWTH INITIATION, ELONGATION AND CELL DIVISION

Wild type bdellovibrio growth and division are host dependent. In the absence of host cells, the bdellovibrios are capable of incorporating amino acids into protein, uracil into RNA, and inorganic phosphate into phospholipids, as well as of elongating to a limited degree (up to 2 to 3 times their original size) (Varon, unpublished). However, the cells remain flagellated, DNA is apparently not synthesized (Horowitz *et al.*, 1974), the cells show low competence for transformation (F. Guerrini, personal communication) and bdellophage development does not take place (Varon and Levisohn, 1972). Only when a bdellovibrio infects a suitable host is the normal developmental cycle initiated: the flagellum is lost, and after a certain lag period (which may last 30–60 min) the cell begins elongating, synthesizing its own DNA. The length attained by the intraperiplasmically growing bdellovibrio, the number of progeny produced and the time of fission were found to be correlated to host cell size; the number of progeny obtained was 3 to 4 in small *E. coli* cells and as high as 80 to 90 in filamentous, multinucleate *E. coli* (Kessel and Shilo, 1976). When bdellovibrios were grown in continuous culture with filamentous *E. coli* suspended in buffer, the bdellovibrio yield per infected host cell was two orders of magnitude higher than that obtained with normal sized *E. coli* (Varon and Seijffers, unpublished). Elongation of the bdellovibrios proceeded until the host protoplast had been devoured and only then did multiple fission take place.

It has been possible to stimulate normal bdellovibrio development by the addition of suitable host extract. Its presence triggered the sequence of morphological events as they occur inside the host bacterium, and allowed thymidine incorporation (Reiner and Shilo, 1969; Horowitz *et al.*, 1974) and bdellophage development (Toppel *et al.*, 1973). Under these conditions, the bdellovibrios elongated until the active component in the host extract was exhausted, or removed by centrifugation or dilution. Consecutive addition of host extract led to further elongation and postponed the time of division (Horowitz *et al.*, 1974).

All the evidence obtained from both the *in vivo* and *in vitro* experiments advocates that initiation of growth, growth itself, as well as onset of bdellovibrio division are controlled by components of the host. The great variability in length attained by bdellovibrios growing under various conditions suggests that there is separate control of elongation and division. This conclusion is also supported by the inhibitory effect of virginiamycin S (Varon *et al.*, 1976), which prevents division of bdellovibrios whether growing intraperiplasmcially or axenically, without affecting their elongation.

Recent evidence indicates that additional to the control exerted by host components, a division-triggering mechanism is operative in which the host plays an indirect role. This mechanism involves a low molecular weight substance, produced by the bdellovibrio just prior to the onset of division. Using axenically growing mutants, Eksztejn and Varon (1977) showed that this substance was released from the cells and accumulated in the medium; division occurred only after it had reached a certain threshold concentration. This would account for the delayed division of axenically growing bdellovibrios at low cell densities. In a 2-membered culture, the developmental cycle is unaffected by cell density and division occurs even in highly diluted cell suspensions (Varon and Shilo, 1969b). If a certain critical concentration of the division factor is indeed necessary for normal divison, its local accumulation would be ensured within the bdelloplast, and its dilution below the effective level prevented.

2.4 SUPERIORITY OF HOST-DEPENDENT BDELLOVIBRIOS OVER MUTANTS CAPABLE OF AXENIC GROWTH

The vulnerability of the free-living stage of the host-dependent (HD) bdellovibrio to external conditions is also shared by mutants growing axenically. New isolates of such mutants grow poorly in axenic culture and require a special environment for more efficacious growth. Upon serial transfers, they become hardier and show increased capacity for axenic growth (Varon and Seijffers, 1975). This is often accompanied by the acquisition of yellow pigmentation (Stolp and Starr, 1963; Seidler and Starr, 1969b) which in turn may play a role in protection against photo-oxidation (Friedberg, 1977). Still, only in 2-membered cultures do these mutants show an optimal growth rate (Varon and Seijffers, 1977) and even then they do not successfully compete with their wild type parent. This has been demonstrated in repeatedly transferred batch culture (Varon and Seijffers, 1975) as well as in a continuous flow system (a 2-stage chemostat similar to that described by Veldkamp and Jannasch, 1972) in which the substrate was *E. coli* suspended in dilute nutrient medium. Under such growth conditions a facultative mutant of *Bdellovibrio* 109J was washed out when grown together with its wild type parent strain. When grown alone, the mutant population was soon displaced by HD revertants which had lost the capacity for axenic growth (Varon and Westra, 1976).

A possible explanation for the disadvantage of the facultative bdellovibrios stems from their capacity of axenic growth; as they start growing, the flagella necessary for the establishment of the bdellovibrio–host association (see section 4) are shed. This curtails their ability of attachment to the host. Since growth of the bdellovibrio outside the periplasmic space is less efficient than inside it, the facultative bdellovibrios will not be able to compete successfully with the host-dependent bdellovibrios, which retain their flagella until they have secured their intraperiplasmic development, or until (in the absence of host) they have exhausted their reserve materials. Only in a milieu rich in organic material but poor in suitable host bacteria would a facultative mutant have any potential advantage over the wild type (HD) form. It is thus conceivable that the HD form is the only form capable of existing in nature, the host-independent bdellovibrios being limited to life under laboratory conditions. Indeed, all the bdellovibrio strains isolated from nature until now have been of the HD type, despite the fact that most of them are capable of throwing off host-independent mutants.

3 Survival of *Bdellovibrio* in nature

The survival of *Bdellovibrio* in nature poses problems intrinsic to its unique bi-phasic life-cycle. Whereas, during intraperiplasmic growth, the cell inhabits a physically protective and nutrient-rich environment, in its free-living state, it is directly exposed to the external environment. Moreover, incapable of axenic growth, it is under stress of starvation even in a milieu rich in organic substances, unless bacteria of suitable nature and concentration are present. The survival capacity of the free-living form is thus of utmost importance for the existence of the species. The question is to what degree does the high sensitivity of the free-living stage limit the existence of *Bdellovibrio* in nature.

Having lysed its prey, the bdellovibrio is faced with starvation conditions and soon dies (Varon and Shilo, 1968; Burger *et al.*, 1968). Hespell *et al.* (1974) showed that the viability of cell suspensions of *Bdellovibrio* 109J and 109D shaken in buffer at 30°C decreased by approximately 50 per cent within 10 h; a third strain, A3-12, was only slightly more resistant. A marine bdellovibrio strain behaved similarly whether shaken in flasks containing buffer (Marbach *et al.*, 1976) or placed in dialysis sacs immersed in sea water (Varon and Shilo, unpublished), both at 25°C. This signifies that there is a very limited time span during which the bdellovibrio must find its host and attach to it.

In spite of the laboratory findings, there is evidence that bdellovibrio populations are maintained in nature for relatively long periods: Fry and Staples (1974) showed that indigenous bdellovibrios in river water could remain viable for long periods of time (at least 70 h) in the absence of added host, and Daniel (1969) proved that in suitable milieu bdellovibrios could survive for at least 96 h.

The maintenance of bdellovibrio populations in nature in environments poor in bacteria or otherwise unsuitable for the intraperiplasmic growth of the bdellovibrios could be explained in several possible ways.

a. *Genotypic and phenotypic heterogeneity of the bdellovibrio population* The degree of sensitivity of the bdellovibrios to starvation, as well as to other adverse conditions, may differ among the various strains. Table 1 demonstrates this genetic heterogeneity in respect to starvation. The strains prevalent in nature could be of the relatively stable type.

In addition to genetic heterogeneity, phenotypic variability is also common among bdellovibrio populations. Such variability has been shown for the bdellovibrio capacity to attach to bacterial hosts of the smooth (S) type (Varon and Shilo, 1969a) and the ability to attach at different temperatures (Varon and Shilo, 1968). There is also heterogeneity of survival potential, as indicated in Table 1. Fry and Staples (1974) described a similar phenomenon: after an initial mortality of 90 per cent during the first day, the survivors retained their viability for at least 5 days. Keya and Alexander (1975a) showed that bdellovibrios added to soil declined in number within a few hours, but the surviving population was thereafter constant for 24 days. The phenotypic variability in resistance to starvation is not a feature unique for *Bdellovibrio*; Novitsky and Morita (1977) showed that although

TABLE 1

The survival of different bdellovibrio strains under non-growing conditions (Varon, unpublished)

Bdellovibrio strain	Bdellovibrio number (PFU) after storage (months) 0	3		6		9	
		DNB	Lysate	DNB	Lysate	DNB	Lysate
UKi2[1]	5×10^9	1×10^6	$<10^3$	<100	<100		
109J[2]	5×10^9	3×10^8	1×10^5	<100	<100		
6-5-S[3]	5×10^9	6×10^7	1×10^3	9×10^5	<100	2×10^5	<10
GD[4]	5×10^9	6×10^4	5×10^3	3×10^4	6×10^2	2×10^2	<10
GB[4]	5×10^9	3×10^7	6×10^6	5×10^6	6×10^6	7×10^5	5×10^5
109D[2]	5×10^9	2×10^7	2×10^7	6×10^6	1×10^7	1×10^6	4×10^5

Each of the six strains were grown with *E. coli* in a 2-membered culture in DNB medium (Varon and Shilo, 1969b) until complete lysis of the host had occurred. The bdellovibrios were then centrifuged and resuspended in the respective supernatant fluid or in fresh medium and stored at 10°C. The viable number (i.e. plaque forming units, PFU) was determined after 3, 6 and 9 months by the agar double-layer technique.

For description of strains see Diedrich *et al.* (1970) (1), Rittenberg (1972) (2), Simpson and Robinson (1968) (3), Varon and Shilo (1969a) (4).

50 per cent of starved marine vibrio cells died within 6 to 7 weeks, a proportion remained viable for more than 1 year.

b. *Protection from starvation* Hespell *et al.* (1974) showed that upon starvation, bdellovibrio death is accompanied by release of the cell carbon as CO_2. The cells apparently use up their own cell constituents as energy source. Nutrients, such as glutamate (Hespell *et al.*, 1974) or ribonucleoside monophosphate (Hespell and Mertens, 1978) added to the medium are utilized by the bdellovibrios, thereby sparing their own cell carbon, and prolonging viability. A similar phenomenon affected by glutamate or an amino acid mixture was noted for marine bdellovibrios (Marbach *et al.*, 1976). Table 1 shows that the sparing effect (in this case of a dilute nutrient solution) is more pronounced in the case of strains highly sensitive to starvation than in less sensitive ones.

The rapid self-destruction of bdellovibrios has been attributed by Hespell *et al.* (1974) to the cells' high rate of endogenous respiration imposed by the high energy requirement for motility. Their experiments as well as those of Marbach *et al.* (1976) were all performed at relatively high temperatures (30°C and 25°C, respectively). Lowering the metabolic activity of the bdellovibrios should prolong their life-span. Indeed, the experiments of Fry and Staples (1974), demonstrat-

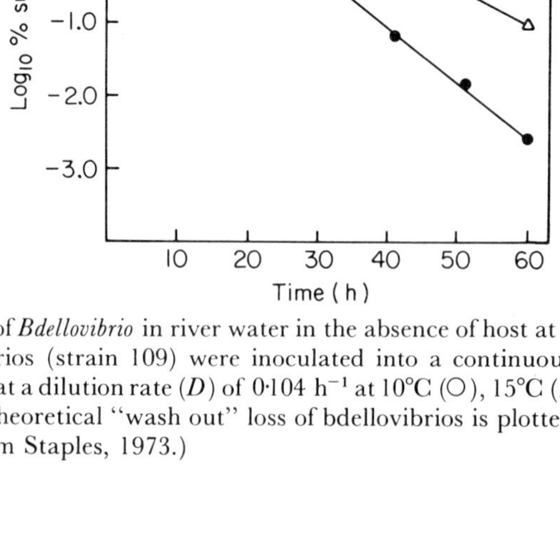

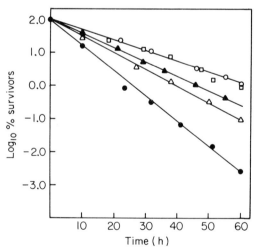

Fig. 2. Survival of *Bdellovibrio* in river water in the absence of host at various temperatures. Bdellovibrios (strain 109) were inoculated into a continuous-culture growth vessel operating at a dilution rate (D) of 0·104 h^{-1} at 10°C (○), 15°C (▲), 20°C (Δ) and 30°C (●). The theoretical "wash out" loss of bdellovibrios is plotted for comparison (□). (Taken from Staples, 1973.)

ing survival of bdellovibrios for at least 70 h, were done at a mean temperature of 13°C. Staples (1973) showed that bdellovibrios maintained in a "chemostat" in the absence of prey bacteria survived starvation if kept at 10°C. They surivived less well at 15°C and 20°C, and an even higher loss was observed at 30°C (Fig. 2). The relatively slow death rate reported in Table 1 could also be attributable to the low temperature (10°C) at which the cells were kept. Longer survival at low temperatures could be of importance for the maintenance of bdellovibrio populations in nature, especially in the marine environment where the prevalent temperatures are usually low.

c. *Bdellocyst formation* Certain bdellovibrio strains are able to form cysts—"bdellocysts" (Burger *et al.*, 1968; Hoeniger *et al.*, 1972; Mishustin and Nikitina, 1973; Tudor and Conti, 1977a, b). The bdellocysts are more resistant than vegetative cells to the effects of elevated temperatures, uv radiation, sonication, dessication and detergents (Tudor and Conti, 1977a). However, of possibly greater importance from the ecological point of view is their increased ability to survive under starvation conditions. Tudor and Conti have shown that bdellocysts retained over 95 per cent viability for up to 15 days under starvation conditions which caused vegetative cells to lose more than 80 per cent viability after only 48 h. The bdellocysts' low rate of endogenous respiration may be an important factor in their survival capacity during starvation.

Bdellocyst formation (as studied in *Bdellovibrio* strain W; Fig. 3(a)) may occur within a variety of Gram-negative bacteria. In fact, all bacteria which can support the growth of this strain allow encystment. No cysts are formed in the absence of prey bacteria. What determines the fate of the penetrating bdellovibrio to grow and multiply or become encysted is an intriguing question. How can a bdellovibrio upon penetration into a host cell, which should suffice to allow completion of its life-cycle, gauge that at the end of the normal intraperiplasmic development its progeny will face starvation? One possibility is that the bdellovibrio can sense the starved condition of the penetrated host, this "sensing" triggering encystment. Another possibility may be that it is the starved condition of the bdellovibrio itself which could trigger encystment even inside a "healthy" host. Indeed, both explanations are in accordance with the findings that encystment is optimal under starvation conditions (Tudor and Conti, 1977a).

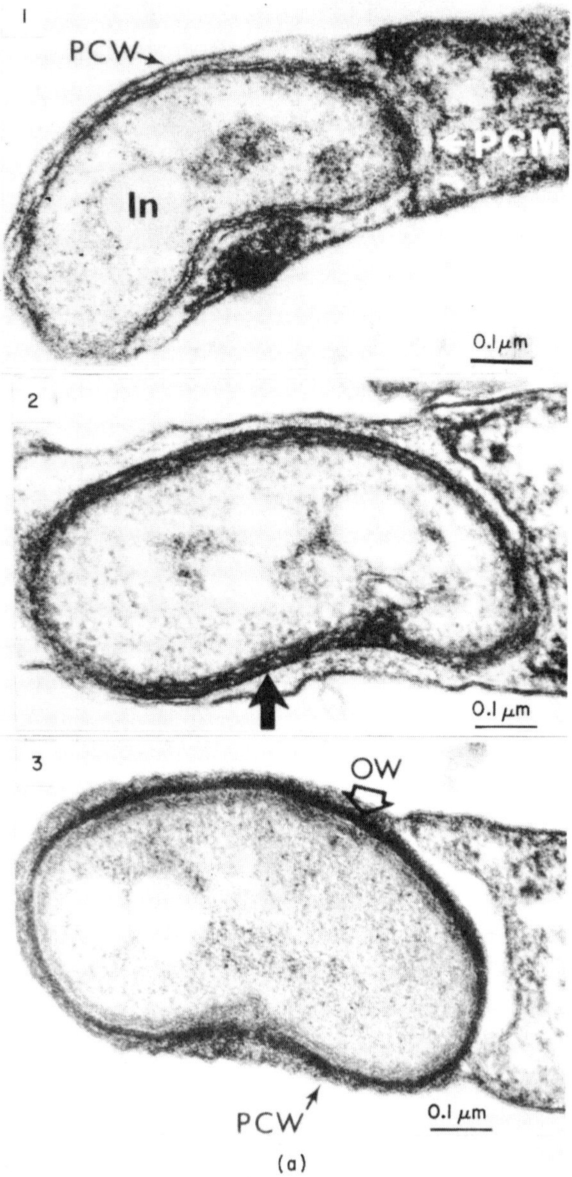

Fig. 3. (a) Electron micrographs of thin sections of *Bdellovibrio* strain W during encystment in *R. rubrum*. (1) A bdellovibrio containing inclusion bodies (In) inside the periplasmic space of its host. (2) An intermediate stage in the process of encystment; a thin layer of amorphous, electron-dense material has been laid down around the encysting cell (arrow). (3) The final stages of encystment: the outer wall (OW) has increased in thickness and is intimately associated with the host cell wall (PCW).

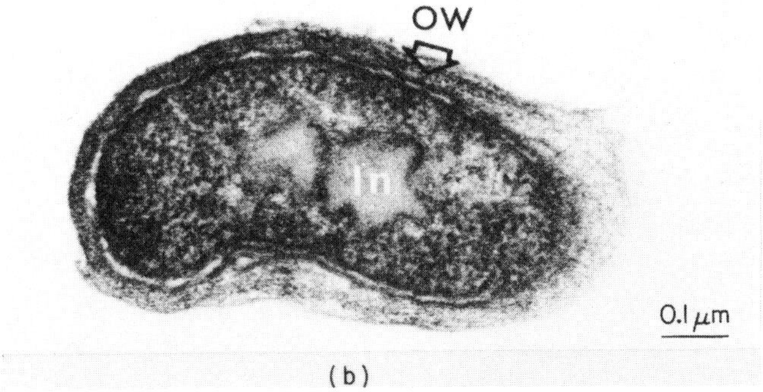

(b) A germiniating bdellocyst: the outer wall (OW) has expanded and become fibrous and less dense. Inclusion material (In) has become irregular in shape. (From Tudor and Conti, 1977b.)

Germination of bdellocysts of strains W (Fig. 3(b)) can be induced by amino acids, as well as by K^+ and NH_4^+; the best single inducer is glutamine though superior germination was obtained in a medium containing a combination of germinants (Tudor and Conti, 1978). Thus, germination is not directly elicited by host cells, but possibly by conditions which allow the development of a dense population of the potential host.

Bdellocysts have until now been demonstrated in only two bdellovibrio strains, though many more have been screened (Tudor and Conti, 1977a). Thus, the extent of the cyst's role in the survival of *Bdellovibrio* in nature is still not clear.

4 Conditions for interaction of *Bdellovibrio* with other bacteria in nature

Except for rare occasions such as the presence of inhibitors (Varon and Shilo, 1968), or conditions which might affect the integrity of the bdelloplast (Seidler and Starr, 1968a; Castro e Melo, 1975; Marbach, 1977), attachment of a bdellovibrio to a suitable host ensures its normal development cycle. Because of this and because the survival of the free bdellovibrio is time-limited, the crucial steps in the life of a bdellovibrio are those leading to its attachment. The prerequisite for the establishment of a stable association between the bdellovibrio and its host is a physical encounter between the two, which is motility and density dependent. It must be followed by a specific process leading to irreversible attachment.

4.1 THE ROLE OF MOTILITY IN ATTACHMENT

Bdellovibrios are equipped with a long, sheathed flagellum and manifest exceptionally rapid motility, about 100 $\mu m\ s^{-1}$ (Stolp, 1967a, b). Impairment of motility or flagellar sheath integrity, induced by any of several procedures (Stolp and Starr, 1963; Varon and Shilo, 1968; Dietrich *et al.*, 1970; Abram and Davis, 1970; Dunn *et al.*, 1974) results in total inability of the bdellovibrio to establish stable contact with its potential host. The high sensitivity of the bdellovibrio flagellum to relatively mild mechanical agitation (2 min on a Vortex mixer, Varon, unpublished) makes it likely that in nature the exertion of shearing forces caused by currents and turbulence in the aquatic environment could impair motility and, thereby, attachment.

The presence of motility inhibitors, as well as the absence of factors required for bdellovibrio motility, should also affect the bdellovibrio–host interaction. Chet and Mitchell (1976) described the inhibition of bacterial chemotaxis by oil pollutants and other compounds in the marine environment. Marbach (1977) showed that a marine bdellovibrio required K^+ for its usual high velocity; this requirement was specific and could not be fulfilled by either Na^+, Ca^{2+} or Mg^{2+}, or any combination of these cations.

Microscopic observations of attaching bdellovibrios led Stolp and Starr (1963) to suggest that the bdellovibrio drills its way into the host's envelope and therefore motility must play a direct role in the process of attachment. However, drilling may not be necessary for attachment; the gyratory motion of an attaching bdellovibrio may be attributable to a physical reaction resulting from the attachment of the proximal pole of the cell, the flagellum at the distal pole providing the motive power and creating the observed spinning motion. Indeed, the same kind of motion is seen when marine pseudomonads (Marshall *et al.*, 1971) or bdellovibrios attach to a glass surface.

Obviously, motility increases the probability of an encounter between a bdellovibrio and its host. Experiments done in our laboratory suggest that only a small proportion (3 per cent) of the total number of encounters result in the formation of stable assocations with the host (Varon and Zeigler, 1978). Possibly, motility only serves to increase the number of chance encounters, and so indirectly increases the probability of irreversible attachment.

The dependence of *Bdellovibrio* on prey organisms and the low probability of chance collision in natural environments would make any mechanism for sensing individual host cells or host-rich niches, greatly advantageous for its survival. Straley and Conti (1974) were the first to find indications for a chemotactic response in bdellovibrios. Recently, LaMarre *et al* (1977), establishing optimal conditions for bdellovibrio chemotaxis, found that bdellovibrios responded (though weakly in comparison to other bacteria) to a number of amino acids such as L-asparagine, L-cysteine, glycine, L-histidine, L-lysine, L-threonine and L-glutamine. However, there was no significant chemotactic response to host cells such as *E. coli* or *P. fluorescens* (even at high densities of 10^7 cells ml^{-1}), their exudates or their lysates (Straley and Conti, 1977). LaMarre *et al.* (1977) suggested that the attraction towards amino acids may be a means for detecting ecological niches rich in prey organisms. Another possibility proposed by these authors was that the attractants to which the bdellovibrios respond are such that can be used for cellular maintenance, thus serving to alleviate starvation.

4.2 THE EFFECT OF HOST DENSITY ON THE BDELLOVIBRIO–HOST INTERACTION

The chances for collision between a bdellovibrio and its potential host are dependent on the cell concentration of both. It was shown by Varon and Shilo (1968) that the attachment efficiency of *Bdellovibrio* is proportional to cell density. Direct studies of the effect of cell density on the attachment efficiency were, however, limited to relatively high bacterial numbers. Information on interactions at lower cell densities such as those found in nature has been recently obtained by following bdellovibrio growth. Fry and Staples (1974) observed that a bdellovibrio culture (containing about 5×10^3 cells ml^{-1}) incubated in buffer in a shaken flask did not increase in number during 5 days in the presence of $7{\cdot}6 \times 10^5$ *E. coli* cells ml^{-1} or lower concentrations, but multiplied at higher host densities. A similar effect was observed when river water with its indigenous bdellovibrio population (10 bdellovibrios ml^{-1}) was transferred into dialysis sacs containing various concentrations of *E. coli*, and immersed in the river (Fry and Staples, 1974).

Keya and Alexander (1975a) observed a similar phenomenon for a soil bdellovibrio strain preying on rhizobia: the bdellovibrio titre increased only when the number of rhizobia in the culture had reached a level of 10^8 cells ml^{-1}. The indigenous bdellovibrio population increased in soil to which $4 \cdot 2 \times 10^8$ rhizobia g^{-1} were added, whereas little or no response could be detected in soil to which $1 \cdot 4 \times 10^5$ rhizobia g^{-1} were added. A similar host-density effect was observed by Westra (1976) and Guélin and Cabioch (1974). The latter introduced four different-sized bdellovibrio inocula into *E. coli* suspensions with initial cell concentrations of 10^4, 10^6 and 10^8 cells ml^{-1}. Only the *E. coli* suspension with a density of 10^8 cells ml^{-1} supported growth of bdellovibrios, while lower densities did not allow an increase in bdellovibrio titre.

A classical method for studying population dynamics at low cell densities is the use of the continuous culture technique (Jannasch and Mateles, 1974). Whitby (1977) followed the pattern of growth of *B. bacteriovorus* 6-5-S and its prey *Aquaspirillum serpens* in continuous cultures operating at various dilution rates. He found that at dilution rates of $0 \cdot 1$, $0 \cdot 2$ and $0 \cdot 3$ h^{-1} the number of both organisms oscillated during prolonged periods of incubation, the bdellovibrio titre increasing as the prey titre decreased, followed by a decrease in the number of bdellovibrios as the prey number increased back to the original level. At the low dilution rate of $0 \cdot 5$ h^{-1} the two organisms went through one full cycle, after which the numbers of the prey and the bdellovibrios remained at 10^3 and 10^8 cells ml^{-1} respectively. The interpretation of these data and their significance as to the density effect should take into consideration whether the prey is homogeneously dispersed in the growth vessel; wall growth could be of great significance in this respect, especially at low cell densities.

Veldkamp and Jannasch (1972) suggested the use of a two-stage "chemostat" for the cultivation of *Bdellovibrio*: the host cells, growing in a chemostat, limited by nutrients flowing in from the reservoir, flow out and into a second fermentor in which they are immediately attacked by the bdellovibrios. The bdellovibrio population achieves a steady state in the second fermentor. Using a similar type of "chemostat", operating at a dilution rate of $0 \cdot 18$ h^{-1}, we found that the steady state of the bdellovibrios maintained on 10^8 *E. coli* cells ml^{-1} was rapidly disturbed when the inflowing host concentration dropped below 10^7 cells ml^{-1}; the bdellovibrio number progressively decreased until they were washed out and the number of host cells concomitantly increased (Varon and Seijffers, unpublished).

A more detailed study of the effect of cell density on the bdellovibrio–host interaction was recently made by Varon and Zeigler (1978). The system chosen consisted of a marine bdellovibrio and its luminous host bacterium, *Photobacterium leiognathi*, and the parameter for following the interaction was the extinction of the bioluminescence of the host cell caused by the penetrating bdellovibrio. The decay rate of the bioluminescence of the population was a function of the number of fertile encounters between the bdellovibrios and their potential hosts. This system is particularly suitable for experimental analysis since it enables direct and continuous measurement of the bdellovibrio–host interaction under various conditions with no need for intervention or separation of the interacting components. Figure 4 shows that the bioluminescence decay rate was dependent on both the predator/prey ratio and the host cell density; upon dilution of the host suspension, a higher predator/prey ratio was required in order to obtain the same decay ratio. Analysis of the results indicated that under given circumstances, the rate of decay of light emission was solely dependent on the concentrations of both bacteria.

Hespell *et al.* (1974), assuming that in the absence of host the half-life

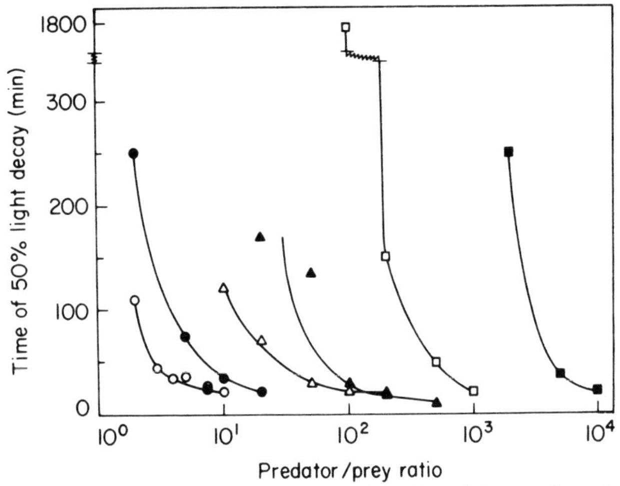

Fig. 4. The rate of bioluminescence decay as a function of the predator/prey ratio at various cell densities. The prey densities were 10^8 cells ml^{-1} (○), 10^7 cells ml^{-1} (●), 10^6 cells ml^{-1} (Δ), 10^5 cells ml^{-1} (▲), 10^4 cells ml^{-1} (□) and 10^3 cells ml^{-1} (■). (Varon and Zeigler, 1978.)

time of the bdellovibrio population is 10 hours, calculated that a minimal density of $1 \cdot 5 \times 10^5$ host cells ml^{-1} would be required to sustain *Bdellovibrio* in nature. Taking into account that only a small proportion (3 per cent) of the total encounters between the bdellovibrios and their hosts leads to a stable association, Varon and Zeigler (1978) calculated that the minimal density required would be approximately 10^6 host cells ml^{-1} while 10^7 cells ml^{-1} would be required for population equilibrium. These values would vary somewhat depending on cell size, since the latter determines the number of bdellovibrio progeny released (see section 2). The amount and composition of nutrients in the natural environment would, obviously, affect the density, the cell size as well as the composition of the bacterial population and, hence, the chances for maintenance of *Bdellovibrio*. The probability of survival in an environment of low bacterial density would be greater if the bdellovibrios could take advantage of a "search" mechanism, such as positive chemotaxis, for the detection of niches rich in host cells, or if the free bdellovibrios could remain viable longer than has been assumed (see section 3).

4.3 ATTACHMENT TO HOST CELLS

Bdellovibrios attach stably and irreversibly only to potential host bacteria, all of which are Gram-negative. A structure found only on the surface of Gram-negative bacteria must, therefore, provide the complementary molecular structure necessary for the specific interaction. Varon and Shilo (1969a) have implicated the lipopolysaccharide (LPS) layer of the cell's outer membrane in this role. The LPS consists of a basal core region, the R antigen, common to most Gram-negative bacteria, and variable polysaccharide side chains which impact the somatic (O) antigenic specificity (Lüderitz *et al.*, 1966). Varon and Shilo showed that rough mutants of *Salmonella* (*S. typhimurium* LT2 and *S. minnesota*) and *E. coli* containing the complete R antigen but lacking the O antigen (chemotype Ra) were the best receptors for bdellovibrio. Deficiencies in the R antigen, as in chemotype Rb (lacking *N*-acetyl-glucosamine) and in chemotype Re (lacking all hexoses and heptoses) reduced receptor activity. The conclusion that the R antigen plays a role in bdellovibrio attachment gained further support from the findings of Houston *et al.* (1974) who showed that R antigen prepared from cells of *S. typhimurium* inhibited the attachment of bdellovibrios to intact host cells, the degree of inhibition being proportional to the amount of R antigen added.

The similarity of the core in most of the Gram-negative bacteria explains the broad host range of *Bdellovibrio*. The importance of the R antigen in maintaining cell integrity may explain the fact that no bdellovibrio-resistant mutants have as yet been found.

The fact that rough mutants were better receptors for bdellovibrio than the smooth bacteria from which they were derived (Varon and Shilo, 1969a) indicated a possible masking of the receptor sites or steric hindrance of attachment. A similar effect was described by Buckmire (1971) for *Spirillum serpens*; an additional external layer (proteinaceous in nature, Buckmire and Murray, 1970) protects this bacterium from attack by bdellovibrios, and its removal by mutation, EDTA, SDS, or heat treatment allows bdellovibrio attachment (Starr and Huang, 1972). The composition of the bacterial envelope is, of course, greatly affected by the composition of the growth medium, e.g. the formation of levan and dextran capsules is dependent on the presence of sucrose (Beijerinck, 1921). The composition of the medium could, therefore, affect the receptor capacity of the host for bdellovibrios.

In the absence of appropriate receptor sites, unstable, reversible attachment may occur. Such attachment has been observed to non-host bacteria such as *Bacillus* (Dunn *et al.*, 1974), *Neisseria gonorrhoaea* (Drutz, 1976) as well as to glass surfaces and organic or inorganic particles. This non-specific adsorption removes the bdellovibrios from the suspension and may reduce their chances of finding suitable host cells.

The attachment of a bdellovibrio to the host cell is known to be affected by many factors including composition and pH of the medium, oxygen tension, and incubation temperature (Varon and Shilo, 1968). It is not known, however, to what extent these factors affect the attachment process itself or other related functions such as motility.

5 Distribution of *Bdellovibrio* in aquatic ecosystems and their role in nature

5.1 PROBLEMS IN THE METHODOLOGY OF ENUMERATION

Studies on *Bdellovibrio* distribution have till now been mostly of a qualitative nature. A major difficulty in the quantitative determination lies in the lack of suitable methods and in the fact that various methods now in use detect but a portion of the population. The mode of sampling, the way of separating the bdellovibrios from other microorganisms present in the sample, the conditions used for cultivation of the bdellovibrios and the criteria used for their enumeration, namely, plaque number, rate of lysis of host suspensions etc., all markedly affect the ultimate numbers obtained. An evaluation of the literature is also complicated by the fact that various investigators have used different enumeration techniques, which cannot be easily intercalibrated nor the results equated.

The procedure first used for isolating bdellovibrios (Stolp and Petzold, 1962; Stolp and Starr, 1963) involved centrifugation of suspensions of soil, sewage, etc., and differential filtration of the supernatant fluid through a consecutive series of cellulose ester membrane filters of decreasing pore size ranging from 3 μm to 0·45 μm. The final filtrate was plated with the prospective host cells in semi-solid agar using the same double-layer technique as for bacteriophages (Adams, 1959). The method, although removing most other bacteria from the sample, involves a considerable loss of bdellovibrios as well. Therefore, though suitable for isolation of bdellovibrios, it is unreliable as a method for quantitative estimation of their number in natural environments (Stolp, 1973). A technique involving repeated centrifugation, used by Dias and Bhat (1965), also suffers from the same drawback, as shown by Klein and Casida (1967). Varon and Shilo (1970) have suggested differential centrifugation through a linear Ficoll gradient or single filtration through 1·2 μm pore-sized membranes. The latter method removes 90–95 per cent of the larger bacteria such as *E. coli* or *Pseudomonas* and allows nearly complete recovery of the bdellovibrios. A combination of the two techniques was recommended for samples rich in bacteria. Eremenko and Mardashev (1972) have recommended the use of phasic distribution in a system of dextran–polyethylene glycol in the presence of 0·3 M NaCl. In this system, about 97 per cent of the bdellovibrios are concentrated in the intermediary layer while 95 per cent of the *E. coli* cells accumulate in the upper phase.

For samples in which the relative number of bdellovibrios is high, the following two methods may be preferable. The first involves direct plating (Klein and Casida, 1967; Parker and Grove, 1970; Staples, 1973) sometimes following mechanical or chemical deflocculation of clumps (Staples, 1973). According to Staples, this technique yields the highest bdellovibrio counts. However, it presents difficulties when the relative number of bdellovibrios in the sample is low, since their plaques are likely to be covered by colonies of other microorganisms. Swarming bacteria, abundant in marine sample, may cover the bdellovibrio plaques even if there are numerous bdellovibrios present. The second technique is the dilution to extinction method (Guélin *et al.*, 1967; Lambina *et al.*, 1974). This method may lead to underestimates of their quantity since bdellovibrios may not grow out when in low numbers. The phenomenon of "population-dependent growth" has been clearly established for some bdellovibrio strains (Shilo and Bruff, 1965).

A variety of media have been used for bdellovibrio enumeration. Dilute media recommended for cultivation of isolated strains (Stolp and Petzold, 1962; Shilo and Bruff, 1965) have also been found most suitable for enumeration purposes (Parker and Grove, 1970; Williams *et al.*, 1977). Staples and Fry (1973) systematically compared the efficacy of various media for bdellovibrio enumeration and found a very dilute medium (NB/500) to be optimal. Guélin *et al.* (1967) used resting host cells suspended in distilled water containing salts (KCl, $CaCl_2$, $MgSO_4$ and NcCl) but no nutrients, thus avoiding growth of bacteria other than *Bdellovibrio.* Using this principle, Varon and Levisohn (1972) were able to isolate bdellophages against bdellovibrios growing in a 2-membered culture with *E. coli* without interference by coliphages.

A special difficulty encountered in the enumeration of *Bdellovibrio* from the marine environment stems from the use of media containing low salt concentrations (Daniel, 1969) or media supplemented with NaCl (Varon and Shilo, 1970; Enziger and Cooper, 1976) but lacking sufficient concentrations of other salts present in the marine environment. Marbach *et al.* (1976) showed that some Mediterranean bdellovibrio strains required K^+, Ca^{2+} and Mg^{2+} in addition to NaCl. Miyamoto and Kuroda (1975) found that even in artificial sea water (Herbst solution), the bdellovibrio yields were ten times lower than in natural sea water. This would indicate that additional and as yet unknown factors present in the sea water may be of importance for bdellovibrio development.

The concentration of agar-agar in the plates used for bdellovibrio enumeration was found to affect plaquing efficiency as well as plaque size (Guélin *et al.*, 1967; Huang, 1969; Staples and Fry, 1973) and time of plaque appearance (Marbach *et al.*, 1976). The optimal concentration of Difco agar was found to be 0·6 per cent for the upper layer and 1 per cent for the bottom layer (Varon, 1968).

An important factor in the enumeration of bdellovibrios from natural environments is the temperature of incubation. The optimal temperature as well as the range of temperature permitting activity and growth vary for bdellovibrios from the different ecosystems. Temperatures favourable for marine strains were found by Marbach *et al.* (1976) to be lower than those described for several soil, sewage and river bdellovibrios isolated from temperate regions (Seidler and Starr, 1969a; Staples and Fry, 1973). Some marine bdellovibrios showed a narrow temperature range, their plaquing efficiency being significantly higher at 15–20°C than at higher or lower temperatures and nil at 10°C. Other strains are even capable, albeit slowly, of forming plaques at 5°C, the yields being comparable to those at 15–30°C (Miyamoto and Kuroda, 1975).

Many different hosts have been used for bdellovibrio enumeration including species of Enterobacteriaceae (Stolp and Petzold, 1962; Dias and Bhat, 1965; Guélin *et al.*, 1967; Klein and Casida, 1967), *Pseudomonas* (Stolp and Starr, 1963), *Rhizobium* (Parker and Grove, 1970) and various marine bacteria (Taylor *et al.*, 1974; Miyamoto and Kuroda, 1975; Marbach *et al.*, 1976). The nature of the host has a major effect on the bdellovibrio count, since the plaquing efficiency of the bdellovibrios on various host organisms may vary. For example, Burger *et al.* (1968) found that the plaquing efficiency of *Bdellovibrio* W on *Proteus vulgaris* was 10 times higher than on *E. coli* or *Agrobacterium*. Staples (1973) found that plaque yields on *Achromobacter, E. coli* and *Proteus vulgaris* were four times higher than on *Spirillum* and twenty times higher than on *Serratia marcescens. Achromobacter* which enabled the most rapid and highest plaque formation was proposed as the host organism of choice for enumerating river and sewage bdellovibrios.

A special case in which the choice of the host may be critical is in environments such as estuaries or polluted ocean coastal waters in which true marine bdellovibrios and terrestrial, sewage or fresh water bdellovibrio types may co-exist. Since many of the marine bdellovibrios are preferentially active on marine host bacteria, the use of terrestrial host organisms such as *E. coli* (Paoletti, 1970; Ferro *et al.*, 1970; Campanile *et al.*, 1970; Bell and Latham, 1975; Enziger and Cooper, 1976) for estimating the number of bdellovibrios present in these ecosystems may result in numbers possibly lower than the actual. Miyamoto and

Kuroda (1975) found that *Bdellovibrio* which they had isolated from Osaka Bay did not form plaques on *E. coli*; it did form plaques on a number of marine vibrio strains, the plaquing efficiency being 2 to 3 times higher on *Vibrio parahaemolyticus* than on the other marine vibrio species. The use of an array of different hosts (as well as growth conditions) may, therefore, be essential for true and total enumeration of bdellovibrio. Corberi and Solaro (1971) in their isolation of bdellovibrios from 30 different soil loci concluded that only by the use of three different host organisms (*E. coli*, *S. marcescens* and *P. vulgaris*) could they ensure the isolation of the maximal number of bdellovibrio types.

Another problem in bdellovibrio enumeration is the differentiation of bdellovibrio plaques from those caused by lytic bacteria such as myxobacteria or bacteriophages. Myxobacterial plaques are often recognizable by their peculiar, crater-like morphology (Shilo, 1970; Miyamoto *et al.*, 1976), while bacteriophage plaques are differentiated on the basis of their early appearance and their cessation of growth after 24 h. This, however, may often be misleading since in the dilute media used for bdellovibrio enumeration, bacteriophage plaques may appear relatively late. The sensitivity of bdellovibrios to chloroform could be an additional useful criterion (Staples, 1973). Protozoa create another difficulty: their large and irregular plaques often cover sizeable areas of the plates used for bdellovibrio enumeration, making the count impossible. In order to overcome this problem Staples and Fry (1973) attempted the incorporation of anti-protozoal drugs such as cyclohexamide, nystatin, griseofluvin or metronidazole into the agar medium. However, these agents also reduced the number of bdellovibrio plaques. The least harmful was the trypanocidal agent, antricide (Staples, 1973).

5.2 DISTRIBUTION

Bdellovibrios are widely distributed in different aquatic ecosystems ranging from the tropics to the polar region. They have been isolated from the coastal water of seas and oceans: the Mediterranean (Guélin *et al.*, 1967, 1970; Shilo, 1969; Daniel, 1969; Paoletti, 1970; Ferro *et al.*, 1970; Campanile *et al.*, 1970; Marbach *et al.*, 1976); the Pacific Ocean near Hawaii (Taylor *et al.*, 1974) and Osaka Bay (Miyamoto and Kuroda, 1975); the Atlantic Ocean near Scotland (Bell and Latham, 1975) and the eastern coast of the United States (Shilo, 1966; Mitchell *et al.*, 1967; Williams *et al.*, 1977), and the Indian Ocean at the Gulf of Elat (Varon and Shilo, unpublished). Bdellovibrios have been found in a hypersaline solar lake (Hirsch, 1978) and were isolated from brackish and freshwater fish ponds (Varon and Shilo, unpublished), irrigation water (Uematsu and Wakimoto, 1970), and rivers and estuaries in Britain (Fry and Staples, 1974, 1976), France (Guélin *et al.*, 1967, 1970; Daniel, 1969; Finance, 1976), Italy (Ferro *et al.*, 1970), the USSR (Lambina *et al.*, 1974), the United States (Hendricks, 1974; Enziger and Cooper, 1976) and the Canadian Rockies (Robinson, personal communication).

Bdellovibrios are a consistent component of the microbial communities in sewage systems and oxidation ponds (Dias and Bhat, 1965; Guélin *et al.* 1967, 1970; Varon, 1968; Maré, 1972; Fry and Staples, 1976; Westergaard, 1977). Staples (1973) demonstrated the presence of *Bdellovibrio* in all aerobic stages of sewage treatment, including primary and secondary settling stages and the final effluent. Relatively high numbers of bdellovibrios were found in the effluents of the percolator filters, suggesting their multiplication within the filter's zoogleal film (Staples, 1973; Fry and Staples, 1976).

Table 2 summarizes our knowledge of the distribution of bdellovibrios in different aquatic ecosystems, and the methods used in the different investigations. Data for soil bdellovibrios are included for comparison.

Several investigators (Staples, 1973; Hendricks, 1974; Lambina *et al.*, 1974; Fry and Staples, 1976; Westergaard, 1977) studying the distribution of *Bdellovibrio* in rivers, found that the number of bdellovibrios correlated to the degree of pollution. Westergaard (1977) showed that bdellovibrios were absent from pure well waters. Lambina *et al.* (1974) found that the number of bdellovibrios increased from 10 cells ml^{-1} at the river's source to 3500 cells ml^{-1} at an urbanized section further along the river. Similarly, Finance (1976), analysing river waters at different seasons and locations, found a correlation between the bacteriolytic activity of the water and the number of coliform bacteria. Staples (1973) found that, whereas no bdellovibrios could be detected in pure rivers (true mountain stream water from various river head-waters), bdellovibrios were present in a polluted river (the Ely River) and their number was correlated to the distance from the sewage outfall site, to the *E. coli* count and to the total number of Gram-negative bacteria in the water. When the Ely River was in flood, the number of both bdellovibrios and potential host bacteria markedly increased due to overflows from the sewage works causing the discharge of untreated sewage into the river. The relation of bdellovibrio number in the river waters to the quality of the water is demonstrated in Fig. 5.

An analysis of the number of bdellovibrios in river sediments also showed correlation to the degree of pollution (Fry and Staples, 1976); unpolluted river sediments contained fewer bdellovibrios than sediment from polluted rivers, the cell number varying from 55 to $2 \cdot 9 \times 10^4$ g^{-1}. The correlation did not hold true for the deeper layers of the sediment in which bdellovibrios were low in number or absent in spite of high concentrations of heterotrophic bacteria. This could be explained by the strict aerobic nature of *Bdellovibrio*, and the facultative-aerobic nature of most of the heterotrophs.

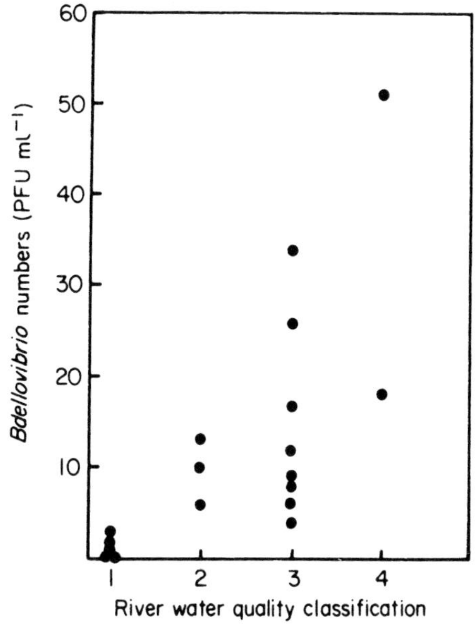

Fig. 5. Number of bdellovibrios in river water in relation to its quality. Bdellovibrio plaque-forming units (PFU) ml^{-1} in various polluted and unpolluted streams and rivers plotted against the water quality of the sampling sites: (1) unpolluted; (2) mildly polluted; (3) badly polluted; (4) grossly polluted. Water quality was classified according to the Report (1970) River Pollution Survey of England and Wales, vol. I, HMSO. (From Staples, 1973.)

TABLE 2

Distribution of *Bdellovibrio* in nature

Source	Isolation technique	Host	Bdellovibrio number l^{-1} or g^{-1}‡	Reference
Coastal sea water, Hawaii	Filtration, 1·2 μm	*B. natriegens*	$9{\cdot}7 \times 10^1$	Taylor *et al.* (1974)
Coastal sea water, Japan	—	*V. parahaemolyticus*	2×10^4 (December) – $1{\cdot}5 \times 10^6$ (July)	Miyamoto (personal communication)
Coastal sea water, France	Sequential filtration, 5·0–0·45 μm	Various Gram-negatives	$0 - 4 \times 10^3$	Daniel (1969)
Coastal sea water, Israel (Mediterranean)	Gradient centrifugation	*B. harveyi* W18	$4 \times 10^4 - 5 \times 10^4$	Shilo (1966)
Coastal sea water, Israel (Red Sea)	Filtration, 1·2 μm	*P. leiognathi*	$1 \times 10^3 - 2 \times 10^3$	Varon and Shilo (unpublished)
Coastal sea water, Israel (Red Sea)	Direct-plating	*P. leiognathi*	$3 \times 10^3 - 2 \times 10^4$	Varon and Shilo (unpublished)
Hypersaline Solar Lake (Gulf of Elat)	Direct microscopy on Henrici slides		1–5/field	Hirsch (1978)
Spring water, USSR	Dilution to extinction	*E. coli*	$10^4 - 3{\cdot}5 \times 10^6$	Lambina *et al.* (1974)
River water, Britain				
(a) unpolluted	Dilution to extinction	*E. coli*	$0 - 3 \times 10^3$	Fry and Staples (1976)
(b) grossly polluted	Dilution to extinction	*E. coli*	$1{\cdot}8 \times 10^4 - 5{\cdot}1 \times 10^4$	Fry and Staples (1976)
Polluted water, USSR	Dilution to extinction	*E. coli*	$0 - 3{\cdot}5 \times 10^5$	Lambina *et al.* (1974)
Polluted water, France	Sequential filtration, 5·0–0·45 μm	Various Gram-negatives	$0 - 6 \times 10^5$	Daniel (1969)

Polluted water, France	—	*Salmonella*	10^8	Guélin *et al.* (1967)
Sewage systems, Israel	Gradient centrifugation	*E. coli*	10^6	Shilo (1966)
Sewage systems, Britain	Filtration, 1·2 μm (+ pretreatment)	*E. coli*	$0 - 4{\cdot}8 \times 10^4$	Fry and staples (1976)
Sewage systems, Britain	Filtration, 1·2 μm (+ pretreatment)	*Achromobacter*	2×10^5	Fry and Staples (1976)
Sewage systems, USA	Stepwise filtration, 5–0·45 μm	*E. coli*	$8 \times 10^4 - 2 \times 10^6$	Stolp and Starr (1963)
Sewage systems, India	Repeated centrifugation	Various Gram-negatives	$0 - 8{\cdot}6 \times 10^5$	Dias and Bhat (1965)
Activated sludge, India	Repeated centrifugation	Various Gram-negatives	$0 - 2{\cdot}6 \times 10^5$	Dias and Bhat (1965)
Municipal waste water, USA	Sequential filtration, 5·0–0·45 μm	*E. coli*	$0 - 1{\cdot}2 \times 10^4$	Westergaard (1977)
Fish pond, Israel	Gradient centrifugation	*E. coli*	$4 \times 10^4 - 5 \times 10^4$	Shilo (1966)
Fish pond, Israel	Filtration, 1·2 μm	*E. coli*	6×10^3	Varon and Shilo (unpublished)
Fish pond, Israel	Direct plating	*E. coli*	3×10^4	Varon and Shilo (unpublished)
Ponds and streams, USA	Sequential/filtration, 5–0·45 μm	*E. coli*	$0 - 10^3$	Westergaard (1977)
Soil, Australia	Sequential filtration, 5–0·45 μm	*Rhizobium*	$2 - 1{\cdot}2 \times 10^3$	Parker and Grove (1970)
Soil, USA	Sequential filtration, 5–0·45 μm	Various Gram-negatives	$4 \times 10^1 - 2 \times 10^2$	Stolp and Starr (1963)
Soil, USA	Direct plating	*E. coli*	$<1 \times 10^3 - 9 \times 10^4$	Klein and casida (1967)

‡Zero signifies that no bdellovibrios were detected by the method used or in the volume tested.

5.3 INDIGENOUSNESS OF THE AQUATIC BDELLOVIBRIOS

The experimental findings described above raised the question whether the bdellovibrios found in aquatic ecosystems are autochthonous or are carried over from soil and sewage. The presence of autochthonous bdellovibrio populations in unpolluted waters has been regarded as doubtful also because the bacterial density considered necessary for their maintenance (see section 4) is rarely encountered in such ecosystems. The total number of bacteria in the non-polluted sea was reported to be in the range of 10^3 cells ml^{-1}, and in rivers and lakes in the range of 10^5–10^6 cells ml^{-1} (Rheinheimer, 1974). Even though the vast majority of the bacteria in the aquatic ecosystems are Gram-negative (see review by Caldwell, 1977), and therefore potential hosts for the bdellovibrios, their number is still lower than necessary for the maintenance of *Bdellovibrio*.

However, there is little doubt as to the autochthonous nature of the marine bdellovibrios: they clearly differ from the terrestrial, sewage and freshwater types in their absolute and specific requirement for Na^+, K^+, Ca^{2+} and Mg^{2+}, as well as in their DNA composition (Taylor *et al.*, 1974; Marbach *et al.*, 1976). There is also evidence that bdellovibrios may be indigenous to river waters: Robinson (personal communication) isolated bdellovibrios from non-polluted waters of the north Saskatchewan River in the Canadian Rockies only 2 miles distant from the river source (glaciers) although, as was to be expected, their number was considerably lower than in more urbanized sections of the river. Obviously, pollution of the waters by bacteria or by organic substances capable of supporting the growth of potential host bacteria will result in the multiplication of the autochthonous bdellovibrio population. Pollution could also introduce bdellovibrios originating from soil or sewage into rivers or seas, thereby increasing their total number. Analysis of bdellovibrio populations from estuaries and polluted coastal water indeed reveals the presence of non-marine bdellovibrios (salt-sensitive) in addition to the marine types (Daniel, 1969).

In explaining the existence of *Bdellovibrio* in aquatic environments two points must be considered: one, the actual number of bacteria in these ecosystems may be considerably higher than that reported; and two, the average numbers given may be irrelevant, since the bacteria are not randomly dispersed but are concentrated in special niches which allow high local densities.

Bacteria are usually enumerated by collection on a filter followed by fixation, staining and counting. However, the use of this procedure underestimates the size of bacterial populations since many bacteria are attached to particles, are clumped, or have an unusual morphology, making it difficult to distinguish them from detritus (Caldwell, 1977). Fluorescent stains, such as acridine orange, which distinguish between microorganisms and detritus, can be used to obtain a better count. However, even then the procedure tends to provide underestimates due to lysis of the bacteria during sample preparation (Straka and Stokes, 1957).

Another practice extensively used is plate counts. However, all plating media are selective, as are also the incubation conditions used. Therefore, the colonies appearing on the plates do not represent the total number of bacteria in the sample, nor do they necessarily represent a true qualitative picture. Recently, Akagi *et al.* (1977) using a special, low organic-carbon medium found that a sample of sea water contained 100-fold more oligotrophic bacteria than heterotrophic ones. Standard media would completely miss the former. Hoppe (1976) showed that the number of colony-formers in samples from the Bay of Kiel was 1·2 per cent of the number of metabolically active heterotrophic bacteria (as detected by autoradiography) and only 0·5 per cent of the total number of fluorescently stained bacteria. The proportion of colony-formers was even smaller (0·01 per cent) in a sample of water from the Pluss-See.

However, the key to the maintenance of *Bdellovibrio* populations in aquatic ecosystems could well be in the unequal and non-homogenous distribution of the inhabitant bacteria in time and space.

The time factor involved in the enrichment of such systems with potential host organisms has been noted by several investigators who showed that blooms of Gram-negative bacteria in lakes accompanied or followed those of cyanobacteria (Caldwell, 1977).

Most of the bacteria in lakes (Caldwell, 1977) and oceans (Sieburth, 1971, 1976) are attached to the surfaces of suspended matter, live as epiphytes on planktonic algae, or are concentrated on the surfaces of aquatic plants and animals. Many microorganisms have evolved mechanisms that favour growth in the attached state, thereby enabling them to take advantage of nutrients accumulating on the surface. Heukelekian and Heller (1940), for example, found that the total mass of *E. coli* incubated with glass beads was 50 times higher than in their absence and was equal to the cell mass of a culture grown without beads and containing a substrate concentration 100 times greater. Jannasch and Pritchard (1972) demonstrated that in batch and continuous culture, the addition of natural silt or particles consisting of chitin, clay or cellulose increased the utilization of substrate at low concentrations. The bacteria were shown to adhere to the particles when the concentration of substrates was low; no bacteria adhered in the absence of substrate or when it was present in high concentrations.

According to Fenchel and Jørgensen (1977) most microbial activity in the aquatic ecosystem is associated with the surfaces of particles. On the basis of primary production data and metabolic considerations, they estimated the total number of active bacteria in a temperate, coastal region to be 5×10^{12} cells m^{-2}, the majority of these apparently concentrated in the detritus-rich sediment layer. Potter (1963) and Keeny *et al.* (1971) also state that the overall number and activity of heterotrophic, benthic bacteria exceed those of planktonic bacteria by several orders of magnitude, and Focht and Verstraete (1977) explain the greater denitrification and nitrification activity of the sediments by this concentration of bacteria. Complex microbial films on stone surfaces (Marshall, 1976), in trickling filters in sewage systems (Staples, 1973) and on activated sludge flocs (Pike, 1975) are examples of periphytic bacterial concentrations.

High local concentrations of heterotrophic bacteria have been shown to accumulate in the thermocline between the epilimnion and hypolimnion of lakes and in oceans (Rheinheimer, 1974), as well as in the layer above the pycnocline or near the air–water interface (Sieburth *et al.*, 1976; Marshall, 1976). According to Sieburth *et al.* (1976), the surface microlayers of the sea can yield extremely high numbers of bacteria since this layer is enriched to the level of laboratory media.

It is, therefore, possible that although the average bacterial number may be too low to sustain a constant bdellovibrio population, *Bdellovibrio* may persist in restrictive niches, multiplying when and where there is an increased host population. An analogous situation exists in the localization of predaceous protozoa on microbial films in nature (Skerman, 1956).

5.4 THE ROLE OF *BDELLOVIBRIO* IN NATURE

Several investigators have suggested that bdellovibrios play a role in the determination of biological equilibria in various ecosystems. This conclusion was mainly based on the autopurification of polluted rivers and sea waters (Mitchell *et al.*, 1967; Mitchell and Morris, 1968; Guélin *et al.*, 1967, 1968a, 1969; Paoletti, 1970). The non-establishment of *E. coli* as a common soil organism was also explained in part by the widespread distribution of bdellovibrios in soil (Klein and Casida, 1967). However, Enziger and Cooper (1976) have shown that the removal of *E. coli* from estuarine waters was associated with an increase in the number of protozoa; when indigenous protozoa were removed by filtration, the destruction of the coliform population was negligible, while prevention of growth of the indigenous bacterial population did not affect coliform mortality. The bdellovibrios' role in autopurification of streams and rivers has also been doubted by Hendricks (1974) and Fry and Staples (1974).

Another approach for studying the problem was applied by Westergaard and Kramer (1977a). Using a small-scale, laboratory waste water treatment system they examined the ability of *Bdellovibrio* to lyse coliform bacteria in domestic waste water, and in waste water filtrates inoculated with *E. coli* and bdellovibrios. On the basis of their results, they concluded that *Bdellovibrio* is of only minor importance in the biological autopurification of water.

The seeming controversy in the conclusions drawn from all these observations is undoubtedly due to the fact that various factors, such as density of the host, its physiological state and its growth rate as well as environmental factors may all affect the *Bdellovibrio*–host interaction. Keya and Alexander (1975a), studying the *Bdellovibrio–Rhizobium* interaction in soil, suggested that only when the host is present in large concentrations will bdellovibrios multiply and evoke a significant decrease in the number of host cells.

Miyamoto and Kuroda (1975) stated that the physiological condition of the host and its growth rate may be of importance for the predator–prey balance. They based their conclusion on the fact that although bdellovibrios were prevalent in the sea water at Osaka Bay throughout the year, *V. parahaemolyticus* disappeared only in the winter. The authors postulated that bdellovibrios play a significant role only in winter, at low water temperatures at which *V. parahaemolyticus* does not proliferate rapidly. An indication that the physiological condition of the host may be of ecological importance was also found by Keya and Alexander (1975b) in their study of soil rhizobia attack by bdellovibros. Rhizobia that had been stored were more rapidly lysed than freshly harvested cells.

The above authors also showed that environmental factors affected the *Bdellovibrio*: host interaction; various colloids such as clay fractions; and soil organic matter markedly protected rhizobia from bdellovibrio attack. The mechanism of this protection is unclear. A similar phenomenon was described by Roper and Marshall (1974) who demonstrated that *E. coli* could be protected from bacteriophage attack by an envelope of colloidal material adsorbed to the cells.

In spite of the widespread occurrence of the bdellovibrios, and in spite of the large number of bacteria susceptible to them, our understanding of the role of *Bdellovibrio* in affecting population dynamics in nature is still fragmentary, sporadic and open to question. This is mainly due to the inherent difficulties of projecting conclusions from laboratory data on to conditions prevalent in nature, as well as to problems associated with collecting valid physiological field data. It may well be that the function of *Bdellovibrio* in nature is restricted to limited sets of conditions, and to specific time intervals and loci.

5.5 POTENTIAL USE OF BDELLOVIBRIOS FOR BIOLOGICAL CONTROL

The broad host range of *Bdellovibrio*, which encompasses most of the Gram-negative bacteria including many human and plant pathogens, their ability to lyse host organisms in stationary as well as in logarithmic growth phase, and the absence of host mutants resistant to attack, all render *Bdellovibrio* a powerful potential tool for the biological control of harmful bacteria (Shilo, 1966).

Protection of experimental animals against model infections was first described by Nakamura (1972) who found that treatment with bdellovibrios was effective against keratoconjunctivitis induced in rabbits by *Shigella flexneri*; bdellovibrios prevented the infection or markedly reduced the duration and severity of disease. They were also active in preventing the pathogenic symptoms caused by *S. flexneri* in the ligated, intestinal loops of rabbits; when inoculated jointly with bdellovibrios the number of shigellas in the loops was drastically reduced. However, experiments to forcefeed animals with bdellovibrios with the aim of obtaining large bdellovibrio population in the intestine, thereby reducing the number of pathogens, have been unsuccessful (Nakamura, 1972; Westergaard and Kramer, 1977b). This may be due to the inability of the bdellovibrios to grow in the intestine's anaerobic environment.

Experiments by Scherff (1973) showed that when the bdellovibrios were inoculated at a very high ratio together with *Pseudomonas glycinea* they prevented the development of local and systemic symptoms of bacterial blight evocable in soybean plants by the pseudomonads.

Venosa (1975) studied the possibility of controlling *Sphaerotilus natans*, which proliferates abundantly in flowing polluted streams, seriously impairing their recreational and economic utilization. Bdellovibrios were found, in this case, to be of limited use since they attacked only the swarmer stage of *Sphaerotilus*, a phase comprising only a small part of the latter's life-cycle.

The extensive damage caused by cyanobacterial blooms in different aquatic ecosystems led Burnham *et al.* (1976) to study the possible use of *Bdellovibrio* in their control. In laboratory experiments, bdellovibrios did kill and lyse cyanobacteria. However, they did not penetrate the periplasmic space of the cyanobacteria, and the activity observed must have been due to components excreted by the bdellovibrios into the medium. Therefore, the potential use of *Bdellovibrio* in control of cyanobacterial blooms is unlikely.

In summary, in spite of some encouraging preliminary results, it is impossible at this stage to predict whether and to what extent *Bdellovibrio* could be used for biological control.

6 *Bdellovibrio*-like organisms

Many bacteria lytic for other bacteria have been described in the literature. These include Myxobacteria (Stewart and Brown, 1969; Shilo, 1970; Miyamoto *et al.*, 1976), *Achromobacter* and *Pseudomonas* species (Postgate, 1967) as well as a whole group of predatory bacteria (*Dictyobacter, Cyclobacter, Trigonobacter, Streptobacter, Desmobacter* and *Teratobacter* observed *in situ* by the use of fine capillaries (Perfil'ev and Gabe, 1969).

In a study of the bactericidal power of sea water, Guélin and coworkers (Guélin and Cabioch, 1972; Guélin *et al.*, 1968b, 1976a, 1976b, 1977) found a group of "micropredators" enriched by the addition of a suspension of potential prey bacteria such as *E. coli, Salmonella* or *Clostridium.* They appeared to attack Gram-positive as well as Gram-negative bacteria. Many of them did not penetrate the wall of the prey but some were seen attached to the prey surface (Guélin and Maillet, 1976). Others grew and multiplied without any direct contact with the prey bacteria (Guélin *et al.*, 1977).

Hirsch (personal communication) has also observed the lysis of marine bacteria by extracellular bacterial parasites which adhere longitudinally to the outer surface of the prey and often form rosette-like structures (Fig. 6). On the basis of their morphology, he suggested the name "*Bdellomonas*" for these bacteria.

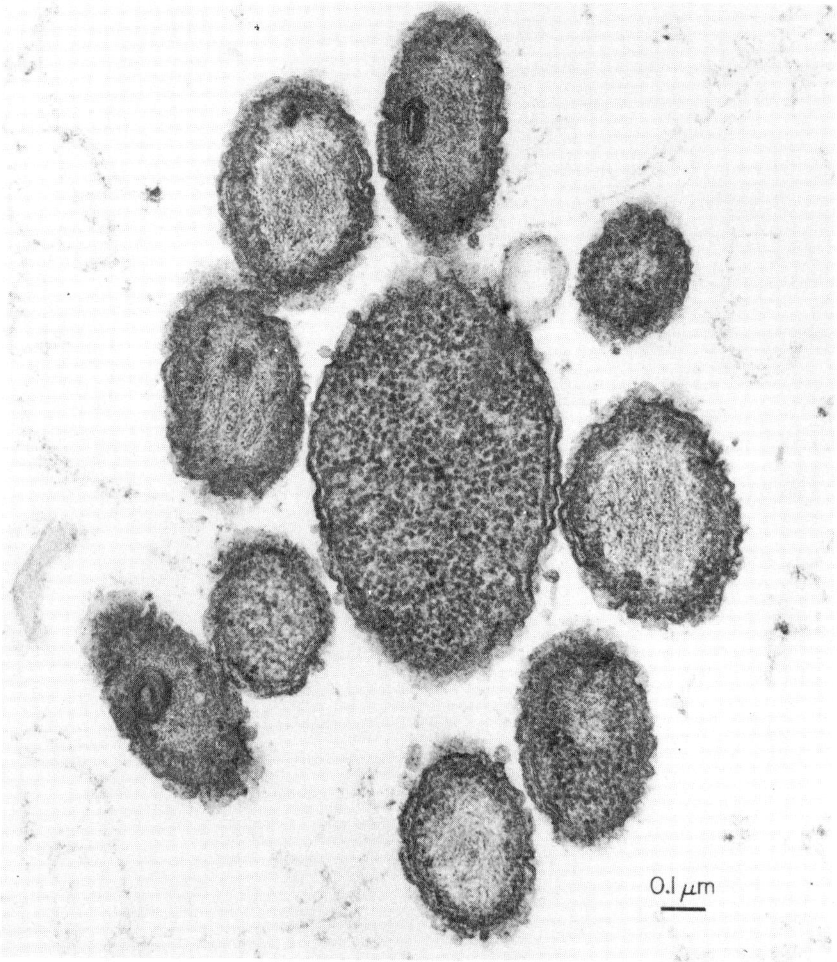

Fig. 6. "*Bdellomonas*" cells surrounding *Pseudomonas fluorescens* cell in a rosette-like formation. Some damage can be seen in areas of contact. (Micrography by P. Hirsch, Unpublished.)

Extracellular lysis has also been shown to occur in cultures of *Chlorella vulgaris* (Gromov and Mamkaeva, 1972). The lytic organism involved in this case, *Bdellovibrio chlorellavorus*, resembles *Bdellovibrio* but has an unsheathed flagellum. It is found in contact with *Chlorella* cells

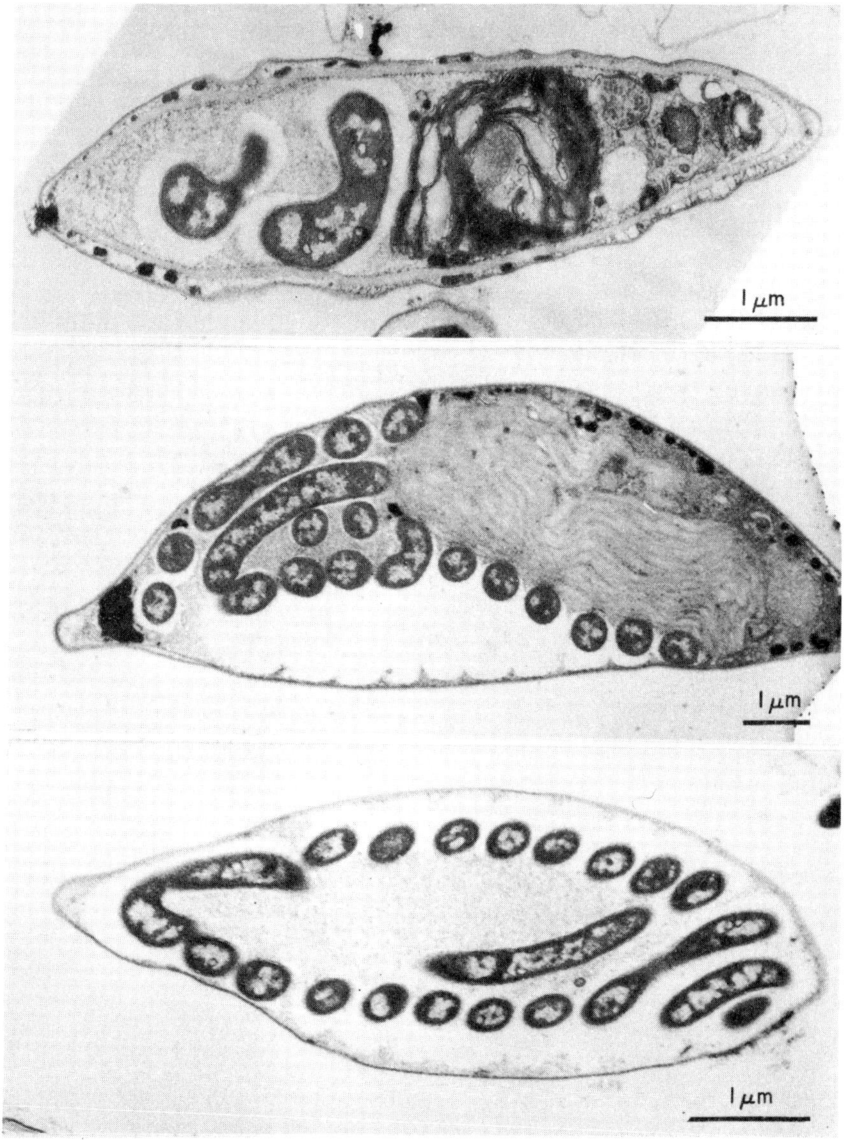

Fig. 7. Cells of *Scenedesmus acutus* infected with a spirillum-like bacterium (from Schnepf *et al.*, 1974).

but never enters the living cell. Other bacteria are, however, capable of penetrating the viable cells of green algae and proliferating there (Schnepf *et al.*, 1974). One of them, a spirillum, is morphologically similar to *Bdellovibrio* (Fig. 7).

All the *Bdellovibrio*-like organisms resemble *Bdellovibrio* in one general property or another. However, they are distinguishable from *Bdellovibrio* in their independence of other bacteria for growth. They do not share the bdellovibrio's ability for intraperiplasmic growth nor do they divide by multiple fission.

Acknowledgements

This research was supported, in part, by a grant from the United States–Israel Binational Science Foundation (BSF), Jerusalem, Israel, and the Deutsche Forschungsgemeindschaft. We are greatly obliged to S. Conti, R. B. Hespell, P. Hirsch, B. V. Gromov, A. Guerrini, S. Miyamoto, S. C. Rittenberg, J. Robinson and H. N. Williams who generously shared their findings with us prior to publication. We are grateful to S. Conti, P. Hirsch, E. Schnepf and D. G. Staples for allowing us to reproduce their figures and photographs. We appreciate very much the aid of M. Isaacs and A. Mahler in the preparation of this manuscript.

References

Abram, D. and Davis, B. K. (1970). Structural properties and features of parasitic *Bdellovibrio bacteriovorus*. *Journal of Bacteriology*, **104**, 948–965.

Abram, D., Castro e Melo, J. and Chou, D. (1974). Penetration of *Bdellovibrio bacteriovorus* into host cells. *Journal of Bacteriology*, **118**, 663–680.

Adams, M. H. (1959). "Bacteriophages". Interscience Publishers, Inc., New York.

Akagi, Y., Taga, N. and Simidu, V. (1977). Isolation and distribution of oligotrophic marine bacteria. *Canadian Journal of Microbiology*, **23**, 981–987.

Bauchop, T. and Elsden, S. R. (1960). The growth of microorganisms in relation to their energy supply. *Journal of General Microbiology*, **23**, 457–469.

Beijerinck, M. W. (1921). Die durch Bakterien aus Rohrzucker erzeugten schleimigen Wandstoffe. *In* "Verzamelde Geschriften" (Ed. M. W. Beijerinck), vol. 5, pp. 89–111, Delft.

Bell, R. G. and Latham, D. J. (1975). Influence of NaCl, Ca^{2+} and Mg^{2+} on the growth of a marine *Bdellovibrio* sp. *Estuarine and Coastal Marine Science*, **3**, 381–384.

"Bergey's Manual of Determinative Bacteriology" (1974). Eighth Edition (Eds R. E. Buchanan and N. E. Gibbons). Williams and Wilkins, Baltimore, Maryland.

Buckmire, F. L. A. (1971). A protective role for a cell wall protein layer of *Spirillum serpens* against infection by *Bdellovibrio bacteriovorus Bacteriological Proceedings*, G122, p. 43.

Buckmire, F. L. A. and Murray, R. G. E. (1970). Studies on the cell wall of *Spirillum serpens*. 1. Isolation and partial purification of the outermost cell wall layer. *Canadian Journal of Microbiology*, **16**, 1011–1022.

Burger, A., Drews, G. and Ladwig, R. (1968). Wirtskeries und Infektionscyclus eines neu isolierten *Bdellovibrio bacteriovorus*-Stammes. *Archiv für Mikrobiologie*, **61**, 261–279.

Burnham, J. C., Hashimoto, T. and Conti, S. F. (1968). Electron microscopic observations on the penetration of *Bdellovibrio bacteriovorus* into gram-negative bacterial hosts. *Journal of Bacteriology*, **96**, 1366–1381.

Burnham, J. C., Stetak, T. and Locher, G. (1976). Extracellular lysis of the blue green alga *Phormidium luridum* by *Bdellovibrio bacteriovorus*. *Journal of Phycology*, **12**, 306–313.

Caldwell, D. E. (1977). The planktonic microflora of lakes. *CRC Critical Reviews in Microbiology*, **5**, 305–370.

Campanile, E., Ferro, V., de Simone, E., Grasso, S. and de Fusco, R. (1970). Souches bacteriolytiques dans l'eau de mer. *Revue Internationale d'Océanographie Médicale*. **18–19**, 117–124.

Castro e Melo, J. (1975). Early stages of *Bdellovibrio*–host cell interaction. Ph.D. Thesis, University of Pittsburgh.

Chet, I. and Mitchell, R. (1976). Ecological aspects of microbial chemotactic behavior. *Annual Reviews of Microbiology*, **30**, 221–239.

Corberi, E. and Solaro, M. L. (1971). Richerche sulla presenza in diversi terreni coltivati di un microorganismo predatore de batteri: il *Bdellovibrio bacteriovorus*. *Annali di Microbiologia*, **21**, 123–133.

Corberi, E. and Solaro, M. L. (1972). *Bdellovibrio bacteriovorus* e diserbanti chimici selettivi. Società Italiana di Microbiologia. Atti del 16 Congresso Nazionale di Microbiologia. 423–428.

Crothers, S. F. and Robinson, J. (1971). Changes in the permeability of *Escherichia coli* during parasitization by *Bdellovibrio bacteriovorus*. *Canadian Journal of Microbiology*, **17**, 689–697.

Daniel, S. (1969). Etude de l'influence de *Bdellovibrio bacteriovorus* dans l'auto-épuration marine. *Revue Internationale d'Océanographie Médicale*. **15–16**, 61–102.

DeBary, A. (1879). "Die Erscheinung der Symbiose". Verlag von Karl J. Trüber, Strassburg.

Dias, F. F. and Bhat, J. V. (1965). Microbial ecology of activated sludge. II. Bacteriophages, Bdellovibrio, Coliforms, and other organisms. *Applied Microbiology*, **13**, 257–261.

Diedrich, D. L., Denny, C. F., Hashimoto, T. and Conti, S. F. (1970). Facultative parasitic strain of *Bdellovibrio bacteriovorus*. *Journal of Bacteriology*, **101**, 989–996.

Drucker, E. (1969). Leakage of intracellular components from *Escherichia coli* B infected with *Bdellovibrio bacteriovorus* 109. M.Sc. Thesis, Hebrew University, Jerusalem.

Drutz, D. J. (1976). Response of *Neisseria gonorrhoeae* species. *Infection and Immunity*, **13**, 247–251.

Dunn, J. E., Windom, G. E., Hansen, K. L. and Seidler, R. J. (1974). Isolation and characterization of temperature-sensitive mutants of host-dependent *Bdellovibrio bacteriovorus* 109D. *Journal of Bacteriology*, **117**, 1341–1349.

Eksztejn, M. and Varon, M. (1977). Elongation and cell division in *Bdellovibrio bacteriovorus*. *Archives of Microbiology*, **114**, 175–181.

Enzinger, R. M. and Cooper, R. C. (1976). Role of bacteria and protozoa in the removal of *Escherichia coli* from estuarine waters. *Applied and Environmental Microbiology*, **31**, 758–763.

Eremenko, V. V. and Mardashev, S. R. (1972). Method of concentration and separation of *Bdellovibrio bacteriovorus* from host cells. *Zhurnal Mikrobiologii, epidemiologii*, Issue N1, 126–130.

Fenchel, T. M. and Jørgensen, B. B. (1977). Detritus food chains of aquatic ecosystems: The role of bacteria. *In* "Advances in Microbial Ecology" (Ed. M. Alexander), vol. 1, pp. 1–58. Plenum Press, New York and London.

Ferro, V., de Simone, E., Campanile, E., de Fusco, R. and Grasso, S. (1970). Recherche du *Bdellovibrio bacteriovorus* dans la mer, les fleuves et les eaux d'égout. *Revue Internationale d'Océanographie Médicale,* **18–19**, 109–115.

Finance, C. (1976). Contribution a l'étude de la bactériolyse spontanée dans les rivières Moselle et Meurthe. Ph.D. Thesis, Université de Nancy, France.

Focht, D. D. and Verstraete, W. (1977). Biochemical ecology of nitrification and denitrification. *In* "Advances in Microbial Ecology" (Ed. M. Alexander), vol. 1, pp. 135–214. Plenum Press, New York and London.

Friedberg, D. (1977). Effect of light on *Bdellovibrio bacteriovorus. Journal of Bacteriology,* **131**, 399–404.

Fry, J. C. and Staples, D. G. (1974). The occurrence and role of *Bdellovibrio bacteriovorus* in a polluted river. *Water Research,* **8**, 1029–1035.

Fry, J. C. and Staples, D. G. (1976). Distribution of *Bdellovibrio bacteriovorus* in sewage works, river waters, and sediments. *Applied and Environmental Microbiology,* **31**, 469–474.

Gromov, B. V. and Mamkaeva, K. A. (1972). Electron microscope examination of *Bdellovibrio chlorellavorus* parasitism on cells of the green alga *Chlorella vulgaris. Tsitologiya,* **14**, 256–260.

Guélin, A. (1976). L'intensité du pouvoir bactericide des eaux marines et leur teneur en microvibrions. *Comptes Rendus de l'Académie des Sciences* (Paris), **282**, 397–400.

Guélin, A. and Cabioch, M. L. (1972). Bactériolyse spontanée et pouvoir bactéricide des eaux douces et marines; isolement d'un nouveau microprédateur. *Comptes Rendus de l'Académie des Sciences* (Paris), **274**, 3317–3319.

Guélin, A. and Cabioch, M. L. (1974). Charactères dynamiques de l'interaction entre le microprédateur *Bdellovibrio bacteriovorus* et la bactérie-hôte en fonction de leur densités initiales respectives. *Comptes Rendus de l'Académie des Sciences* (Paris), **278**, 1293–1296.

Guélin, A. and Maillet, P.-L. (1976). Observations sur l'ultrastructure de microprédateurs de Bacilles Gram-positif et la dégénerescence de *C. perfringens. Comptes Rendus de l'Académie des Sciences* (Paris), **283**, 1675–1678.

Guélin, A., Lépine, P. and Lamblin, D. (1967). Pouvoir bactéricide des eaux polluées et role de *Bdellovibrio bacteriovorus. Annales de l'Institut Pasteur,* **113**, 660–665.

Guélin, A., Lépine, P. and Lamblin, D. (1968a). Sur l'autoépuration des eaux. *Revue Internationale d'Océanographie Médicale*, **10**, 221–227.

Guélin, A., Lépine, P., Lamblin, D. and Sisman, J. (1968b). Isolement d'un parasite bactérien actif sur les germes Gram-positifs à partir d'echantillons d'eau polluée. *Comptes Rendus de l'Académie des Sciences* (Paris), **266**, 2508–2509.

Guélin, A., Lépine, P. and Lamblin, D. (1969). Microorganisms responsables du pouvoir bactéricide des eaux. *Verhand Internationale Vereinigung Limnologie*, **17**, 744–746.

Guélin, A., Bychovskaja, I., Lépine, P. and Lamblin, D. (1970). Distribution des germes parasites des bactéries pathogenes dans les eaux mondiales. *Revue Internationale d'Océanographié Medicale,* **18–19**.

Guélin, A., Michoustina, I. E., Goulevskaya, S. A., Petchnikov, N. V. and Ledova, L. A. (1977). Étude sur les microvibrions marins de Roscoff (*Microvibrio marinus roscoffensis). Comptes Rendus de l'Académie des Sciences* (Paris), **284**, 2171–2174.

Gunsalus, I. C. and Shuster, C. W. (1961). Energy-yielding metabolism in bacteria. *In*

"The Bacteria" (Eds I. C. Gunsalus and R. Y. Stanier), vol. 2, pp. 1–58. Academic Press, New York and London.

Hatch, T. P. (1976). Utilization of exogenous thymidine by *Chlamydia psittaci* growing in thymidine kinase-containing and thymidine kinase-deficient L cells. *Journal of Bacteriology*, **125**, 706–712.

Hendricks, C. W. (1974). *Bdellovibrio bacteriovorus—Escherichia coli* interactions in the continuous culture of river water. *Environmental Letters*, **7**, 311–319.

Hespell, R. B. and Mertens, M. (1978). Effects of nucleic acid compounds on viability and cell composition of *Bdellovibrio bacteriovorus* during starvation. *Archiv für Mikrobiologie*, **116**, 151–159.

Hespell, R. B., Rosson, R. A., Thomashow, M. F. and Rittenberg, S. C. (1973). Respiration of *Bdellovibrio bacteriovorus* strain 109J and its energy substrates for intraperiplasmic growth. *Journal of Bacteriology*, **113**, 1280–1288.

Hespell, R. B., Thomashow, M. F. and Rittenberg, S. C. (1974). Changes in cell composition and viability of *Bdellovibrio bacteriovorus* during starvation. *Archiv für Mikrobiologie*, **97**, 313–327.

Hespell, R. B., Miozzari, G. F. and Rittenberg, S. C. (1975). Ribonucleic acid destruction and synthesis during intraperiplasmic growth of *Bdellovibrio bacteriovorus*. *Journal of Bacteriology*, **123**, 481–491.

Heukelekian, H. and Heller, A. (1940). Relation between food concentration and surface for bacterial growth. *Journal of Bacteriology*, **40**, 547–558.

Hirsch, P. (1978). Microbial mats in a hypersaline Solar Lake: Types, composition and distribution. *In* "Environmental Biogeochemistry and Geomicrobiology" (Ed. W. E. Krumbein), vol. 1, pp. 189–201. Ann Arbor Science Publishers, Ann Arbor, Michigan.

Hoeniger, J. F. M., Ladwig, R. and Moor, H. (1972). The fine structure of "resting bodies" of *Bdellovibrio* sp. strain W developed in *Rhodospirillum rubrum*. *Canadian Journal of Microbiology*, **18**, 87–92.

Hoppe, H.-G. (1976). Determination and properties of actively metabolizing heterotrophic bacteria in the sea, investigated by means of microautoradiography. *Marine Biology*, **36**, 291–302.

Horowitz, A. T., Kessel, M. and Shilo, M. (1974). Growth cycle of predaceous bdellovibrios in a host-free extract system and some properties of the host extract. *Journal of Bacteriology*, **117**, 270–282.

Houston, K. J., Aldridge, K. E. and Magee, L. A. (1974). The effect of R antigen on the attachment of *Bdellovibrio bacteriovorus* to *Salmonella typhimurium*. *Acta Microbiologia Polonica*, **6**, 253–255.

Huang, J. C.-C. (1969). Parasitization of *Spirillum serpens* by *Bdellovibrio bacteriovorus*. Ph.D. Thesis, University of Western Ontario, London, Canada.

Jannasch, H. W. and Mateles, R. I. (1974). Experimental bacterial ecology studied in continuous culture. *Advances in Microbial Physiology*, **11**, 165–212.

Jannasch, H. W. and Pritchard, P. (1972). The role of inert particulate matter in the activity of aquatic microorganisms. Cited by Fenchel and Jørgensen, 1977.

Keeney, D. R., Herbert, R. A. and Hollding, A. J. (1971). Microbiological aspects of the pollution of fresh water with inorganic nutrients. *In* "Microbial Aspects of Pollution" (Eds G. Sykes and F. A. Skinner), pp. 181–200. Academic Press, London and New York.

Kessel, M. and Shilo, M. (1976). Relationship of *Bdellovibrio* elongation and fission to host cell size. *Journal of Bacteriology*, **128**, 477–480.

Keya, S. O. and Alexander, M. (1975a). Regulation of parasitism by host-density: The *Bdellovibrio-Rhizobium* interrelationship. *Soil Biology and Biochemistry*, **7**, 231–237.

Keya, S. O. and Alexander, M. (1975b). Factors affecting growth of *Bdellovibrio* on *Rhizobium*. *Archives of Microbiology*, **103**, 37–43.

Klein, D. A. and Casida, Jr., L. E. (1967). Occurrence and enumeration of *Bdellovibrio bacteriovorus* in soil capable of parasitizing *Escherichia coli* and indigenous soil bacteria. *Canadian Journal of Microbiology*, **13**, 1235–1241.

Kramer, T. T. and Westergaard, J. M. (1977). Antigenicity of *Bdellovibrio*. *Applied and Environmental Microbiology*, **33**, 967–970.

Kuenen, J. G. and Rittenberg, S. C. (1975). Incorporation of long-chain fatty acids of the substrate organism by *Bdellovibrio bacteriovorus* during intraperiplasmic growth. *Journal of Bacteriology*, **121**, 1145–1157.

LaMarre, A. G., Straley, S. C. and Conti, S. F. (1977). Chemotaxis toward amino acids by *Bdellovibrio bacteriovorus*. *Journal of Bacteriology*, **131**, 201–207.

Lambina, V. A., Chuvilskaya, N. A., Ledova, L. A., Afinogenova, A. V. and Averburg, I. V. (1974). Quantitative characteristics of distribution of *Bdellovibrio bacteriovorus* in river waters. *Microbiologia*, **42**, 715–720.

Lüderitz, O., Staub, A. M. and Westphal, O. (1966). Immunochemistry of O and R antigens of *Salmonella* and related *Enterobacteriaceae*. *Bacteriological Reviews*, **30**, 192–255.

Marbach, A. (1977). The isolation and characterization of marine bdellovibrios and the interaction with their host. Ph.D. Thesis, Hebrew University.

Marbach, A., Varon, M. and Shilo, M. (1976). Properties of marine bdellovibrios. *Microbial Ecology*, **2**, 284–295.

Maré, I. J. (1972). The isolation and characterization of a strain of *Bdellovibrio bacteriovorus*. *South Africa Medical Journal*, **46**, 68–71.

Marshall, K. C. (1976). "Interfaces in Microbial Ecology". Harvard University Press, Cambridge, Massachusetts, and London, England.

Marshall, K. C., Stout, R. and Mitchell, R. (1971). Mechanism of the initial events in the sorption of marine bacteria to surfaces. *Journal of General Microbiology*, **68**, 337–348.

Matin, A. and Rittenberg, S. C. (1972). Kinetics of deoxyribonucleic acid destruction and synthesis during growth of *Bdellovibrio bacteriovorus* strain 109D on *Pseudomonas putida* and *Escherichia coli*. *Journal of Bacteriology*, **111**, 664–673.

Mishustin, E. N. and Nikitina, E. S. (1973). Parasitic soil microorganisms. *Bulletin Ecological Research Committee* (Stockholm), **17**, 37–44.

Mitchell, R. and Morris, J. C. (1968). The fate of intestinal bacteria in the sea. Fourth International Conference on Water Pollution Research. 2–6 September, 1968. Pergamon Press, Oxford.

Mitchell, R., Yankofsky, S. and Jannasch, H. W. (1967). Lysis of *Escherichia coli* by marine microorganisms. *Nature, London*, **215**, 891–893.

Miyamoto, S. and Kuroda, K. (1975). Lethal effect of fresh sea water on *Vibrio parahaemolyticus* and isolation of *Bdellovibrio* parasitic against the organism. *Japanese Journal of Microbiology*, **19**, 309–317.

Miyamoto, S., Kuroda, K., Hanaoka, M. and Okada, Y. (1976). Isolation of a small rod with lytic activity against *Vibrio parahaemolyticus* from fresh sea water. *Japanese Journal of Microbiology*, **20**, 517–527.

Nakamura, M. (1972). Alteration of Shigella pathogenicity by other bacteria. *American Journal of Clinical Nutrition*, **25**, 1441–1451.

Novitsky, J. A. and Morita, R. Y. (1977). Survival of a psychrophilic marine vibrio under long-term nutrient starvation. *Applied and Environmental Microbiology*, **33**, 635–641

Paoletti, A. (1970). Facteurs biologiques d'autoépuration des eaux de mer: points clairs

et points obscurs d'une question discutée. *Revue Internationale d'Océanographic Médicale*, **18–19**, 33–68.
Parker, C. A. and Grove, P. L. (1970). *Bdellovibrio bacteriovorus* parasitizing *Rhizobium* in western Australia. *Journal of Applied Bacteriology*, **33**, 253–255.
Perfil'ev, B. V. and Gabe, D. R. (1969). "Capillary Methods of Investigating Micro-organisms". Oliver and Boyd, Edinburgh.
Pike, E. B. (1975). Aerobic bacteria. *In* "Ecological Aspects of Used-Water Treatment" (Eds C. R. Curds and H. A. Hawkes), vol. 1, pp. 1–63. Academic Press, London and New York.
Postgate, J. R. (1967). Soil bacteria parasitic on Azotobacteriaceae. *Antonie van Leeuwenhoek Journal of Microbiology and Serology*, **33**, 113–120.
Potter, L. F. (1963). Planktonic and benthic bacteria of lakes and ponds. *In* "Principles and Application in Aquatic Microbiology" (Eds H. Heukelekian and N. C. Dondero), pp. 148–166. John Wiley, New York.
Pritchard, M. A., Langley, D. and Rittenberg, S. C. (1975). Effects of methotrexate on intraperiplasmic and axenic growth of *Bdellovibrio bacteriovorus*. *Journal of Bacteriology*, **121**, 1131–1136.
Reiner, A. M. and Shilo, M. (1969). Host-independent growth of *Bdellovibrio bacterivorus* in microbial extracts. *Journal of General Microbiology*, **59**, 401–410.
Rheinheimer, G. (1974). "Aquatic Microbiology". John Wiley and Sons, London.
Rittenberg, S. C. (1972). Nonidentity of *Bdellovibrio bacteriovorus* strains 109D and 109J. *Journal of Bacteriology*, **109**, 432–433.
Rittenberg, S. C. and Hespell, R. B. (1975). Energy efficiency of intraperiplasmic growth of *Bdellovibrio bacteriovorus*. *Journal of Bacteriology*, **121**, 1158–1165.
Rittenberg, S. C. and Langley, D. (1975). Utilization of nucleoside monophosphates *per se* for intraperiplasmic growth of *Bdellovibrio bacteriovorus*. *Journal of Bacteriology*, **121**, 1137–1144.
Rittenberg, S. C. and Shilo, M. (1970). Early host damage in the infection cycle of *Bdellovibrio bacteriovorus*. *Journal of Bacteriology*, **102**, 149–160.
Roper, M. M. and Marshall, K. C. (1974). Modification of the interaction between *Escherichia coli* and bacteriophage in saline sediment. *Microbial Ecology*, **1**, 1–13.
Schelling, M. E., Anderson, C. S. and Conti, S. F. (1977). Serotyping of bdellovibrios by agglutination and indirect immunofluorescence. *American Society for Microbiology Annual Meeting* (Abstract).
Scherff, R. H. (1973). Control of bacterial blight of soybean by *Bdellovibrio bacteriovorus*. *Phytopathology*, **63**, 400–402.
Scherff, R. H., DeVay, J. E. and Carroll, T. W. (1966). Ultrastructure of host–parasite relationships involving reproduction of *Bdellovibrio bacteriovorus* in host bacteria. *Phytopathology*, **56**, 627–632.
Schnepf, E., Hegewald, E. and Soeder, C.-J. (1974). Electron microscopic observations on parasites of *Scenedesmus* mass culture. 4. Bacteria *Archiv für Microbiologie*, **98**, 133–145.
Seidler, R. J. and Starr, M. P. (1969a). Factors affecting the intracellular parasitic growth of *Bdellovibrio bacterivorus* developing within *Escherichia coli*. *Journal of Bacteriology*, **97**, 912–923.
Seidler, R. J. and Starr, M. P. (1969b). Isolation and characterization of host-independent *Bdellovibrios*. *Journal of Bacteriology*, **100**, 769–785.
Seidler, R. J., Mandel, M. and Baptist, J. N. (1972). Molecular heterogeneity of the bdellovibrios: Evidence of two new species. *Journal of Bacteriology*, **109**, 209–217.
Serban, A. and Levisohn, R. (1973). Isolation of bacteriophages for *Bdellovibrio starii:*

Implications about the host. *First International Congress for Bacteriology*, (Abstract), **2**, 142.

Shilo, M. (1966). Predatory bacteria. *Science Journal*, **2**, 33–37.

Shilo, M. (1969). Morphological and physiological aspects of the interaction of *Bdellovibrio* with host bacteria. *Current Topics in Microbiology and Immunology*, **50**, 174–204.

Shilo, M. (1970). Lysis of blue-green algae by myxobacter. *Journal of Bacteriology*, **104**, 453–461.

Shilo, M. (1973). Rapports entre *Bdellovibrio* et ses hôtes. Nature de la dépendance. *Bulletin de l'Institut Pasteur*, **71**, 21–31.

Shilo, M. (1974). *Bdellovibrio bacteriovorus* as a model for study of bacterial endoparasitism. *In* "Dynamic Aspects of Host–Parasite Relationships" (Eds A. Zuckerman and D. W. Weiss), vol. 1, pp. 1–12. Academic Press, New York and London.

Shilo, M. and Bruff, B. (1965). Lysis of gram-negative bacteria by host-independent ectoparasitic *Bdellovibrio bacteriovorus* isolates. *Journal of General Microbiology*, **40**, 317–328.

Sieburth, J. McN. (1971). Distribution and activity of oceanic bacteria. *Deep-Sea Research*, **18**, 1111–1121.

Sieburth, J. McN. (1976). Bacterial substrates and productivity in marine ecosystems. *Annual Reviews of Ecology and Systematics*, **7**, 259–285.

Sieburth, J. McN., Willis, P. J., Johnson, K. M., Burney, C. M., Lavoie, D. M., Hinga, K. R., Cavon, D. A., French, F. W., Johnson, P. W. and Davis, P. G. (1976). Dissolved organic matter and heterotrophic microneuston in the surface microlayers of the North Atlantic. *Science, New York*, **194**, 1415–1418.

Simpson, F. J. and Robinson, J. (1968). Some energy producing systems in *Bdellovibrio bacteriovorus*, strain 6-5-S. *Canadian Journal of Biochemistry*, **46**, 865–873.

Skerman, T. M. (1956). The nature and development of primary films on surfaces submerged in the sea. *New Zealand Journal of Science and Technology, B.*, **38**, 44–57.

Snellen, J. E. and Starr, M. P. (1974). Ultrastructural aspects of localized membrane damage in *Spirillum serpens* VHL early in its assocation with *Bdellovibrio bacteriovorus* 109D. *Archiv für Mikrobiologie*, **100**, 179–195.

Snellen, J. E. and Starr, M. P. (1976). Alterations in the cell wall of *Spirillum serpens* VHL early in its assocation with *Bdellovibrio bacteriovorus* 109D. *Archives of Microbiology*, **108**, 55–64.

Staples, D. G. (1973). The ecology and physiology of *Bdellovibrio bacteriovorus*. Ph.D. Thesis, University of Wales Institute of Science and Technology, Cardiff, Wales.

Staples, D. G. and Fry, J. C. (1973). Factors which influence the enumeration of *Bdellovibrio bacterivorus* in sewage and river water. *Journal of Applied Bacteriology*, **36**, 1–11.

Starr, M. P. (1975). *Bdellovibrio* as symbiont: The associations of bdellovibrios with other bacteria interpreted in terms of a generalized scheme for classifying organismic associations. *Symposium of the Society for Experimental Biology*, **29**, 93–124.

Starr, M. P. and Baigent, N. L. (1966). Parasitic interaction of *Bdellovibrio bacteriovorus* with other bacteria. *Journal of Bacteriology*, **91**, 2006–2017.

Starr, M. P. and Huang, J. C.-C. (1972). Physiology of the bdellovibrios. *In* "Advances in Microbial Physiology" (Eds A. R. Rose and D. W. Tempest), vol. 8, pp. 215–265. Academic Press, London and New York.

Starr, M. P. and Seidler, R. J. (1971). The bdellovibrios. *Annual Reviews of Microbiology*, **25**, 649–678.

Stewart, J. R. and Brown, R. M. (1969). Cytophaga that kills or lyses algae. *Science, New York*, **164**, 1523–1524.

Stolp, H. (1976a). Film E-1314. Institut für den wissenschaftlichen Film, Göttingen.

Stolp, H. (1967b). Film C-972. Institut für den wissenschaftlichen Film, Göttingen.
Stolp, H. (1973). The bdellovibrios: bacterial parasites of bacteria. *Annual Review of Phytophathology*, **11**, 53–76.
Stolp, H. and Petzold, H. (1962). Untersuchungen über einen obligat parasitischen Mikroorganismus mit lytischer Aktivitat für Pseudomonas Bakterien. *Phytopathogische Zeitschrift*, **45**, 364–390.
Stolp, H. and Starr, M. P. (1963). *Bdellovibrio bacteriovorus* gen. et sp. n., a predatory, ectoparasitic, and bacteriolytic microorganism. *Antonie van Leeuwenhoek Journal of Microbiology and Serology*, **29**, 217–248.
Stouthamer, A. H. (1973). A theoretical study on the amount of ATP required for synthesis of microbial cell material. *Antonie van Leeuwenhoek Journal of Microbiology and Serology*, **39**, 545–565.
Straka, R. P. and Stokes, J. L. (1957). Rapid destruction of bacteria in commonly used dilutents and its elimination. *Applied Microbiology*, **5**, 21–25.
Straley, S. C. and Conti, S. F. (1974). Chemotaxis in *Bdellovibrio bacteriovorus*. *Journal of Bacteriology*, **120**, 549–551.
Straley, D. C. and Conti, S. F. (1977). Chemotaxis by *Bdellovibrio bacteriovorus* toward prey. *Journal of Bacteriology*, **132**, 628–640.
Taylor, V. I., Baumann, P., Reichelt, J. L. and Allen R. D. (1974). Isolation, enumeration, and host range of marine bdellovibrios. *Archives of Microbiology*, **98**, 101–114.
Toppel, J., Hagilaadi, A. and Levisohn, R. (1973). An assay for the substance that supports host free growth of *Bdellovibrio bacteriovorus* using bacteriophages. *First International Congress for Bacteriology* (Abstract), **2**, 144.
Tudor, J. J. and Conti, S. F. (1977a). Characterization of bdellocysts of *Bdellovibrio* sp. *Journal of Bacteriology*, **131**, 314–322.
Tudor, J. J. and Conti, S. F. (1977b). Ultrastructural changes during encystment and germination of *Bdellovibrio*. *Journal of Bacteriology*, **131**, 323–330.
Tudor, J. J. and Conti, S. F. (1978). Characterization of germination and activition of *Bdellovibrio* bdellocysts. *Journal of Bacteriology*, **133**, 130–138.
Uematsu, T. and Wakimoto, S. (1970). Biological and ecological studies on *Bdellovibrio*. 1. Isolation , morphology, and parasitism of *Bdellovibrio*. *Annals of the Phytopathological Society of Japan*, **36**, 48–55.
Varon, M. (1968). Interaction of *Bdellovibrio bacteriovorus* and other bacteria. Ph.D. Thesis, The Hebrew University, Jerusalem, Israel.
Varon, M. and Levisohn, R. (1972). Three-membered parasitic system: a bacteriophage *Bdellovibrio bacteriovorus*, and *Escherichia coli*. *Journal of Virology*, **9**, 519–525.
Varon, M. and Seijffers, J. (1975). Symbiosis-independent and symbiosis-incompetent mutants of *Bdellovibrio bacteriovorus* 109J. *Journal of Bacteriology*, **124**, 1191–1197.
Varon, M. and Seijffers, J. (1977). Osmoregulation in symbiosis-independent mutants of *Bdellovibrio bacteriovorus*. *Applied and Environmental Microbiology*, **33**, 1207–1208.
Varon, M. and Shilo, M. (1968). Interaction of *Bdellovibrio bacteriovorus* and host bacteria. I. Kinetic studies of attachment and invasion of *Escherichia coli* B by *Bdellovibrio bacteriovorus*. *Journal of Bacteriology*, **95**, 744–753.
Varon, M. and Shilo, M. (1969a). Attachment of *Bdellovibrio bacteriovorus* to cell wall mutants of *Salmonella* spp. and *Escherichia coli*. *Journal of Bacteriology*, **97**, 977–979.
Varon, M. and Shilo, M. (1969b). Interaction of *Bdellovibrio bacteriovorus* and host bacteria. II. Intracellular growth and development of *Bdellovibrio bacteriovorus* 109 in liquid cultures. *Journal of Bacteriology*, **99**, 136–141.
Varon, M. and Shilo, M. (1970). Methods for separation of *Bdellovibrio* from mixed bacterial population by filtration through Millipore filters or by gradient differential centrifugation. *Revue Internationale d'Océanographie Médicale*, **18–19**, 145–152.

Varon, M. and Westra, H. (1976). Competition between symbiosis-dependent and symbiosis-independent bdellovibrios in continuous cultures. *Proceedings 7th Scientific Conference of the Israel Ecology Society.*

Varon, M. and Zeigler, B. P. (1978). Bacterial predator–prey interaction at low prey density. *Journal of Applied and Environmental Microbiology,* **36**, 11–17.

Varon, M., Drucker, I. and Shilo, M. (1969). Early effects of *Bdellovibrio* infection on the synthesis of protein and RNA of host bacteria. *Biochemical and Biophysical Research Communications,* **37**, 518–525.

Varon, M., Dickbuch, S. and Shilo, M. (1974). Isolation of host-dependent and nonparasitic mutants of the facultative parasitic *Bdellovibrio* UKi2. *Journal of Bacteriology,* **119**, 635–637.

Varon, M., Cocito, C. and Seijffers, J. (1976). Effect of virginiamycin on the growth cycle of *Bdellovibrio. Antimicrobial Agents and Chemotherapy,* **9**, 179–188.

Veldkamp, H. and Jannasch, H. W. (1972). Mixed culture studies with the chemostat. *Journal Applied Chemistry and Biotechnology,* **22**, 105–123.

Venosa, A. D. (1975). Lysis of *Sphaerotilus natans* swarm cells by *Bdellovibrio bacteriovorus. Applied Microbiology,* **29**, 702–705.

Wehr, N. B. and Klein, D. A. (1971). Herbicide effects on *Bdellovibrio bacteriovorus* parasitism of a soil *pseudomonad. Soil Biology and Biochemistry,* **3**, 143–149.

Westergaard, J. M. (1977). Biological and ecological studies on the interaction of *Bdellovibrio* and *Enterobacteriaceae.* Ph.D. Thesis, Auburn University.

Westergaard, J. M. and Kramer, T. T. (1977a). The role of *Bdellovibrio* in biologic purification of domestic wastewater. *American Society for Microbiology, Annual Meeting* (Abstract).

Westergaard, J. M. and Kramer, T. T. (1977b). *Bdellovibrio* and the intestinal flora of vertebrates. *Applied and Environmental Microbiology*, **34**, 506–511.

Westra, H. (1976). Some ecological aspects of *Bdellovibrio bacteriovorus* in continuous culture, M. Sc. Thesis, The University of Groningen, Groningen, The Netherlands.

Whitby, G. E. (1977). *Bdellovibrio bacteriovorus* 6-5-S and *Aquaspirillum serpens* VHL in continuous culture. Ph.D. Thesis, The University of Western Ontario.

Williams, H. N., Falkler, Jr., W. A. and Shay, D. E. (1977). Growth of marine *Bdellovibrio* in various media. *American Society for Microbiology Annual Meeting* (Abstract).

Winkler, H. H. (1976). Rickettsial permeability. An ADP-ATP transport system. *Journal of Biological Chemistry,* **251**, 389–396.

Yagil, E. and Beacham, I. R. (1975). Uptake of adenosine 5′-monophosphate by *Escherichia coli. Journal of Bacteriology,* **121**, 401–405.

Recent advances in the area of barotolerant protein synthesis in bacteria and implications concerning barotolerant and barophilic growth

J. V. LANDAU and D. H. POPE

Department of Biology, Rensselaer Polytechnic Institute, Troy, NY, USA

1 Introduction

The utilization of high hydrostatic pressure as an experimental tool has yielded data which have provided considerable insight into the mechanisms underlying a variety of life processes. Such phenomena as cell division (Marsland and Landau, 1954), ciliary movement (Pease and Kitching, 1939), maintenance of cellular form (Landau *et al.*, 1954), protein and nucleic acid synthesis (Landau, 1966), enzyme activity (Haight and Morita, 1962; Hockachka, 1971; Low and Somero, 1975a, b), bacterial luminescence (Brown *et al.*, 1942) and growth in microbial systems (ZoBell and Johnson, 1949; ZoBell and Oppenheimer, 1950) have been investigated. A reasonable coverage of all but the most recent studies may be found in the volumes edited by Zimmerman (1970), MacDonald (1975) and Brauer (1972).

The rationale for the use of this physical parameter as an experimental tool and the analysis of the resulting data on the basis of reaction rate theory has been superbly treated by Johnson *et al.* (1954) and we will not devote any major portion of this paper to such a discussion.

The scope of this presentation will be restricted to pressure experimentation in the area of the biosynthesis of protein. It is this area in which our essential expertise lies and to which we have devoted the bulk of our laboratory effort. Hopefully, we will be able to have the reader trace our experimental paths, follow and concur in our scientific logic, understand and accept our conclusions, and at least appreciate and share in our enthusiasm and optimism concerning the future of such experimentation.

Over the years there has been a great deal written concerning the environmental conditions of high-pressure, low-temperature, high-salinity and low-nutrient concentration experienced by organisms that are known to inhabit the ocean depths. In most cases, the inaccessability of the organisms, the lack of proper collection and transfer methods, and the inability to maintain *in situ* conditions, have been cited as major limitations in dealing with these organisms. A limited interpretation of a limited number of experiments, along with a carefully delineated extrapolation of the data, has often been presented. We feel this to be a valid approach provided there is strict distinction between what is known to be and what is thought to be. There are, also, unfortunately, examples where through broad or inaccurate citation and re-citation, tentative hypothesis has become substantiated theory, limited data from inadequately designed experiments have reappeared as solid and substantial experimental evidence, and extrapolations have been presented as hard fact. We would hope to at least partially rectify the situation.

2 Definition of terms

As will soon be made clear, the application of hydrostatic pressure results, in general, in an inhibition of the rate of protein synthesis in both *in vivo* (whole cell) and *in vitro* (cell-free extract) preparations. Some bacteria show a stimulation of protein synthesis at specific temperatures upon application of pressures up to 400 atmospheres (atm). Although it has been suggested that this "lower" pressure stimulation is the result of at least two oppositely affected reactions, no solid experimental analysis has yet been made. We have concerned ourselves primarily with the region in which protein synthesis is inhibited by pressure and particularly with the level of synthesis that can be achieved at 680 atm. *Escherichia coli* K-12 (streptomycin sensitive) has been utilized as our standard. At 680 atm, protein synthesis in preparations of this organism is totally inhibited and the inhibition of the rate of synthesis at pressures between 400 and 600 atm is fairly steep (at 15°C and 37°C). The inhibition corresponds to the effect upon a reaction having a volume increase of activation (ΔV^*) of 100 $cm^3 mol^{-1}$. We have designated this type of response to pressure as *barosensitive* and the process, therefore, as barosensitive protein synthesis.

Pseudomonas fluorescens, on the other hand, continues to synthesize protein at about 35 per cent and 20 per cent of its 1 atm rate at 680 and 800 atm, respectively. Also, the inhibition of the rate of protein synthesis corresponds to the effect on a reaction undergoing a ΔV^* of about 50 $cm^3 mol^{-1}$, only half that of *E. coli.* This type of response we have designated, in relative terms, as *barotolerant* and we describe such synthesis as barotolerant protein synthesis.

It must be made emphatically clear that the ability of an organism to synthesize protein, however slowly, under conditions of high pressure does not classify that organism as barotolerant with respect to growth. Certainly, barotolerant protein synthesis is a necessary requirement for growth under environmental conditions of high pressure but for such growth many other life processes would also have to evidence barotolerance. Therefore, while barotolerant protein synthesis would be a requirement for barotolerant growth, the fact of such synthesis does not, in itself, indicate the existence of barotolerant growth.

3 Technical aspects

The instrumentation utilized in our high-pressure studies is relatively simple and has been described in detail by Landau and Thibodeau (1962) Schwarz and Landau (1972b) and Shen and Berger (1974). Caution must be taken that all air space is eliminated from the experimental vessel, temperature must be critically controlled, and organic buffers used to prevent all but negligible shifts in pH during pressure treatment.

Of course, the biochemical analyses utilized in our experiments are not carried out under conditions of high hydrostatic pressure. The development of techniques which either allowed for "fixation" of the reaction within the chamber just prior to pressure release or which permitted an absolute minimum time period for sampling after the release of pressure was of major importance. The chamber of Landau and Thibodeau (1962) is suitable for the larger volumes used in experiments with whole cell preparations and that of Schwarz and Landau

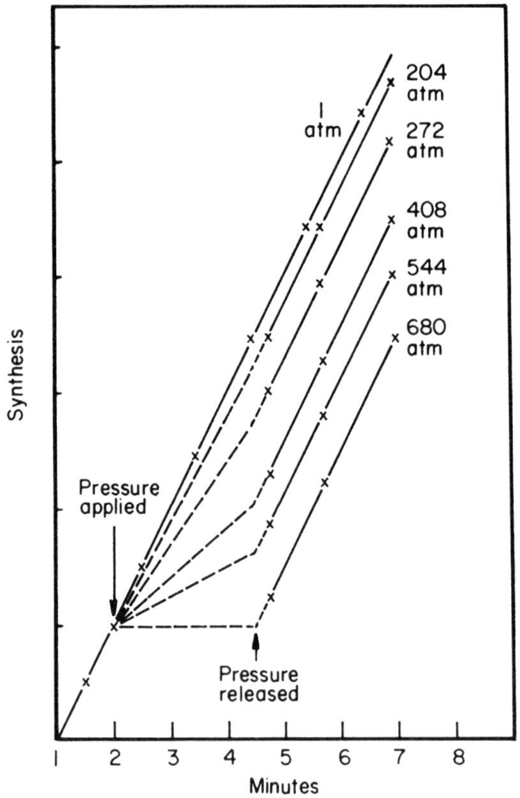

Fig. 1a. A representation of the data obtained from a typical series of pressure experiments.

(1972b) for the smaller volumes of cell-free extract preparations. In both systems specified additions or "fixations" can be performed under pressure conditions. The most convenient apparatus is that developed by Shen and Berger (1974). Although pressure must be released for sampling, the time period at atmospheric pressure between release and fixation for analysis can be held to about 8 seconds. However, when assaying the effect of pressure on a relatively rapid process such as protein synthesis, even this short duration of time at 1 atm must be accounted for in analysing the data. A typical plot of the data obtained is shown in Fig. 1a. Note the importance of the extrapolation to the time of pressure release. The lack of consideration of this time interval could obviously lead to a quantitative error in determining the effect of pressure on synthesis. Data in support of such extrapolation are obtained by control experiments involving "fixation" at the time of pressure release. Protein synthesis is generally measured by the incorporation of labelled amino acids into the acid precipitable fraction. Figure 1b demonstrates how the data of Fig. 1a can be used to characterize the inhibition of the rate of synthesis as a function of pressure.

At the onset of this experimentation it became evident that the effect of pressure was immediate and that upon release of pressure the original or atmospheric rate of synthesis was restored. Considering our primary objective to be the determination of the direct effect of pressure on the rate of protein synthesis, we applied the following conditions to our experimental methods: (1) a rate of synthesis at atmospheric pressure

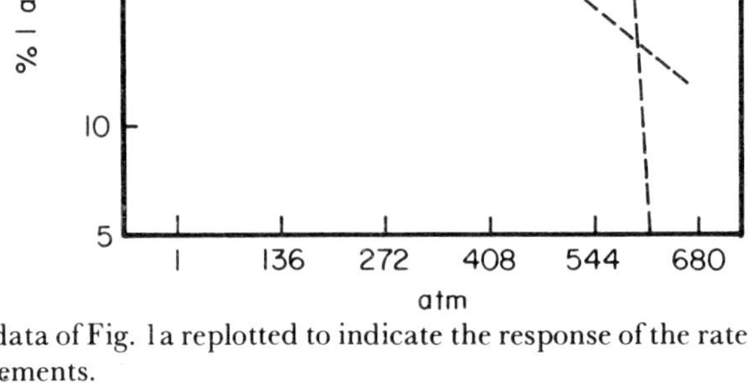

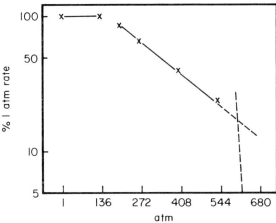

Fig. 1b. The data of Fig. 1a replotted to indicate the response of the rate of synthesis to pressure increments.

C

must be determined; (2) pressure, in most cases, was to be applied after the start of the reaction; (3) the rate of synthesis at atmospheric pressure must be linear throughout the course of the experiment; and (4) upon release of pressure the atmospheric rate of synthesis must resume promptly.

4 Effects of Pressure on protein synthesis

4.1 *ESCHERICHIA COLI*

The early work of Landau (1966), in which *E. coli* whole cells were utilized, established the fact that high pressure inhibited protein synthesis to a greater extent than ribonucleic acid (RNA) synthesis. Further, the extent to which the rate of protein synthesis was inhibited could be related to the level of pressure, and such inhibition was quickly reversed upon decompression. The question then arose as to whether the amino acids were incorporated at high pressure into functional protein or into non-functional polypeptide. Therefore, studies were undertaken to determine the effect of pressure on the synthesis of β-galactosidase (Landau, 1967). At about the same time Pollard and Weller (1966) reported on similar experiments. The results obtained by both groups were substantially the same, although slightly different techniques were used. The rate of synthesis of a functional protein was found to be inhibited by pressure in a precise quantitative manner and at 680 atm no synthesis was detectable. Release of pressure resulted in an immediate return to the atmospheric rate. More recently, Marquis and Keller (1975) reported that β-galactosidase synthesis continued at a reduced but persistent rate over an extended period of time at 680 atm. An analysis of the experimental technique, however, would tend to negate their interpretation of the data. It would seem that pressure was applied for a given period of time and released, the pressure chamber dismantled and a sample of the cell suspension taken, the chamber reassembled and pressure reapplied. This procedure was repeated several times during the experiment. Considering the type of pressure chamber used, each sampling must have resulted in several minutes' exposure to atmospheric pressure. Under such conditions, the curve drawn by those authors and interpreted by them as indicating continuing β-galactosidase synthesis really indicates, we feel, a stepwise increase in β-galactosidase during the periods of time at which the sample was at 1 atm. Our analysis of their methods and data indicates that it is unlikely that any β-galactosidase synthesis occurred during the time at pressure.

The results of our experiments on β-galactosidase were readily interpretable on the basis of a primary effect of pressure on the translation phase of protein synthesis. The data also indicated the possibility that the effect could be due to the inhibition of a single rate-limiting process. Could we identify more specifically this postulated pressure-sensitive rate-limiting reaction by utilizing the techniques of molecular biology?

The first step was to determine whether amino acid permeability, formation of amino-acyl transfer RNAs or polysome integrity were affected at 680 atm sufficiently to cause an immediate cessation of protein synthesis. Schwarz and Landau (1972a) showed that whole cells of *E. coli* exposed to 680 atm accumulated labelled amino acids, formed amino-acyl-tRNA, and retained a polysomal profile similar to that of the atmospheric pressure controls. Each of these phenomena were relatively unaffected over short time periods although protein synthesis was totally inhibited. Arnold and Albright (1971) had meanwhile reported a pressure inhibition of aa-tRNA binding to ribosomes in an isolated, non-synthesizing system. The accumulated evidence allowed us to present a tentative hypothesis indicating that the rate-limiting reaction of protein synthesis primarily affected by pressure involved the binding of aa-tRNA to the ribosome and/or the translocation step of translation (also see Hildebrand and Pollard, 1972). We then proceeded to test this hypothesis through the use of cell-free synthesizing systems (Schwarz and Landau, 1972a, b).

The inhibition by pressure of protein synthesis in cell-free systems, whether directed by a synthetic messenger such as polyuridylic acid (poly-U) or natural messenger (MS-2 viral mRNA), was identical to that observed in the intact cell. Furthermore, it was shown that peptide bond formation *per se* was not prevented at 680 atm. Additional evidence was accumulated which indicated that the binding of aa-tRNA to the ribosome was affected in a manner quantitatively identical to that of overall protein synthesis. Hardon and Albright (1974) subsequently reported that all the reactions of translation were inhibited by pressure and indicated that no single reaction could possibly be delineated as rate limiting. Their experiments involved relatively extended time periods during which amino acid transport and aa-tRNA integrity seemed to be affected. It is difficult to understand how aa-tRNA integrity could be the primary factor limiting protein synthesis at 680 atm, when at this pressure protein synthesis was immediately inhibited but the total amount of aa-tRNA breakdown, even after several minutes, was only a minor fraction of the amount available. Similarly, how can the inhibition of amino acid transport through the membrane be a major factor in the inhibition of protein synthesis if whole cell and cell-free systems are inhibited identically by pressure? All of our experimental evidence indicated that the primary and immediate effect of pressure involved an inhibition of some ribosome function. Such an inhibition of a ribosomal function could readily cause a subsequent reversal or disruption of a series of interrelated reactions during extended periods at pressure.

Of particular interest was the indication by Arnold and Albright (1971) that pressure disrupted polysome integrity but in a completely reversible manner, i.e. ribosomes bound to poly-U would be released under pressure and then would rebind immediately upon restoration to atmospheric pressure. Since it was our contention that at pressures totally inhibiting protein synthesis the polysome configuration was held as such without ribosome run-off, it was apparent that further investigation was necessary. The key to our experimental design was supplied by the report of Pato *et al.* (1973). They showed, quantitatively, that mRNA could be protected from degradation in the intact cell as long as the polysomal configuration was maintained, i.e. naked messenger would degrade rapidly while that which had ribosomes bound to it would be less prone to RNAase activity. We reasoned, therefore, that if pressure caused a release of ribosomes the polysome mRNA, now stripped of ribosomes, would rapidly be degraded, provided that RNAase could function at high pressure. On the other hand, if 680 atm caused a retention of the polysome configuration by "freezing" the ribosomes in place on the mRNA, any degradation of the mRNA should be effectively retarded during the period at pressure. Pope *et al.* (1975a) showed conclusively that RNAase activity *per se* was unaffected by pressure and that the polysome mRNA was protected against degradation during the period of pressure application.

Several workers have investigated the effect of hydrostatic pressure generated during centrifugation on the ribosomes of cell-free systems (Subramanian and Davis, 1971; Infante *et al.*, 1971a, b, c; Nieuwenhuysen *et al.*, 1975). In general, they report a change in the polysome-ribosome-ribosomal subunit equilibrium at high pressure. However, where aa-tRNA-mRNA-ribosome complexes were already formed and functional, no structural disruption was noted. It must be noted that Arnold and Albright (1971) did not utilize aa-tRNA in their experiments and that their data probably represented the effect of pressure on non-specific, non-functional binding of ribosomes or ribosomal subunits to poly-U.

4.2 *PSEUDOMONAS BATHYCETES* AND *P. FLUORESCENS*

Having established an experimental base for the investigation of the effect of pressure on protein synthesis with *E. coli* preparations, we were ready to proceed with other bacteria. The question was posed as to whether a deep-sea isolate would react in the same manner as *E. coli*.

P. bathycetes, an isolate from the sediment of the Marianas Trench, was graciously supplied to us by Dr R. Colwell. Experiments on intact cells showed a distinct resistance to pressure (Swartz *et al.*, 1974). The inhibition of protein synthesis by pressure was decidedly less than that seen for *E. coli* and a significant rate of synthesis could be maintained at 1000 atm. The organism was classified as being capable of barotolerant protein synthesis. This barotolerant feature was lost, however, when a cell-free synthesizing system was utilized. The response of the cell-free system was identical to that of *E. coli*, i.e. barosensitive. As will be discussed later, this loss of barotolerance resulted from the absence of specific ions at particularly high concentrations in the reaction mixtures.

P. fluorescens, a "terrestrial" pseudomonad, was then selected for comparison with *P. bathycetes*. In this case, both intact cells and cell-free systems showed highly barotolerant protein synthesis. These results were unexpected but happily accepted since they now afforded us with the opportunity to interchange the various components of two cell-free synthesizing systems, one barotolerant and the other barosensitive, and thus attempt a delineation of the pressure-sensitive (or tolerant) component.

4.3 HYBRID PROTEIN-SYNTHESIZING SYSTEMS

The experiments of Pope *et al.* (1975b), and Smith *et al.* (1975) clearly showed that the original source of the supernatant factors (tRNA, soluble factors G and T, termination factors, amino-acyl-tRNA synthetases, and other factors) was not a critical element affecting the nature of the pressure response. Similarly, it was later shown by Landau *et al.* (1977) that the source of the initiation factors was also not critical. The component governing the relative sensitivity to pressure was first found to be the ribosome and then further delineated as the 30S ribosomal subunit. The 50S subunit from either *E. coli* or *P. fluorescens* appeared to be functionally equivalent and seemingly played no role in modifying the pressure effect.

Figure 2 shows the results of a typical subunit exchange experiment. The inhibition of protein synthesis by pressure is identical for whole cell systems, homologous cell-free systems directed by synthetic or natural

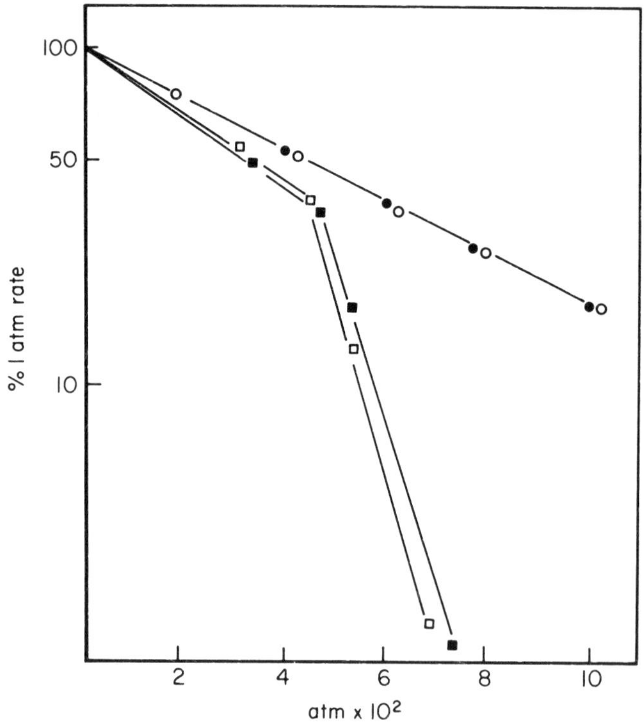

Fig. 2. Effect of hydrostatic pressure on the rate of (^{14}C) phenylalanine incorporation into polyphenylalanine by polyuridylate-directed cell extract systems utilizing reaggregated ribosomes at 25°C. Symbols: ○, *P. fluorescens* S-100 plus *P. fluorescens* 30S and 50S ribosomal subunits; □, *E. coli* S-100 plus *E. coli* 30S and 50S ribosomal subunits; ●, *P. fluorescens or E. coli* S-100 plus *P. fluorescens* 30S ribosomal subunit and *E. coli* 50S subunit; ■, *P. fluorescens* or *E. coli* S-100 plus *E. coli* 30S ribosomal subunit and *P. fluorescens* 50S subunit. (From Smith *et al.*, 1975.)

messenger, and hybrid cell-free systems similarly directed, if the 30S ribosomal subunit is of the same cellular origin. Here then is definitive evidence that the 30S ribosomal subunit is a component of the translation process which exhibits a specific activation volume increase. Logically, one may consider that the binding of aa-tRNA and/or translocation involves a conformational change of the 30S subunit with a concomitant increase in volume. The nature or inherent structure of the subunit would then dictate the magnitude of the volume increase.

4.4 SPECIFIC ION REQUIREMENTS

The loss of barotolerance in cell-free systems of *P. bathycetes* posed a very interesting problem. We felt that the best approach was to attempt to modify the pressure response through modification of ionic conditions. Several earlier reports influenced this approach. Palmer and Albright (1970) had reported that increased extracellular Na^+ concentration extended the maximum pressure at which *Vibrio marinus* could grow. Low and Somero (1975a) demonstrated the importance of maintaining ionic conditions close to that *in situ* for valid determinations of the effects of pressure on enzyme reactions. Takacs *et al.* (1964) had provided essential data on the internal ion concentrations of a marine pseudomonad.

Specific concentrations of Mg^{2+} and Na^+ were found to be critical for barotolerant synthesis in *P. bathycetes* cell-extract systems. Figure 3 indicates the effect of pressure on synthesis in *P. bathycetes*, *P. fluorescens* and *E. coli* systems as a result of modification of specific ion concentrations in the reaction mixtures. Hybrid systems, using all the possible combinations of supernatant fluids and ribosomes from the three organisms, were then tested for barotolerance under conditions of low (16 mM Mg^{2+} plus 0 mM Na^+) and high (60 mM Mg^{2+} plus 150 mM Na^+) ion concentrations. The results are given in Table 1 and reveal that the ion-regulated barotolerance is a characteristic determined by the ribosome. Certain concentrations of specific ions will increase barotolerance in those cases where the ribosomes are derived from barotolerant bacteria. The same ions will not alter the response of barosensitive ribosomes. The supernatant fluid of *E. coli* is obviously incapable of supporting synthesis at high ion concentrations but the ribosomes remain fully functional in the presence of supernatant fluid obtained from either of the pseudomonads.

Following our previous pattern of experimentation, the next step was to interchange ribosomal subunits. However, isolation of the *P. bathycetes* ribosomal subunits had not yet been attempted. We found that specific ions play a critical role during such isolation (Landau *et al.*, 1977). *P. bathycetes* 30S subunits isolated in a buffer solution containing 0 mM NaCl and 0 mM KCl were non-functional at any pressure, those isolated in the presence of 150 mM NaCl and 0 mM KCl were functional at 1 atm but barosensitive, and those isolated in the presence of 0 mM NaCl and 150 mM KCl retained the ion-mediated barotolerance characteristic of *P. bathycetes* protein synthesis. The 50S subunit remained functional regardless of the method of isolation and had no effect on pressure sensitivity. The data are shown in Table 2 and clearly indicate, through a variety of recombinations, that the component controlling relative sensitivity of protein synthesis to pressure is the 30S subunit.

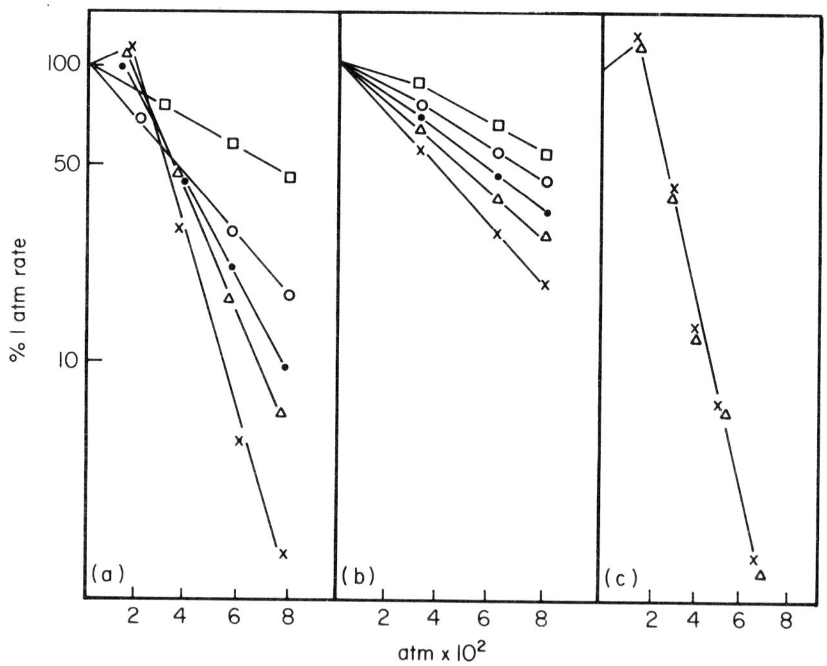

Fig. 3. Effect of hydrostatic pressure on the rate of phenylalanine incorporation into polyphenylalanine by poly U-directed cell-extract preparations of (a) *P. bathycetes*, (b) *P. fluorescens*, and (c) *E. coli* at various ion concentrations. (X) 16 mM Mg^{2+}, 0 mM K^+, 0 mM Na^+; (Δ) 16 mM Mg^{2+}, 0 mM Na^+, 50 mM K^+; (●) 60 mM Mg^{2+}; (○) 16 mM Mg^{2+}, 150 mM Na^+; (□) 60 mM Mg^{2+}, 150 mM Na^+. (From Smith *et al.*, 1976.)

TABLE 1

Effect of specific ion concentrations on the relative rates of polyphenylalanine synthesis by homologous and hybrid cell extract systems at 1 and 680 atm (from Smith *et al.*, 1976)

Source of ribosomes	Source of S-100	1 atm, low[b]	1 atm, high[c]	680 atm, low	680 atm, high
Escherichia coli	*E. coli*	100 ± 8	0	6 ± 3[a]	0
	P. bathycetes	100 ± 8	65 ± 6	6 ± 3	6 ± 3
	P. fluorescens	100 ± 8	65 ± 6	6 ± 3	6 ± 3
Pseudomonas bathycetes	*E. coli*	100 ± 8	0	12 ± 8	0
	P. bathycetes	100 ± 8	90 ± 6	12 ± 6	65 ± 6
	P. fluorescens	100 ± 8	90 ± 6	12 ± 6	65 ± 8
Pseudomonas fluorescens	*E. coli*	100 ± 8	0	35 ± 6	0
	P. bathycetes	100 ± 8	80 ± 6	35 ± 5	65 ± 5
	P. fluorescens	100 ± 8	80 ± 6	35 ± 5	65 ± 5

[a]Relative rate is given as the percentage of the 1 atm rate at low ion concentration.
[b]Low ion concentration is 16 mM Mg^{2+} and 0 mM Na^+.
[c]High ion concentration is 60 mM Mg^{2+} and 150 mM Na^+.

TABLE 2

Effect of specific ion concentrations on the relative rates of polyphenylalanine incorporation by systems utilizing homologous and hybrid reconstituted ribosomes at 1 and 680 atm and 25°C (from Landau *et al.*, 1977)

Source of 30S ribosomal subunit	Source of 50S ribosomal subunit	Rate at 1 atm and low ion concentration[a] dpm min^{-1} mg^{-1} ribosomes (± 6%)	Relative rate[c] at 680 atm and low ion concentration	Relative rate at 680 atm and high ion concentration[b]
P. bathycetes I[d]	*E. coli* I	0	—	—
P. bathycetes I	*P. fluorescens* I	0	—	—
P. bathycetes I	*P. bathycetes* I, II, III	0	—	—
E. coli I	*P. bathycetes* I, II, III	230	6%	6%
P. bathycetes II	*E. coli* I	220	11%	13%
P. bathycetes II	*P. fluorescens* I	230	11%	13%
P. bathycetes II	*P. bathycetes* I, II, III	220	11%	13%
P. fluorescens I	*P. bathycetes* I, II, III	280	36%	65%
P. bathycetes III	*E. coli* I	300	12%	66%
P. bathycetes III	*P. fluorescens* I	240	11%	67%
P. bathycetes III	*P. bathycetes* I, II, III	270	12%	65%

[a]Low ion concentration is 16 mM Mg^{2+}.
[b]High ion concentration is 60 mM Mg^{2+}, 150 mM Na^{+}.
[c]Relative rates expressed as percentage of the 1 atm rate at low ion concentration.
[d]I, isolation in 0 mM NaCl and 0 mM KCl; II, isolation in 150 mM NaCl and 0 mM KCl; III, isolation in 0 mM NaCl and 150 mM KCl.

The evidence is now rather substantial that protein synthesis involves an increase in activation volume, the magnitude of which may be modified by specific ions or temperature (Pope *et al.*, 1975b), and that the capability to undergo this volume change is inherent within the structural or conformational state of the specific 30S subunit.

4.5 OTHER BACTERIA

Was it possible that barotolerant protein synthesis might be a characteristic of organisms able to grow at low temperature or of pseudomonads as a group? A number of organisms representative of several taxonomic and physiological groups were tested for the ability to synthesize protein at high pressure (Pope *et al.*, 1976). The resultant data are shown in Table 3. The ability to synthesize protein at 680 atm does not seem to be related to environmental origin, physiologic type or taxonomic group.

Is there any possible indication of a common factor that would allow for barotolerant synthesis? Five of the eleven bacterial types tested show such synthesis. In *P. fluorescens* and *P. bathycetes* barotolerant synthesis has been related to the 30S ribosomal subunit. *E. coli* K-12 S^R (streptomycin-resistant mutant) contains an altered 30S ribosomal subunit (Wittman and Wittman-Liebold, 1974) and *Bacillus stearothermophilus* is known to have a functionally different S-12 ribosomal protein from that of *E. coli* K-12 S^S (Nomura and Held, 1974). No comparable information is available for *Halobacterium salinarium*, although it is known that the ribosome composition of some halophiles is different from that of *E. coli* (Bayley, 1966). In very broad terms, we can still postulate that the key to barotolerant synthesis in a wide range of bacterial types lies within the inherent structure of the ribosome. The barosensitive nature of the isolates from the gut of deep-sea fish is perplexing from an ecological viewpoint on gross analysis, but considering how little we know of the host organism, of the bacterium itself, and of their interrelationship, it cannot be thought of as a critical bit of information at this time.

5 The barotolerant bacterium: growth at high pressure

Our discussion thus far has been primarily concerned with the effects of pressure on only one facet of growth, i.e. protein synthesis. However, in

TABLE 3

Effect of 680 atm pressure on protein synthesis by various bacteria (from Pope *et al.*, 1976)

Organism	Growth at 5°C	Source	Date of isolation[a]	Temperature at which bacteria were grown and tested (°C)	(Rate of protein synthesis at 680 atm)/rate of protein synthesis at 1 atm) × 100 (%)
E. coli K-12S^S	−	Faeces	?	37	0
E. coli K-12S^R	−	Mutant strain of above S^S strain	?	37	35
P. fluorescens	+	Prefilter tanks	1959	25	35
P. bathycetes	+	Deep-sea sediments	1968	25	35
P. aeruginosa	−	Faeces	?	37	0
Pseudomonas 412 (Szer)	+	Peat	?	25	0
V. fischeri PS-207	+	Pacific cod	1964	25	0
B. stearothermophilus	−	Thermal pool	1971	62	60
H. salinarium	−	Salted fish	1966	37	80
Deep-sea isolate (Berger, no. 3)[b]	+	Deep-sea fish	1973	25	0
Deep-sea isolate (Berger, no. 6)[b]	+	Deep-sea fish	1973	25	0

[a]Dates are approximate. ?, Date of original isolation is unknown.

[b]Isolated from intestinal contents of a fish captured at a depth of 1 mile off the coast of the island of Hawaii.

order for an organism to grow at high pressure, no matter how slowly, an entire series of cellular processes must be at least minimally functional. Within this series we would place (1) transcription and translation, (2) replication of the genome, (3) membrane transport, (4) energy conversions, (5) enzyme activity, and (6) phenomena of assembly.

The search for bacteria capable of barotolerant growth has, with some logic, centred on the deep-sea environment. For the purposes of this discussion, however, we will, in a limited sense, define a barotolerant bacterium as one which will divide and grow under conditions of high pressure. The combined effects of the other environmental conditions of the deep sea will temporarily be put aside.

P. bathycetes is an example of a barotolerant bacterium that has been recovered from the deep sea. We also have preliminary data indicating that some thermophiles and halophiles are barotolerant. *P. fluorescens* exhibits barotolerant protein synthesis but we do not as yet know whether it is a barotolerant organism. *E. coli* is not a barotolerant organism and the same can probably be said concerning many other bacteria. What do we actually know about the effect of pressure on each of the previously listed life processes? It is possible that in assaying the barotolerance (or barosensitivity) of an organism, certain of these processes are of major importance while others are relatively inconsequential.

The remainder of this section will be devoted to a discussion of the known effects of pressure on each of the listed processes. We will attempt to point out those areas in which hard data exist, and those areas where there is little available but inference and assumption. Paths for further investigation will be suggested.

5.1 PROTEIN SYNTHESIS

The effect of pressure on protein synthesis has already been described in detail. This process has been perhaps more intensively investigated than most of the others listed. There are, however, many significant areas for further investigation: (1) it should be possible to disassemble the components of the 30S ribosomal subunits and reconstitute hybrid subunits for test of barotolerant capability. Such experiments would help delineate the role of specific subunit components in barotolerant protein synthesis; (2) possible conformational differences between barotolerant and barosensitive subunits should be studied; (3) the overall experimental approach should be broadened to include a wider range of bacterial types; (4) further study is needed on the role of temperature and specific ion concentration in modifying the pressure response of the protein synthesizing apparatus; (5) the stimulation of the rate of synthesis in some bacteria at the lower pressure range should be investigated, and (6) the effect of pressure on transcription must be investigated in greater detail.

5.2 REPLICATION

To date there has been very little work measuring the direct effect of pressure on DNA replication. The major data available are for the barosensitive organism, *E. coli* (Pollard and Weller, 1966; Yayanos and Pollard, 1969; and Yayanos, 1975). These data show that DNA synthesis (thymidine incorporation) continues at increased pressure for a short period of time and then ceases. The investigators have interpreted their data as an indication that the activity of the DNA polymerases is affected by pressure to a lesser extent than is protein synthesis and they further suggest that the process of replication is probably blocked by high pressure at its initiation or termination steps. It seems possible, however, that the inhibition of protein synthesis (and, therefore, DNA polymerase synthesis) by pressure might be a major factor contributing to the failure to initiate or terminate rounds of DNA replication. An experimental approach is needed in which each of the various phases of DNA synthesis may be analysed in terms of direct pressure effects and a comparison of such effects on the replication process of barotolerant and barosensitive organisms is essential.

5.3 MEMBRANE TRANSPORT

There has been little direct investigation on the effect of pressure on the process of membrane transport. Most experimental design has been limited to overall uptake studies or relatively extended time periods with subsequent inference concerning the transport phenomenon (for example see Baross *et al.*, 1974). It seems obvious that a primary inhibition of protein synthesis, RNA synthesis, enzyme activity, or energy conversion would eventually manifest itself as an alteration in the magnitude or rate of total uptake. Such uptake measurements can only assist in an indirect or inferential evaluation of any direct pressure effect on transport.

A more valid approach to the problem has been taken by Shen and Berger (1974). A non-metabolizable glucose analogue was utilized and its transport during short periods of time under pressure was studied. Pressure apparently has little or no effect on the rate of influx but the rate of efflux is increased, resulting in a reduction of the steady-state level in the cell. Unfortunately, there is no evidence indicating whether the reduction in steady-state level has a proportionate bearing on a possible inhibition of growth.

Although we have shown that some amino acids can enter the cell at pressures which inhibit protein synthesis and that the presence or absence of a membrane does not alter the effect of pressure on protein synthesis, our methods have not involved direct measurements of the kinetics of transport. There is sufficient methodology available at the present time to allow for a comparative study involving pressure effects on transport kinetics in barotolerant and barosensitive species along with an analysis of membrane components.

5.4 ENERGY CONVERSION

The effect of pressure on lactate production in *Streptococcus faecalis* (Marquis and ZoBell, 1971) and algal photosynthesis (Pope and Berger, 1973a) has been investigated. A comparison of the data on respiration and fermentation with that on growth has been made by Pope and Berger (1973b) and they conclude that, for the several prokaryotic organisms that have been used by various workers, the inhibition of protein synthesis rather than that of energy conversion processes seemed to be limiting to growth. Ehrlich (1974) reported that Mn^{2+} oxidation and MnO_2 reduction by manganese nodule bacteria can occur at pressures and temperatures extant in deep-sea sediments. Jannasch *et al.* (1976) measured the rates at which deep-sea and surface water bacterial populations respired and incorporated various substrates into cellular material. Their results clearly demonstrated that, under deep-sea conditions, respiration is affected much less than incorporation in both types of populations. Kriss *et al.* (1967) studied a marine pseudomonad that was capable of growing at 550 atm. At this pressure its growth rate was significantly reduced and filament formation was observed. However, the amount of glucose utilized per cell at 550 atm was more than twice that utilized at 1 atm. Further, the primary product of the metabolism of glucose at 1 atm was CO_2, while at 500 atm the amount of CO_2 produced was reduced and large amounts of organic acids were produced. They concluded that at high pressures glucose was incompletely metabolized by the cell and the higher rate of glucose utilization reflected an attempt to compensate for an energy deficit. It is obvious that further experimentation can and should be done comparing barosensitive and barotolerant organisms. The ability of bacterial cytochromes to function and the levels of adenosine nucleotides within the cell should be measured as a function of pressure.

5.5 ENZYME ACTIVITY

There have been several reports regarding the effects of pressure on the activity of bacterial enzymes (Morita, 1957; Haight and Morita, 1962; Berger, 1958; Mohankumar and Berger, 1972; also see review by ZoBell and Kim, 1972). Few of these experiments have involved conditions in which the actual time course of the reaction could be followed under pressure. Frequently, only the rates of activity following pressure release were measured. In many cases the effect of pressure on the enzyme activity has not been measured over a range of substrate concentrations. The data, for the most part, do not allow for any firm conclusion regarding the possibility that such inhibition of enzyme activity by pressure is limiting to growth. The exception is the rather thorough study by Mohankumar and Berger (1972) (see also Berger, 1974) on the activity of a malate dehydrogenase of a marine bacterium. They used an optical pressure cell, which allowed them to follow the course of the enzyme reaction under pressure. It was concluded that the apparent ΔV^* of the reaction seemed to be dependent on the concentration of malate, that the affinity of the enzyme for malate increased as the temperature decreased and similarly increased as the pressure increased. The overall indication was that the rate decreased under pressure because of the enhanced affinity of the enzyme for malate.

Rather extensive studies by Hochachka and co-workers (Hochachka, 1971) and Low and Somero (1975a, b), using enzymes from fish and mammals, have demonstrated that the effects of pressure on enzyme activity may be quite different at different combinations of pressure, temperature, and ionic conditions. The data available on these enzymes have led to the conclusion that there are at least two major roles that pressure and temperature may play in relation to enzyme activity; (1) the rate at which equilibrium of the reaction is attained may be affected, and (2) the tertiary or quaternary structure of the enzyme may be altered, therefore modifying its ability to bind substrates and effector molecules. The latter effects presumably involve non-covalent interactions. Baldwin *et al.* (1975) have concluded that the primary problems faced by deep-sea organisms are concerned with the maintenance of high potential reaction rates in an environment of low thermal energy and high pressure and the retention of weak-bond structural interaction. Excellent experimental work has been done in this area and such work should be extended to microbial systems of terrestrial, shallow water and deep-sea origin.

5.6 ASSEMBLY PHENOMENA

The effect of pressure on the phenomena of assembly is an important area in which data on microbial systems is extremely sparse. Lauffer (1962) has discussed the polymerization of tobacco mosaic virus protein in terms of pressure–volume effects. In higher forms, the effect of pressure on the structure and assembly of the mitotic apparatus (Zimmerman and Marsland, 1964) and on the structure of microtubules in Heliozoa (Tilney *et al.*, 1966) has been investigated. Kettman *et al.* (1966) investigated the aggregation of poly-L-valyl ribonuclease as a function of pressure. In all cases, high hydrostatic pressure drastically modifies normal aggregative or assembly processes.

Berger (1959) reported that *E. coli* exhibited filamentous growth at relatively low pressures and found that cross-wall precursor molecules accumulated in the culture medium. Several workers (see previous discussion on ribosomes) have reported on disruptions of ribosome-ribosome subunit equilibria.

Penniston (1971) and Hochachka (1971) have suggested that the aggregation of multimeric proteins is inhibited by pressure such that the equilibrium is shifted toward the monomeric state. Hochachka (1975) has provided a detailed discussion, emphasizing that pressure seems to distort quarternary structure but possibly not secondary or even tertiary structure, since the pressures causing dissociation of the polymeric proteins are at least an order of magnitude less than those required to denature the protein. Unfortunately there has been little, if any, comparable investigation on proteins of bacterial origin and certainly none in which the effects of pressure have been studied utilizing methods which would afford a comparison between those obtained from barosensitive and barotolerant strains.

6 The barophilic bacterium and the ocean environment

In accordance with our previous discussion concerning barotolerant bacteria, we will designate a barophilic bacterium as one which either requires a high-pressure environment for growth or exhibits a greater growth or functional capacity at high pressure than at atmospheric pressure. The search for such organisms has centred on the deep-sea sediments. Under these conditions, of course, the added environmental parameters of low temperature, high salinity and questionable nutrient availability become major problems relative to experimental methods and design.

There have been few reports indicating the existence of barophiles and only a few such obligate barophiles have been found (see ZoBell, 1968). Seki *et al.* (1974) have reported that the amount of bacterial growth in one of two samples collected and incubated *in situ* at 5200 m exceeded that of a 1 atm control sample. Unfortunately no isolates from this material were maintained. In addition, there appears to be some doubt that the collection device functioned properly (see Jannasch *et al.*, 1976). It is apparent that most attempts to recover barophilic organisms from the deep seas have in the past been hampered by instrumentation that could not maintain *in situ* temperatures and pressures and afforded no possibility of transfer without return to atmospheric pressure. Significant progress has recently been made by at least two groups (Colwell *et al.*, personal communication; Jannasch *et al.*, 1976) in achieving instrument design which will allow for the proper collection, transfer and isolation of deep-sea bacteria. The use of such devices should, in the near future, allow for the direct determination of the existence of barophilic bacteria.

Those *in situ* and laboratory studies that have been done with the use of techniques which avoid decompression (see Jannasch *et al.*, 1976) have indicated that bacteria of either deep-sea origin or surface water origin respond similarly to pressure. Deep-sea samples placed at 1 atm yielded metabolic rates comparable to those of surface origin and the metabolic rates of surface samples exposed to deep-sea pressures paralleled those of undecompressed deep-sea samples. In no instance was the metabolic rate enhanced by pressure. These results seem to indicate that there are significant numbers of barotolerant bacteria present in populations from quite diverse origins, but that barophilic bacteria in these same populations may be quite rare.

Schwarz *et al.* (1976) found little difference at 1 or 750 atm in either growth or respiration rates of bacterial samples obtained from the gut contents of amphipods trapped at a depth of 7000 m. Isolates obtained from these cultures, however, did not behave in the same manner. When incubated at 700 atm, the isolates showed an approximate 50 per cent reduction in activity. It may well be that a large portion of the microbial activity in the deep sea occurs in association with higher forms of life. It is possible that when in association with the host the bacteria are capable of achieving rates of metabolism and growth far greater than those which would be achieved outside the host. We know practically nothing concerning the feeding habits or digestive processes of such host animals and any details of host–bacterium interrelationship remain, at this time, in the realm of pure speculation.

Morita (1976) in a recent review has discussed salinity and temperature effects. His speculation concerning the latter would seem to merit further comment. He has presented data on the uptake of glutamic acid by the psychrophile Ant-300 at a variety of different pressure and temperature levels. The data indicated a relationship between pressure and temperature on the basis of their combined effects on uptake at pressures up to 800 atm and temperatures between 3·8 and 15·4°C. Using the calculated relationship, an extrapolation was made to 2·5°C and 1000 atm and this was determined as the equivalent of −28·8°C at 1 atm. Clearly, if such extrapolation is valid, bacteria would be metabolically "frozen" in the deeper portions of the ocean, as he states. Morita quotes Baross *et al.* (1974) in concluding that the primary physical parameter affecting microorganisms in the deep sea is temperature, since metabolically "frozen" bacteria can hardly be further affected by pressure. Such speculation, while intellectually stimulating, should be subjected to test. Strict adherence to Ideal Gas Law over a complete range of temperature and pressure is not generally the case in biological systems. It is undeniable that a reaction must be proceeding in order for pressure to modify its rate. However, it is difficult for us to extend this to a consideration that a life process that is reversibly inactivated by temperature can not be further affected by the application of high pressure. Configurational states of component molecules, at one point or another, may suddenly undergo irreversible changes. Is there sufficient convincing definitive evidence available to state that psychrophilic bacteria as a group display a greater sensitivity to pressure than mesophilic types at low temperature if both reversible and irreversible effects are considered? We think not. Again, while Morita's speculation is stimulating, we wish to point out that it should not be interpreted as actual fact.

Jannasch *et al.* (1976) suggested that, in consideration of standard pressure–temperature relationships, the minimum temperature for growth and incorporation of substrate by an organism might be raised somewhat at increased pressure. In a test of this hypothesis, Wirsen and Jannasch (1975) measured the activities of psychrophilic and mesophilic marine isolates at $-1{\cdot}5$ and 3°C and at pressures up to 400 atm. Their results indicated that the psychrophilic isolates were affected less by pressure, at low temperature, than the mesophilic isolates. They also concluded that their original hypothesis was not substantiated in actual test.

7 Conclusions

It seems obvious that much too little of a definitive nature is known about life as it exists in the deep seas, an area comprising the major portion of our planet. Our own investigative approach has been primarily on the molecular level utilizing comparisons with known functions in terrestrial organisms. We feel strongly that it is insufficient to acknowledge that life exists in the deep seas without asking how it exists. What differences in molecular structure and function must be present to allow for life under the environmental stresses presented? Is it not both possible and probable that molecular considerations can help define and delineate areas of general ecological investigation concerning the oceans?

There is no doubt that information obtained from experiments comparing life functions in the deep sea with those under "terrestrial" conditions in microbes as well as higher forms will eventually lead to a better knowledge of the mechanisms pertinent to the existence of human life.

We are still in the early stages of investigation. Admittedly, the early stages have been quite protracted, but perhaps we are now at least approaching the tunnel even if the light at its end is still obscure.

Note added in proof. A. Yayanos and co-workers at Scripps Institution of Oceanography have recently achieved significant results in the collection, transfer and maintenance of deep-sea organisms (personal communication). They have begun studies on amphipods which seem to function well only at high pressures and on a spirillum which may be a true barophile.

Acknowledgements

The experimental work on protein synthesis has been supported by a series of grants from the National Science Foundation, currently no. PCM 76-15862.

The manuscript of this paper was submitted in February 1977.

References

Arnold, R. M. and Albright, L. J. (1971). Hydrostatic pressure effects on the translation stages of protein synthesis in a cell free system from *Escherichia coli*. *Biochimica et Biophysica Acta*, **238**, 347–354.

Baldwin, J., Storey, K. B. and Hochachka, P. W. (1975). Lactate dehydrogenase M_4 of an abyssal fish: Strategies for function at low temperature and high pressure. *Comparative Biochemistry and Physiology*, **52B**, 19–23.

Baross, J. A., Hanus, F. J. and Morita, R. Y. (1974). Effects of hydrostatic pressure on uracil uptake, ribonucleic acid synthesis and growth of three obligately psychrophilic marine vibrios, *Vibrio alginolyticus*, and *Escherichia coli*. *In* "Effect of the Ocean Environment on Microbial Activities" (Eds R. R. Colwell and R. Y. Morita), pp. 180–202. University Park Press. Baltimore, Maryland.

Bayley, S. T. (1966). Composition of ribosomes of an extremely halophilic bacterium. *Journal of Molecular Biology*, **15**, 420–427.

Berger, L. R. (1958). Some affects of pressure on phenylglycosidase. *Biochimica et Biophysica Acta*, **30**, 522–529.

Berger, L. R. (1959). The effect of hydrostatic pressure on cell wall formation. *Bacteriological Proceedings*, **12**, 129.

Berger, L. R. (1974). Enzyme kinetics, microbial respiration and active transport at increased hydrostatic pressure. *In* "Review of Physical Chemistry of Japan, Proceedings of the Fourth International Conference on High Pressure". Kyoto Shobo, Kyoto.

Brauer, R. W. (1972). "Barobiology and the Experimental Biology of the Deep Sea". North Carolina Sea Grant Program, University of North Carolina.

Brown, D. E., Johnson, F. H. and Marsland, D. A. (1942). The pressure–temperature relations of bacterial luminescence. *Journal of Cellular and Comparative Physiology*, **20**, 151–168.

Ehrlich, H. L. (1974). Response of some activities of ferromanganese nodule bacteria to hydrostatic pressure. *In* "Effect of the Ocean Environment on Microbial Activities" (Eds R. R. Colwell and R. Y. Morita), pp. 208–221. University Park Press. Baltimore, Maryland.

Haight, R. D. and Morita, R. Y. (1962). Interaction between the parameters of hydrostatic pressure and temperature on aspartase of *Escherichia coli*. *Journal of Bacteriology*, **83**, 112–120.

Hardon, M. J. and Albright, L. J. (1974). Hydrostatic pressure effects on several stages of protein synthesis in *Escherichia coli*. *Canadian Journal of Microbiology*, **20**, 359–365.

Hildebrand, C. E. and Pollard, E. C. (1972). Hydrostatic pressure effects on protein synthesis. *Biophysical Journal*, **12**, 1235–1250.

Hochachka, P. W. (1971). Pressure effects on biochemical systems of abyssal fishes: The 1970 Alpha Helix expedition to the Galapagos Archipelago. *American Zoologist*, **11**, 399–576.

Hochachka, P. W. (1975). How abyssal organisms maintain enzymes of the "right" size. *Comparative Biochemistry and Physiology*, **52B**, 39–41.

Infante, A. A. and Baierlein, R. (1971a). Pressure induced dissociation of sedimenting ribosomes: Effect on sedimentation patterns. *Proceedings of the National Academy of Science, USA*, **68**, 1780–1785.

Infante, A. A. and Graves, P. N. (1971b). Stability of free ribosomes, derived ribosomes, and polysomes of the sea urchin. *Biochimica et Biophysica Acta*, **246**, 100–110.

Infante, A. A. and Krauss, M. (1971c). Dissociation of ribosomes induced by centrifugation: Evidence for doubting conformational changes in free ribosomes. *Biochimica et Biophysica Acta*, **246**, 81–89.

Jannasch, H. W., Wirsen, C. O. and Taylor, C. D. (1976). Undecompressed microbial populations from the deep sea. *Applied and Environmental Microbiology*, **32**, 360–367.

Johnson, F. H., Eyring, H. and Polissar, M. J. (1954). "The Kinetic Basis of Molecular Biology". John Wiley, New York and London.

Kettman, M. S., Nishikawa, A. H., Morita, R. Y. and Becker, R. R. (1966). Effect of hydrostatic pressure on the aggregation reaction of poly-L-valyl-ribonuclease. *Biochemical and Biophysical Research Communications*, **22**, 262–267.

Kriss, A. E., Chumak, M. D., Stupakova, T. P. and Kirikova, N. N. (1967). Glucose metabolism in barotolerant bacteria cultivated under high hydrostatic pressure. *Microbiologiya*, **36**, 51–59.

Landau, J. V. (1966). Protein and nucleic acid synthesis in *Escherichia coli:* Pressure and temperature effects. *Science, New York*, **153**, 1273.

Landau, J. V. (1967). Induction, transcription, and translation in *Escherichia coli:* A hydrostatic pressure study. *Biochimica et Biophysica Acta*, **149**, 506–512.

Landau, J. V. and Thibodeau, L. (1962). Micromorphology of *Amoeba proteus* during pressure induced changes in the sol-gel cycle. *Experimental Cell Research*, **27**, 591–594.

Landau, J. V., Zimmerman, A. M. and Marsland, D. A. (1954). Temperature–pressure experiments on *Amoeba proteus*; plasmagel structure in relation to form and movement. *Journal of Cellular and Comparative Physiology*, **44**, 211–232.

Landau, J. V., Smith, W. P. and Pope, D. H. (1977). Role of the 30S ribosomal subunit, initiation factors, and specific ion concentration in barotolerant protein synthesis in *Pseudomonas bathycetes*. *Journal of Bacteriology*, **130**, 154–159.

Lauffer, M. A. (1962). Polymerization–depolymerization of tobacco mosaic virus protein. *In* "The Molecular Basis of Neoplasm". (Eds H. W. Schultz and A. F. Anglemeier). University of Texas Press, Austin, Texas.

Low, P. S. and Somero, G. N. (1975a). Pressure effects on enzyme structure and function *in vitro* and under simulated *in vivo* conditions. *Comparative Biochemistry and Physiology*, **52B**, 67–74.

Low, P. S. and Somero, G. N. (1975b). Activation volumes in enzymatic catalysis: Their sources and modification by low molecular weight solutes. *Proceedings of the National Academic of Sciences, USA*, **72**, 3014–3018.

MacDonald, A. G. (1975). "Physiological Aspects of Deep Sea Biology". Monographs of the Physiological Society. Cambridge University Press, London.
Marquis, R. E. and Keller, D. M. (1975). Enzymatic adaptation by bacteria under pressure. *Journal of Bacteriology,* **122**, 575–584.
Marquis, R. E. and ZoBell, C. E. (1971). Magnesium and calcium ions enhance barotolerance of streptococci. *Archiv für Mikrobiologie,* **79**, 80–92.
Marsland, D. A. and Landau, J. V. (1954). Mechanism of cytokinesis: Temperature–pressure studies on the cortical gel system in various cleaving eggs. *Journal of Experimental Zoology,* **125**, 507.
Mohankumar, K. C. and Berger, L. R. (1972). A method for rapid enzyme kinetic assays at increased hydrostatic pressure. *Analytical Biochemistry,* **49**, 336–342.
Morita, R. Y. (1957). Effect of hydrostatic pressure on succinic, formic and malic dehydrogenases in *Escherichia coli. Journal of Bacteriology,* **74**, 251–255.
Morita, R. Y. (1976). Survival of bacteria in cold and moderate hydrostatic pressure environments with special reference to psychrophilic and barophilic bacteria. *Symposia of the Society of General Microbiology,* **26**, 279–298.
Nieuwenhuysen, P., Clauwaert, J. and Heremaus, K. (1975). Pressure induced dissociation of ribosomes isolated from *Artemia salina* observed by light scattering. *Archives Internationales de Physiologie et de Biochimie,* **83**, 5.
Nomura, M. and Held, W. (1974). Reconstitution of ribosomes: Studies of ribosome structure, function, and assembly. *In* "Ribosomes" (Eds M. Nomura, A. Tissieres and P. Lengyel). *Cold Spring Harbor Symposia,* 1974, 193–223.
Palmer, P. S. and Albright, L. J. (1970). Salinity effects on the maximum hydrostatic pressure for growth of the marine psychrophilic bacterium, *Vibrio marinus. Limnology and Oceanography,* **15**, 343–347.
Pato, M. L., Bennett, P. M. and von Meyenberg, K. (1973). Messenger ribonucleic acid synthesis and degradation in *Escherichia coli* during inhibition of translation. *Journal of Bacteriology,* **116**, 710–718.
Pease, D. C. and Kitching, J. A. (1939). The influence of hydrostatic pressure upon ciliary frequency. *Journal of Cellular and Comparative Physiology,* **14**, 135–142.
Penniston, J. T. (1971). High hydrostatic pressure and enzymic activity: Inhibition of multimeric enzymes by dissociation. *Archives of Biochemistry and Biophysics,* **142**, 322–332.
Pollard, E. C. and Weller, P. K. (1966). The effect of hydrostatic pressure on the synthetic processes in bacteria. *Biochimica et Biophysica Acta,* **112**, 573–580.
Pope, D. H. and Berger, L. R. (1973a). Inhibition of metabolism by hydrostatic pressure: What limits microbial growth? *Archiv für Mikrobiologie,* **93**, 367–370.
Pope, D. H. and Berger, L. R. (1973b). Algal photosynthesis at increased hydrostatic pressure and constant pO_2. *Archiv für Mikrobiologie,* **89**, 321–325.
Pope, D. H., Connors, N. and Landau, J. V. (1975a). Stability of *Escherichia coli* polysomes at high hydrostatic pressure. *Journal of Bacteriology,* **121**, 753–758.
Pope, D. H., Smith, W. P., Swartz, R. W. and Landau, J. V. (1975b). Role of bacterial ribosomes in barotolerance. *Journal of Bacteriology,* **121**, 664–669.
Pope, D. H., Smith, W. P., Ogrinc, M. A. and Landau, J. V. (1976). Protein synthesis at 680 atmospheres: Is it related to environmental origin, physiological type, or taxonomic group? *Applied and Environmental Microbiology,* **31**, 1001.
Schwarz, J. R. and Landau, J. V. (1972a). Hydrostatic pressure effects on *Escherichia coli:* Site of inhibition of protein synthesis. *Journal of Bacteriology,* **109**, 945–948.
Schwarz, J. R. and Landau, J. V. (1972b). Inhibition of cell-free protein synthesis by hydrostatic pressure. *Journal of Bacteriology,* **112**, 1222–1227.

Schwarz, J. R., Yayanos, A. A. and Colwell, R. R. (1976). Metabolic activities of the intestinal microflora of a deep-sea invertebrate. *Applied and Environmental Microbiology*, **31**, 46–48.

Seki, H., Wada, E., Koike, I. and Hattori, A. (1974). Evidence of high organotrophic potentiality of bacteria in the deep ocean. *Marine Biology*, **26**, 1–4.

Shen, J. and Berger, L. (1974). Measurement of active transport by bacteria at increased hydrostatic pressure. *In* "Effects of the Ocean Evnrionment on Microbial Activities" (Eds R. R. Colwell and R. Y. Morita), pp. 173–179. University Park Press, Baltimore, Maryland.

Smith, W., Pope, D. and Landau, J. V. (1975). Role of bacterial ribosome subunits in barotolerance. *Journal of Bacteriology*, **124**, 582–584.

Smith, W., Landau, J. V. and Pope, D. H. (1976). Specific ion concentration as a factor in barotolerant protein synthesis in bacteria. *Journal of Bacteriology*, **126**, 654–660.

Subramanian, A. R. and Davis, B. D. (1971). Release of 70S ribosomes from polysomes in *Escherichia coli*. *Journal of Molecular Biology*, **74**, 45–54.

Swartz, R. W., Schwarz, J. R. and Landau, J. V. (1974). Comparative effects of pressure on protein and RNA synthesis in bacteria isolated from marine sediments. *In* "Effects of the Ocean Environment on Microbial Activities" (Eds R. R. Colwell and R. Y. Morita), pp. 145–149. University Park Press, Baltimore, Maryland.

Takacs, F. P., Matula, T. I. and MacLeod, R. A. (1964). Nutrition and metabolism of marine bacteria. XIII. Intracellular concentrations of sodium and potassium ions in a marine pseudomonad. *Journal of Bacteriology*, **87**, 510–518.

Tilney, L. G., Hiramoto, Y. and Marsland, D. (1966). Studies on the microtubules in heliozoa. III. A pressure analysis of the role of these structures in the formation and maintenance of the axopodia of *Actinosphaerium nucleofilum* (Barrett). *Journal of Cell Biology*, **29**, 77–95.

Wirsen, C. O. and Jannasch, H. W. (1975). Activity of marine psychrophilic bacteria at elevated hydrostatic pressures and low temperatures. *Marine Biology*, **31**, 201–208.

Wittman, H. G. and Wittman-Liebold, B. (1974). Chemical structure of bacterial ribosomal protein. *In* "Ribosomes" (Eds M. Nomura, A. Tissieres and P. Lengyel). *Cold Spring Harbor Symposia*, 1974, 115–140.

Yayanos, A. A. (1975). Stimulatory effect of hydrostatic pressure on cell division in cultures of *Escherichia coli*. *Biochimica et Biophysica Acta*, **392**, 271–275.

Yayanos, A. A. and Pollard, E. C. (1969). A study of the effects of hydrostatic pressure on macromolecular synthesis in *Escherichia coli*. *Biophysical Journal*, **9**, 1464–1482.

Zimmerman, A. M. (1970). "High Pressure Effects on Cellular Processes". Academic Press, New York and London.

Zimmerman, A. M. and Marsland, D. (1964). Cell division: Effects of pressure on the mitotic mechanisms of marine eggs (*Arbacia punctulata*). *Experimental Cell Research*, **35**, 293–302.

ZoBell, C. E. (1968). Bacterial life in the deep sea. *Bulletin of the Misaki Marine Biological Institute, Kyoto University*, **12**, 77–96.

ZoBell, C. E. and Johnson, F. H. (1949). Influence of hydrostatic pressure on the growth and viability of terrestrial and marine bacteria. *Journal of Bacteriology*, **57**, 179–189.

ZoBell, C. E. and Kim, J. (1972). Effects of deep-sea pressures on microbial enzyme systems. *Symposia of the Society for Experimental Biology*, **26**, 125–146.

ZoBell, C. E. and Oppenheimer, C. H. (1950). Some effects of hydrostatic pressure on the multiplication and morphology of marine bacteria. *Journal of Bacteriology*, **60**, 771–781.

Methane cycling in aquatic environments

J. W. M. RUDD* and C. D. TAYLOR

Woods Hole Oceanographic Institution, Woods Hole, Massachusetts, USA

*Present address: Department of Fisheries and the Environment, Freshwater Institute, 501 University Crescent, Winnipeg, Manitoba, Canada.

1 Introduction

Methane has been associated with aquatic environments since its discovery in 1776 by Volta (Söhngen, 1906) who described the formation of "combustible air" in the sediments of several lakes, ponds and streams. Since then pure bacterial cultures have been isolated, usually from aquatic sources, which either produce or oxidize methane. Until very recently, however, almost all of the research on these bacteria has been confined to the laboratory. Therefore very little information has been generated concerning their activities in their native aquatic habitats (e.g. Hutchinson, 1957). During the past approximately ten years, ecological information of this type has begun to emerge. This information, when used in conjunction with laboratory data, has in some cases provided a reasonably clear picture of methane cycling in aquatic environments.

A brief introduction to the aquatic methane cycle[1] in a stratified body of water is given in Fig. 1. Methane is produced by bacteria primarily in the sediments (Rudd and Hamilton, 1978) mainly from acetic acid, carbon dioxide and hydrogen. These substrates are the end products of a series of fermentative reactions which consume plant remains that have settled on to the sediment surface and are the known substrates of methanogenic bacteria. In marine environments it has been proposed that most of the methane produced in the sediments is often oxidized anaerobically by sulphate-reducing bacteria before it leaves the sediments. In freshwater most methane oxidation occurs aerobically at the oxic–anoxic interface either within the sediments or in the water column. Methane that is not oxidized by either of these processes escapes to the atmosphere. The carbon dioxide produced via methane oxidation can be converted back to methane, fixed photosynthetically, escape to the atmosphere or precipitate as insoluble carbonates. Cell material produced by the methane oxidizers and producers may reenter the fermentative processes or be consumed by animal grazers.

The emphasis of this review will be on the activities of methane-

[1]While this is not a cycle in the strictest sense because methane is not present at every point it does provide a useful framework for discussion. A more precise term could be the aquatic methane–carbon cycle.

producing and oxidizing bacteria *in situ*, including the physical and chemical factors controlling their rates of activity and some of the effects these activities have on other members of aquatic ecosystems. Laboratory data will be considered from the point of view of its contribution to the understanding of these bacteria *in situ*.

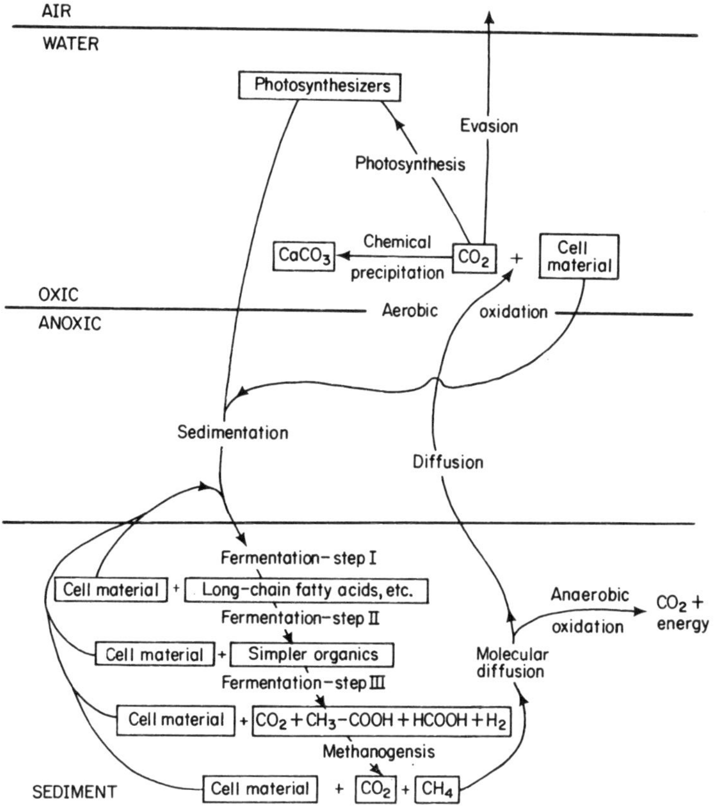

Fig. 1. A schematic representation of methane cycling in a stratified body of water.

Several diverse scientific disciplines have contributed to the study of methane cycling in aquatic environments. In addition to biochemists and microbiologists, important contributions have been made by oceanographers, limnologists, environmental engineers, geochemists and atmospheric chemists. This review will attempt to summarize these efforts and present them within the context of the aquatic methane cycle.

2 Methane production in aquatic environments

2.1 GENERAL PROPERTIES OF METHANE BACTERIA AND PRODUCTION OF METHANE IN ANOXIC FRESHWATER ENVIRONMENTS

An excellent summary of the progress through part of 1977 in the study of the ecology, microbiology, and biochemistry of methane formation is provided in the following review articles: Barker (1956), Stadtman (1967), Wolfe (1971), Hobson *et al.* (1974), Hungate (1975), Zeikus (1977), and Mah *et al.* (1977). Except for areas not previously covered we will not reiterate in detail what has already been discussed.

Methanogenesis is typically observed in anaerobic habitats where polymeric organic material deposited from the aerobic biosphere is being decomposed. Examples of such habitats include (a) the anaerobic muds of freshwater lakes, rivers, marshes, swamps, bogs and flooded soils (Zeikus, 1977; Mah *et al.*, 1977), (b) marine sediments rich in organic matter (Reeburgh, 1972; Martens and Berner, 1974, 1977; Oremland, 1975), (c) sludge digestors (Hobson *et al.*, 1974), (d) gastrointestinal tracts (Hungate, 1966, 1975) and (e) the heartwood of waterlogged hardwood trees (Zeikus and Ward, 1974).

Methane bacteria are a morphologically diverse group of very strict anaerobes which require specialized techniques for their manipulation and culture (see Wolfe, 1971, and Zeikus, 1977, for a description of these techniques). Presently there are eight taxonomically described species of methanogens in pure culture: *Methanobacterium arbophilicum*, *M. formicicum*, *M. ruminantium*, *M. mobile*, *M. thermoautotrophicum*, *Methanospirillum hungatii*, *Methanococcus vannielii*, and *Methanosarcina barkerii*.

Study of the nutritional properties of methanogens (see Wolfe, 1971; Zeikus, 1977; Mah *et al.*, 1977, for details) has revealed that they are nutritionally simple. All use hydrogen as an energy source, reducing CO_2 to methane. All species except *M. arbophilicum*, *M. thermoautotrophicum* and *M. barkerii* will in addition utilize formate. *Methanosarcina barkerii* is the only known species which will form methane from methanol and from acetate. None of the known pure cultures of methanogens will form methane from compounds more complex than acetate. In all organisms tested ammonia is used as a source of nitrogen, phosphate, the source of phosporus, and sulphide (or cysteine), the source of sulphur. One-half of the taxonomically described species of methane bacteria have been shown to grow autotrophically. The rumen species *M. ruminantium* and *M. mobile*, however, require various fatty acids, amino acids, co-factors, and B vitamins for growth. *Methanospirillum hungatii* and *M. vannielii* have not been tested for autotrophy.

Methanogens are terminal organisms in the anaerobic microbial food chain which utilize potentially toxic compounds produced from the anaerobic fermentation or respiration of organic material. The final product of their energy metabolism, methane, is a non-toxic, relatively mobile substance, and an important component of the carbon cycle in these environments. That fraction which is not recycled within the anaerobic zone (see section 3.5) is returned to the aerobic biosphere by diffusion or ebullition.

Recent study has shown that methane bacteria dramatically influence carbon and electron flow in anaerobic habitats by an interaction termed interspecies hydrogen transfer (Iannotti *et al.*, 1973). During fermentation by many chemoorganotrophs reduced pyridine nucleotides generated from the oxidation of substrates are recycled by the reduction of substrate-derived intracellular intermediates. The resultant reduced fatty acids and alcohols are excreted. Certain of these organisms are in addition able to dispose of electrons by pyridine nucleotide-mediated reduction of protons to molecular hydrogen (Wolin, 1976). This reaction is energetically unfavourable, however, and its utility for electron disposal often limited to conditions of very low partial pressures of hydrogen P_{H_2} (Wolin, 1976). Therefore the continued anaerobic oxidation of many of the primary fermentation products via pyridine nucleotide-linked generation of H_2 can occur only at reduced P_{H_2} (Bryant *et al.*, 1967; Wolin, 1976). Possessing a high affinity for hydrogen, methane bacteria are effective at reducing the P_{H_2} in anaerobic habitats and mixed cultures to less than 10^{-3} atm (Hungate, 1967; Winfrey *et al.*, 1977; Bryant *et al.*, 1977), and in essence serve as a sink for electrons. As a result, the stoichiometry of primary fermentation reaction is shifted towards the production of more oxidized products, primarily CO_2 and acetate (Hungate, 1966, 1967; Wolin, 1975; 1976; Weimer and Zeikus, 1977; Latham and Wolin, 1977) and residual primary fermentation products such as ethanol or lactate are amenable to further oxidation (Bryant *et al.*, 1967; Reddy *et al.*, 1972; Wolin, 1975, 1976; Bryant *et al.*, 1977). It has also been shown that interspecies hydrogen transfer will occur when non-methanogenic hydrogen oxidizers are grown in co-culture with hydrogen-producing fermentative microorganisms (Miller and Wolin, 1973; Iannotti *et al.*, 1973). Similar shifts towards more oxidized final products are observed.

Interspecies hydrogen transfer thus increases the efficiency of the anaerobic microbial food chain and would be expected to be of ecological significance in anaerobic habitats. It has in fact been demonstrated that perturbation of methane-producing habitats by termination of methanogenesis with inhibitors (Hobson *et al.*, 1974; Cappenberg, 1974b; Hungate, 1975), or high partial pressures of methane (J. Romesser, unpublished data) results in significant accumulations of hydrogen, acetate, and a variety of fatty acids which were previously undetectable or present at low levels.

In most documented instances 70–75 per cent of the methane produced in sludge digestors and freshwater sediments is derived from acetate (Jeris and McCarty, 1965; Smith and Mah, 1966; Cappenberg and Prins, 1974; Cappenberg, 1976). It should be kept in mind, however, that such values may depend upon the particular environment from which the sample is taken. For example, Belyaev *et al.* (1975a) observed that between 2 and 69 per cent of the methane produced in sediment samples taken from three freshwater lakes was derived from acetate. It is also likely that in sediments the fate of methane precursors will vary with depth (Cappenberg and Jongejan, 1977).

Early isotopic studies have indicated that the methyl group of acetate is converted intact to methane (Buswell and Sollo, 1948; Stadtman and Barker, 1949, 1951; Pine and Barker, 1956; Pine and Vishniac, 1957). Recent isotope studies by Ferry and Wolfe (1976) with a stable consortium which degrades benzoate to methane and carbon dioxide corroborates the earlier observations. There are indications that similar labelling patterns were observed in recent studies with pure cultures of *M. barkerii* (Mah *et al.*, 1977).

It is of interest to note that in freshwater sediments both methane and carbon dioxide are produced from methyl carbon of acetate (Cappenberg and Prins, 1974; Winfrey *et al.*, 1977; Cappenberg and Jongejan, 1977).

Recently two new organisms have been described, *Desulfuromonas acetoxidans* (Pfennig and Biebl, 1976) and *Desulfotomaculum acetoxidans* (Widdel and Pfennig, 1977), which oxidize acetate to carbon dioxide using elemental sulphur and sulphate, respectively, as electron acceptors. Circumstantial evidence suggests that these or physiologically similar organisms may be responsible for the complete oxidation of acetate observed in freshwater sediment:

a. these organisms have been found in sulphide-containing freshwater sediments;
b. Cappenberg and Jongejan (1977) observed from measurements of the distribution of label from 2- ^{14}C-acetate into ^{14}C-methane and ^{14}C-carbon dioxide that the oxidation of acetate was more prevalent in the upper layers of the sediments of Lake Vechten where it was mentioned that elemental sulphur was present from the oxidation of FeS;
c. Winfrey and Zeikus (1977) observed that the addition of 10 mM sulphate to freshwater sediment samples from Lake Mendota prevents methanogenesis from acetate and results almost exclusively in the oxidation of this compound to carbon dioxide.

These results suggest that in the presence of sulphate, acetate-oxidizing organisms effectively outcompete methanogens for acetate. Similar observations have been made with non-acetate-oxidizing sulphate reducers (Bryant *et al.*, 1977). In naturally occurring freshwater sediments acetate-oxidizing, sulphur- and sulphate-reducing anaerobes may be limited by the amount of sulphur which is recycled into forms which can be used as electron acceptors. In the absence of sulphur *Desulfuromonas acetoxidans* will use dithiols, malate, and fumarate as electron acceptors (Pfennig and Biebl, 1976). *Desulfotomaculum acetoxidans* has been shown to use fumarate in place of sulphate (Widdel and Pfennig, 1977). These or similar compounds are likely to also be limiting in the natural environment. In contrast to observations with conventional sulphate reducers (Bryant *et al.*, 1977) there is no positive evidence suggesting that these acetate-oxidizing anaerobes are capable of participating in interspecies hydrogen transfer (Mah *et al.*, 1977).

There remains the possibility that methane produced from the methyl carbon of acetate is oxidized to CO_2 anaerobically (Panganiban and Hanson, 1976; section 3.5 of this review). However, over the time periods in which significant total oxidation of acetate to CO_2 was observed, neither Winfrey and Zeikus (1977) nor Cappenberg and Jongejan (1977) could demonstrate anaerobic methane oxidation in freshwater sediments. Oxidation of the methyl carbon of acetate by methanogens has not been demonstrated to be of major significance (Zeikus *et al.*, 1975; Ferry and Wolfe, 1976).

Cumulative evidence suggests that the balance of methane produced in sediments and digestors from sources other than acetate results primarily from the reduction of CO_2 by hydrogen (Jeris and McCarty, 1965; Smith and Mah, 1966; Smith, 1966; Belyaev *et al.*, 1975a; Ferry and Wolfe, 1976; Zeikus, 1977; Mah *et al.*, 1977).

Formate has not been demonstrated to serve as a direct precursor to methane. In sediments available evidence suggests that this compound is rapidly decomposed to CO_2 and hydrogen which are subsequently converted into methane (Winfrey *et al.*, 1977).

Methanol, although readily converted to CO_2 and methane when introduced into sediment samples (Winfrey *et al.*, 1977) is not thought to be a naturally available substrate (Jeris and McCarty, 1965).

2.2 PRODUCTION OF METHANE IN THE ANOXIC MARINE ENVIRONMENT

It has been shown in numerous instances that the highest levels of methane in anoxic marine sediments are observed in the deep layers where the interstitial sulphate concentration is low (Nissenbaum *et al.*, 1972; Claypool and Kaplan, 1974; Barnes and Goldberg, 1976; Martens and Berner, 1974, 1977; section 3.6 of this review). Several lines of evidence indicate that sulphate reducers play an important role in determining the rates and occurrence of methanogenesis:

a. The rate of sulphate reduction is the greatest in the upper regions of marine sediments (Jørgensen and Fenchel, 1974; Jørgensen, 1978b) where sulphate is high and methane is low.

b. the maximum numbers of sulphate reducers are distributed nearer the sediment surface than is the population maximum of methanogens (Cappenberg, 1974a).

c. Laboratory incubation experiments with marine sediments show that methanogenesis does not occur until sulphate is depleted (Martens and Berner, 1974, 1977; Ferry and Peck, 1977). Likewise, the addition of sulphate to freshwater sediment samples results in a nearly complete inhibition (>90 per cent) of endogenous methanogenesis (MacGregor and Keeney, 1973; Cappenberg, 1974a; Winfrey and Zeikus, 1977).

d. Sulphide increases when inhibitory levels of sulphate are added to freshwater sediments (Winfrey and Zeikus, 1977).

e. Sulphate *per se* does not inhibit methanogens (Bryant *et al.*, 1971; Pine, 1971; Zeikus and Wolfe, 1972).

Cappenberg (1974a, 1975) suggested that this phenomenon is due to the inhibition of methanogens by sulphide generated from the reduction of sulphate. This view, however, has not been corroborated by other workers. Winfrey and Zeikus (1977), for example, observed methanogenesis in sediment samples where the interstitial sulphide concentration was artificially increased approximately two orders of magnitude over that observed in sulphate-inhibited sediments. Martens and Berner (1977) reported that methane formation occurred at interstitial sulphide concentrations an order of magnitude greater than that reported by Cappenberg to be inhibitory (< 1 mM). Similar observations were made by Oremland (1976) and experiments of Bryant *et al.* (1977) showed that *Methanobacterium* strain M.o.H. can readily tolerate 20 mM sulphide.

The present consensus for explaining the suppression of methane formation by sulphate is that methanogens are effectively out-competed for available hydrogen or substrate by sulphate reducers. Winfrey and Zeikus (1977) showed that the addition of exogenous hydrogen or acetate to sulphate-containing freshwater sediment samples significantly reversed the inhibition of methanogenesis. When the supplemental hydrogen of acetate became depleted, inhibition of methanogenesis resumed. Ferry and Peck (1977) observed the same phenomenon in marine sediments. Little or no methanogenesis was observed until sulphate was depleted. On the other hand when incubated in the presence of hydrogen significant methanogenesis occurred and the rate of sulphate reduction was increased. In fact the observed rate of methane formation was found to be proportional to the partial pressure of hydrogen. Upon depletion of sulphate in hydrogen-containing sediment samples methanogenesis increased several fold. It is interesting to note that if the methanogens were allowed to grow in sulphate-depleted, hydrogen-containing sediments, reintroduction of sulphate had little effect upon the rate of methanogenesis even though sulphate reduction resumed. Thus in sufficient numbers methanogens can effectively compete for available hydrogen.

2.3 PRODUCTION OF METHANE IN THE AEROBIC MARINE ENVIRONMENT

Dissolved methane in the upper regions of the oceanic water column has been shown in numerous instances to be present at concentrations higher than would be expected from equilibration with the atmosphere

(Swinnerton *et al.*, 1969; Brooks and Sackett, 1973; Brooks *et al.*, 1973; Lamontagne *et al.*, 1973; Williams and Bainbridge, 1973; Seiler and Schmidt, 1974; Scranton and Brewer, 1977; Scranton and Farrington, 1977). The methane in the upper mixed layers was typically found to be 30 to 70 per cent supersaturated relative to atmospheric equilibrium concentration. Below this region a subsurface maximum associated with the pycnocline was often observed where methane may be two or more fold greater than equilibrium solubility. At greater depths the concentration of methane gradually decreased. Below 400 to 500 meters the waters often became less than 30 per cent saturated.

These phenomena are not likely to have resulted from physical processes such as surface wave injection of air, barometric pressure changes, and temperature changes in the water mass after isolation from the atmosphere. For example the concentration of argon (a conservative gas with solubility properties similar to methane) observed in the water column deviates from calculated equilibrium values by no more than ± 5 per cent (Craig and Weiss, 1968).

Two possible sources of this excess methane are: (a) transport from near-shore anoxic sediments, and (b) *in situ* production. Recent work (Scranton, 1977; Scranton and Brewer, 1977; Scranton and Farrington, 1977) suggests that both mechanisms are operative. The relative contribution of each depends upon the location and hydrography of the water mass.

A conservative estimation of the relative importance of horizontal advection and vertical eddy diffusion in the physical transport of methane from the coastal waters of the Lesser Antilles into the western subtropical North Atlantic some 100 km to the north-east revealed that some of the excess methane could be supplied from this source (Scranton and Brewer, 1977). However, potential losses of methane via biological oxidation and the fact that the currents in this region tend to flow from east to west (away from the region of interest) were purposely excluded from their calculations. In a truer sense a small fraction at most of the excess methane would be expected to have originated via transport from the coast. On the other hand, evidence suggests that *in situ* production is the more probable source of methane supplied to the subsurface maximum. It would be expected, for example, that if the subsurface maximum resulted from horizontal transport, diffusive losses of methane from the maximum into the mixed layer should equal the flux of the gas entering the atmosphere from the mixed layer. This was not found to be the case (Scranton and Brewer, 1977). The rate of diffusion of methane from the maximum into the mixed layer was calculated to be 4·1 nmol $cm^{-2}year^{-1}$ as compared to 84 nmol $cm^{-2}year^{-1}$ leaving the sea surface. This large imbalance not only supports the concept of *in situ* production of methane but also defines the region of production to be within the mixed layer. Evidence in support of this was provided by the observation, in certain instances, of narrow maxima within the mixed layer. Rapid production within this region would be required to maintain such maxima.

Similar studies conducted near Walvis Bay, Nambia (Scranton and Farrington, 1977) revealed that in this region excess methane was supplied by *in situ* production as well as lateral transport from anoxic sediments.

At the other end of the spectrum the presence of high concentrations of methane in the waters of the Murray–Wilkinson Basin in the Gulf of Maine is best explained by advective transport from nearby anoxic sediments (Scranton, 1977).

An interesting question is what is the source of methane produced in the mixed layer? Of high probability is biological production. The well-known biological source of methane, the methane bacteria (Stadtman, 1967; Wolfe, 1971; Zeikus, 1977; Mah *et al.*, 1977) and trace sources such as sulphate reducers of Clostridia (Postgate, 1969) require strict anaerobic conditions. In the well-oxygenated mixed layer production of methane from such sources would require anaerobic microenvironments within, for example, faecal pellets, suspended particles, or within the intestinal tracts of animals.

There is evidence to suggest that faecal pellets may not be an important source of methane supplied to the mixed layer. Data of Bishop (1977) and Turner (1977) indicate that in the equatorial Atlantic the average faecal pellet diameter and density is approximately 100 μm and 1·5 g cm^{-3} respectively. The velocity of settling is on the order of 100 m day^{-1} and the flux through the 400 m horizon is approximately 0·9 mg faecal pellets $cm^{-2}year^{-1}$. Using the growth equation described in Taylor and Jannasch (1976), the rate of methane formation in an exponentially growing culture of *Methanobacterium ruminantium* was calculated from the data of Balch and Wolfe (1976) to be 1·5 fmol $cell^{-1}h^{-1}$ (an absorbance at 600 nm of 0·8 corresponds to a direct cell count of $2{\cdot}5 \pm 0{\cdot}6 \times 10^9$ cells ml^{-1}, W. Balch, personal communication). Assuming that the faecal pellets are completely anoxic and contain numbers of functioning methanogens similar to that of the rumen (10^8 cells cm^{-3}) (Smith and Hungate, 1958; Hungate *et al.*, 1964), the rate of methane supplied to the mixed layer, as calculated by a method similar to that described by Scranton (1977), would be approximately 0·2 nmol $l^{-1}year^{-1}$. The rate of methane production in the mixed layer of the western subtropical North Atlantic required to maintain a flux of 80 nmol $cm^{-2}year^{-1}$ was determined by Scranton (1977) to be 8 nmol $l^{-1}year^{-1}$. It thus appears that under conditions most favourable for methanogenesis that 40-fold more methane is required than can be supplied by faecal pellets. There is in fact some question as to whether the aerobic microorganisms associated with 100 μm faecal pellets will be able to maintain an interior anoxic zone sufficient to support methanogenesis. Calculations of Jørgensen (1977a) suggest that particles of a composition similar to faecal pellets suspended in air-equilibrated water would have a required diameter greater than 500 μm if the centre is to have a redox potential low enough to support processes such as sulphate reduction or methanogenesis. If indeed anoxic microenvironments should exist within faecal pellets precedence would suggest, as discussed earlier in this review, that sulphate reduction would occur largely to the exclusion of methanogenesis.

Methanogens in the interior portions of clusters of microorganisms associated with particulate material would face similar difficulties. Jannasch (1960), for example, used the ability of *Pseudomonas stutzeri* to denitrify under conditions of low oxygen concentration (must be $\leq$15 μM) as an assay for nearly anoxic microenvironments. When the organism was placed in a medium containing inert particulate material the organism grew in clusters (approximately 40–80 μm in diameter (Jannasch, personal communication)) around the particles. Denitrification was not detectable until the oxygen concentration in the surrounding medium was reduced (from an estimated 230 μM) to 130–150 μM. Again, the question becomes whether one can expect stringent anaerobic conditions to exist within such particles when exposed to oxygenated waters typical of the upper regions of the oceanic water column.

The production of methane from the intestinal tracts of marine zooplankton or fish has not been directly examined. In fact, little research has been directed at establishing whether or not anaerobic conditions exist within intestinal tracts of these organisms. Some information exists, however, on the retention time of intestinal contents. It has been shown with some of the smaller pelagic animals that the contents of intestinal tract turns over rapidly. For example, studies on the feeding and digestion in copepods (Marshall and Orr, 1955, pp. 105–106; Roman, 1977) indicate that the residence time of the intestinal contents can be less than one hour. Typical residence times of 3 to 6 hours have been observed in salps and ctenophores (R. Harbinson and L. Madin, personal communication); 3 to 4 hours in the chaetognath *Sagitta hispida* (Cosper and Reeve, 1975; Reeve *et al.*, 1975). Rosenthal and Hempel (1970) observed in sardine larvae an average passage time of 4 to 10 hours with average filling of the gut. Lasker (1970) demonstrated that the Pacific sardine empties its gut within 12 hours. It is thus possible in certain instances that the retention time of the intestinal contents may be insufficient to maintain anaerobiosis. A blanket statement to this effect, however, cannot be made without direct evidence and production of methane from this source remains an open possibility.

Production of methane from aerobic microorganisms is also possible. For example, preliminary growth studies (Scranton, 1977; Scranton and Brewer, 1977) indicate that exponentially growing cultures of the coccolithophore *Coccolithus huxleyi* produces trace amounts of methane at a rate of approximately 10^{-4} fmol cell $^{-1}$ h^{-1}. At typical algal cell densities of approximately 10 cells ml^{-1} (Ortner *et al.*, 1978) it would be difficult to account for more than a small fraction of the required methane production from this source except perhaps in a bloom situation (approximately 1000 cells ml^{-1}). The potential for trace methane production from heterotrophic bacteria has not been investigated using the sensitive techniques developed by Swinnerton *et al.* (1962a, b) and Swinnerton and Linnenbom (1967), and remain possible sources.

As is obvious from the above discussion an unambiguous identification of the source(s) of the methane produced in the aerobic oceanic water column will require further investigation.

2.4 QUANTIFICATION OF THE RATES OF METHANE PRODUCTION IN THE AQUATIC ENVIRONMENT

In regions where active decomposition of organic material is occurring the methane found in the water column almost exclusively originates from the sediments. Significant production in the water column has not been observed (Zeikus and Winfrey, 1976; Rudd and Hamilton, 1978a). One component of the information which is necessary for the understanding of the cycling of methane in these aquatic environments is a quantitative description of the flux of methane entering the water column from the sediment. An important fact which must be realized in the procurement of this information is that the rate at which methane enters the water column is not necessarily equal to the rate at which it is produced within the sediments. In environments where the upper regions of the sediment are aerobic and in anoxic marine sediments, significant amounts of methane may be oxidized before they enter the water column (section 3). Therefore, in these instances only measurements that describe the flux of methane crossing the sediment–water interface will be applicable. On the other hand, if neither oxidation within the sediments nor ebullition is significant, as may occur in certain freshwater systems, methane production within the sediments and the flux of methane diffusing from the sediments will approximate one another.

Whichever approach is determined to be appropriate for a given situation, the measurement of production rates within the sediments or the measurement of the flux of methane entering the water column, great care and attention to detail will be required in order that the data obtained reasonably describe what is occurring *in situ*.

2.4.1 *Direct measurement of rates within the sediments*

An ideal requisite for the determination of the *in situ* rates of a microbial transformation such as methanogenesis is that the sediment system under investigation should not be altered during analysis. However, the very experimental procedures by which quantitative rate measurements are made often requires that a detectable accumulation or diminution of a specific metabolite occur over a specific time interval. The hope is that these alterations can be quantified before the system is significantly altered.

The use of radiotracers in many instances makes the above task easier. Improved sensitivity over most chemical procedures is gained and in instances where high background levels of the substance of interest (e.g. methane) are present within the sediments, measurement of the changes in specific activity can be quite useful. When used, radiotracers must be introduced at levels small enough to prevent alteration of pool sizes and therefore measured rates. This will require the determination of the concentration of pertinent precursors. When the precursors of interest are produced and consumed concomitantly within the sediment the determination of the turnover time of that pool will also be necessary. This is of particular importance when a radiolabelled substrate is transformed via more than one independent pathway, only one of which is quantified. Failure to monitor the resultant specific activity changes within a pool can lead to significant errors in the calculated rates when expressed on a molar basis. The methane precursor acetate is an example where such precautions may be necessary (see, for example, Cappenberg and Prins, 1974; Cappenberg and Jongejan, 1977).

Direct measurements of the rates of processes within the sediments require, by design, that a sample, which was originally a component of an open system, be in some way confined within an experimental vessel. This will result in gradual changes in pool and microbial population sizes. The time at which confinement begins to affect rates will depend upon the system and can only be determined via time course experiments (Hungate, 1966, 1975). It may in fact be necessary to obtain data points over quite close time intervals in order that the initial rates obtained are a true reflection of what is occurring in the sediments. Assume, for example, that an enrichment is established within a confined sample and that an important organism is increasing in numbers with an average generation time of 30 h. The resultant rate measurements may be up to 12 per cent high in 10 h, 27 per cent high by 20 h. This is, of course, an oversimplification since enrichment may or may not be the response observed. The important point is that it may be necessary in some instances to conduct an entire time course experiment in a time period of a few hours.

Manipulation of the sample during rate measurements can dramatically affect rates unless extreme care is taken. In a recent study Jørgensen (1978a) compared various techniques for measuring the *in situ* rate of sulphate reduction. A promising technique was described in which 2 μl of ^{35}S-sulphate was anaerobically injected into the central portion of an undisturbed sediment core and incubated for time periods up to eight hours. The cores were subsequently quick-frozen, sectioned, and analysed for ^{35}S-sulphide. Minimal disturbance of the incubating samples occurred as the radiolabel diffused from the site of injection. It can be theoretically shown (and was experimentally verified) that as long as all of the radiolabel is recovered the decreasing specific activity of ^{35}S-sulphate as diffusion proceeds is exactly compensated by the increasing sphere of sediment that is exposed to radiolabel. The injection technique will thus yield the same calculated rate as would a sediment sample in which radiolabel was homogeneously distributed, provided homogenization does not affect the sample. Results of comparative experiments indicated, however, that sediments were quite sensitive to any changes which did not closely match *in situ* conditions. Any attempt to dilute the sediments resulted in drastic changes in rate measurements. The procedure described by Jørgensen has yet to be implemented for the measurement of methane production rates. However, the technique should prove useful when employed in conjunction with currently available analytical techniques. For example, methane may be stripped from sediment samples and absorbed on to activated charcoal by a procedure analogous to that described by Swinnerton *et al.* (1962a, b) and Swinnerton and Linnenbom (1967). After thermal release from the charcoal, labelled and unlabelled methane may be simultaneously analysed by gas chromatography–gas proportional counting (see Nelson and Zeikus, 1974).

Chemical and biological stratification exists in undisturbed sediments (Cappenberg and Jongejan, 1977; Jørgensen, 1978a) and an overall picture of a microbial transformation such as methanogenesis within the sediment as a whole will require that rate measurements be made in profile. Since the conditions within the sediment may vary drastically over a few centimetres' depth it is particularly important to prevent mixing of sediments from different levels when rate measurements are made (Jørgensen, 1978a). Probable differences in the conditions within the sediments from different regions of a habitat (Zeikus and Winfrey, 1976) and heterogeneity within sediment samples taken only a short distance from one another (Jørgensen, 1978a; Robertson, 1979) make it necessary that many replicate analyses be performed before a statistically significant description of the rate processes occurring within the habitat can be obtained.

It can be seen from the above discussions that to obtain a reasonable approximation of the true *in situ* rates of methanogenesis within sediments is a complex process. However, relative to the alternative method of estimating rates from profiles of the standing concentration of methane and mathematical models, direct rate measurements are probably the more reliable if carefully done. Although it is relatively easy to obtain detailed concentration profiles in sediments (Hesslein, 1976a; Winfrey and Zeikus, 1977) the mathematical relations used for calculation (see Jørgensen, 1978b) contain a number of variables (rates of sediment deposition, rates of diffusion, etc.) which must be determined from separate experiments. The overall error propagated from the uncertainty in the independent variables can become quite large. Likewise, assumptions (e.g. assuming that the sediments are in steady state, defining the shape of the profile of the rate of methanogenesis, etc.) may be difficult or impossible to verify by any method other than comparison of the results with direct rate measurements.

2.4.2 *Determination of the flux of methane leaving the sediments*

The direct measurement of the diffusion of a substance such as methane from the sediments can be made using enclosed traps placed on to the sediment surface. The flux of methane across the sediment–water surface is calculated from the measured rate of accumulation of methane within the trap and the area of the sediment which is enclosed. The constraints and precautions discussed in the above section apply here as well. If concentration measurements are made over as short time intervals as feasible the effects of confinement will be minimized. This is particularly important when the flux of methane is being measured from anoxic sediments whose surface is normally exposed to oxygenated water. The consumption of oxygen is usually high in such systems and the enclosed water mass can quickly become anoxic. This will dramatically influence flux rates by modulating aerobic methane oxidation in the surface layers of the sediments (Hesslein, 1976b). The patterns of mixing over the sediments must also be identified for the region under study and duplicated within the chamber. If this is not done, unnatural gradients will quickly form over the experimental surface and influence the flux of methane leaving the sediments. Variability in parameters such as the input of organic material into the sediments, the physical properties of the sediments, circulation patterns over the sediments, oxygen content in the water at the sediment–water interface, etc., are the rule rather than the exception in most natural systems. If direct flux measurements of the kind described here are to be used to describe an average flux of an entire aquatic system such as a lake, it is obvious that a systematic programme of multiple measurements, both on a geographic and temporal scale, will be necessary.

In certain well-defined freshwater lake systems the flux of dissolved methane from the sediments can be measured by indirect means (see sections 4.5 and 4.6). If (a) the lake possesses a well-defined hypolimnion, (b) methane is conserved within the anaerobic water column, and (c) all of the methane is oxidized aerobically before it reaches the air–water interface, an average flux from the sediments underlying the hypolimnion can be calculated from the rate of accumulation of methane within the hypolimnion and direct measurements of the rate of methane oxidation (see section 3). Relative to any kind of direct measurement of the flux from the sediments which can be made, the determination of methane concentration and methane oxidation rates within the water column are much more straightforward. The sediments are undisturbed and spatial variations in the flux rate are automatically averaged. Since the rate of change of the concentration of methane in the hypolimnion is slow the accuracy of the method is improved by conducting the measurements over extended periods of time (e.g. months). The flux calculations will then also be averaged in a temporal sense.

2.4.3 *Measurement of rates of ebullition*

When the production of a gas within the sediments exceeds the rate of its removal the total gas pressure within the pore waters will ultimately exceed the ambient pressure (sum of the atmospheric and hydrostatic pressures). Supersaturation and thus bubble formation will occur. In anoxic sediments bubble formation is predominantly controlled by methane formation (Hammond *et al.*, 1975; Reeburgh, 1969; Martens and Berner, 1974, 1977; Hesslein, 1976b), and will result in the stripping of other gases from the pore waters in proportion to their partial pressure (Hesslein, 1976b). In fact the measurement of the depletion of a conservative gas such as argon, whose only source into the sediments is by diffusion from the water column, will provide an indication of whether or not ebullition is occurring (Reeburgh, 1969, 1972; Martens and Berner, 1977) in instances where it cannot be determined visually.

Most of the methane lost from the sediments by ebullition is lost from the aquatic environment entirely. It has been shown, for example, that as sediment-derived bubbles traverse a 10 m water column 10 per cent or less of the gas redissolves (Hesslein, 1976b; Robertson, 1979). On the other hand, the amounts of methane which dissolve from bubbles may in certain instances be quite significant relative to the amount which escapes oxidation and diffuses past the oxic–anoxic interface in the sediments or water column (sections 3.4, 3.6, 4.5, 4.6). Thus the measurement of the rates of ebullition can in pertinent instances provide information both on the loss of carbon from the aquatic habitat as well as cycling within the water column.

Rate measurements are usually made by determining the volume of gas collected in inverted funnel-like devices placed over the sediments. Determination of (a) the volume of gas collected at the sediment surface and pertinent positions within the water column (normalized with respect to pressure), and (b) the composition of the retained gas at each position will allow the calculation of the amount of methane which is redissolved into the water column.

Gas which is accumulating within sediments tends to collect in pockets which are heterogeneously distributed within a given area of sediment. Also release of the gas from the sediments will often not occur at regular time intervals. Thus, a statistically significant representation of the rate of ebullition from a given location will require multiple measurements. Likewise an overall picture of ebullition from a given habitat will likely require that a number of measurements be made at different locations since differences in sediment composition and/or hydrostatic pressure will likely influence ebullition rates.

3 Methane oxidation in aquatic environments

3.1 AEROBIC METHANE-OXIDIZING BACTERIA

Several review articles have effectively summarized progress through 1976 in the study of the microbiology and biochemistry of methane-oxidizing bacteria (Leadbetter and Foster, 1958; Fuhs, 1961; Silverman, 1964; Ribbons *et al.*, 1970; Wilkinson, 1971; Quayle, 1972; Wake *et al.*, 1973; Kosaric and Zajic, 1974; Quayle, 1976). We shall not repeat the information previously covered unless it applies directly to the understanding of the ecology of methane-oxidizing bacteria.

Microorganisms were first implicated in the oxidation of methane by Kaserer (1906) in Germany and Söhngen (1906) working in the Netherlands. Söhngen described the disappearance of methane and the appearance of carbon dioxide and organic (cell) material in enrichment cultures obtained from manure, pond water and canal water. From these cultures he isolated a rod-shaped bacterium which he named *Bacillus methanicus*. Although doubts have since been expressed concerning this culture's purity (e.g. Dworkin and Foster, 1956), the same culture has apparently been reisolated and/or renamed three times in the succeeding sixty-four years and is presently known as *Methylomonas methanica* (Whittenbury *et al.*, 1970b). Enrichment cultures of aerobic methane-oxidizing bacteria have since been obtained from a variety of other terrestrial and aquatic sources including soil, fermenting hay, sewage, lakes, pond and marine sediments, and the surface of aquatic plants (e.g. Hutton, 1948; Hutton and ZoBell, 1949; Dworkin and Foster, 1956; Leadbetter and Foster, 1958; Whittenbury *et al.*, 1970b; Naguib, 1971; Hazeu, 1975; Patt *et al.*, 1976a). Their world-wide distribution was demonstrated by Whittenbury *et al.* (1970b) who obtained methane oxidizers from a variety of sources in several countries. Considering the known world-wide distribution of methane-producing bacteria in various types of anoxic environments the widespread distribution of methane oxidizers is not unexpected.

Until recently only three species, in addition to *Methylomonas methanica*, have been isolated in pure culture and described reasonably fully: *Pseudomonas methanitrificans* (Davis *et al.*, 1964), *Methanomonas methanooxidans* (Brown *et al.*, 1964) and *Methylococcus capsulatus* (Foster and Davis, 1966). Difficulties in the isolation of methane-oxidizing bacteria have often been traced to the presence of hydrocarbon impurities in compressed methane and/or the relatively slow growth of methane oxidizers in enrichment culture versus other heterotrophs. In addition, isolation from aquatic sources may be complicated by some other inherent characteristic of the bacteria. For example, methane oxidizers in some cases prefer to grow in close association with other bacteria. Vary and Johnson (1967) obtained several mixed cultures from mud and water but were unable to obtain pure cultures utilizing methane as a sole source of carbon and energy. Sheehan and Johnson (1971) described an association of two different types of bacteria in a methane-oxidizing culture which was stable over a period of years in the absence of aseptic techniques. Other workers (Rudd and Hamilton, 1975a; Rudd *et al.*, 1976) were unable to separate small rod-shaped bacteria from "yeast-like" bacteria resembling the Methylobacter group of Whittenbury *et al.* (1970b). The enrichment cultures of Rudd and co-workers were repeatedly obtained from the eutrophic Lake 227 (Schindler *et al.*, 1973) and were apparently an interdependent association of two bacterial species (A. Furutani, unpublished data). Another difficulty in the isolation of methane-oxidizing bacteria has been their frequent dependence on unidentified growth factors in lake water which could not be easily provided in the culture medium. For example the Lake 227 culture required unidentified heat and uv-labile organic material(s) present only in the lake water. This necessitated the use of at least 5 per cent lake water in the growth medium. Other workers have made similar observations. Söhngen (1906) believed that certain factors present in plant extracts were necessary for growth. Dworkin and Foster (1956) found that calcium panthothenate would substitute for a cold-water agar extract as a growth factor and Naguib (1971) observed that additions of lake water to mixed cultures enhanced methane oxidation. The importance of growth factors present in the natural environment should always be considered when attempting to enrich for the principle bacterial type responsible for any *in situ* activity. In the case of the Lake 227 methane oxidizer the most important methane oxidizer active in the lake was believed to have been obtained in culture by including lake water in the medium and by maintaining other growth parameters (pH, temperature, oxygen and methane concentration) at their *in situ* optima (Rudd *et al.*, 1976).

Despite these difficulties more than 100 strains of methane-oxidizing bacteria have recently been isolated (Whittenbury *et al.*, 1970b; Hazeu and Steenis, 1970; Hazeu, 1975, and Trotsenko, 1976). Whittenbury *et al.* (1970b) have tentatively classified aerobic methane oxidizers into five groups on the basis of fine structure, cell morphology and type of resting stage. However, an unambiguous classification system has not yet been generally accepted (Whittenbury *et al.*, 1976). Cultural characteristics common to all methane oxidizers are:

1. A strict requirement for methane, methanol or dimethyl ether as a source of carbon and energy. A possible biochemical explanation of this is given in Ribbons *et al.* (1970). One exception to this is discussed in section 3.2.
2. They are all Gram-negative and require oxygen for growth.
3. They all possess a complicated internal membrane system (e.g. Proctor *et al.*, 1969; Davies and Whittenbury, 1970; de Boer and Hazeu, 1972). Two structurally different membrane systems desig-

nated type I and type II have been identified and have been used as a basis for dividing the methane oxidizers into two general groups. Recent biochemical studies have shown parallel differences in the pathways of C-1 metabolism as well as certain key enzymes (Lawrence and Quayle, 1970; Ribbons *et al.*, 1970; Quayle, 1972; Davey *et al.*, 1972; Wadzinski and Ribbons, 1975).

4. They do not have absolute requirements for organic growth factors. However this would be expected since these bacteria were isolated in a mineral salts medium. In fact, it is probable that ecologically important methane-oxidizing bacteria which required growth factors were missed in the survey conducted by Whittenbury since no attempt was made to simulate natural conditions in the cultures.

Several different types of methane-oxidizing bacteria which did not fit the descriptions given above have been isolated from Lake Kivu (Jannasch, 1975; Degens *et al.*, 1973). None of these isolates possessed internal membrane systems.

Other groups of microorganisms (e.g. mycobacteria, yeasts and fungi) are apparently incapable of growth on methane despite several reports to the contrary. This point is reviewed in detail by Wake *et al.* (1973).

3.2 SURVIVAL UNDER ADVERSE CONDITIONS

Aerobic methane oxidizers must survive the usual problems of aquatic bacteria such as desiccation and lack of electron acceptors. In addition they seem to lead a very precarious existence because they are principlely dependent on a single substrate, methane, as a sole energy and carbon source. To ensure their survival in the aquatic environment they have developed several specialized properties. Some of these are discussed below.

The majority of methane oxidizers can produce either exospores or cysts (Whittenbury *et al.*, 1970a). The exospores are heat resistant and survive at least eighteen months of desiccation. The three types of cysts observed are not heat resistant but withstand desiccation for several weeks or months. Presumably these resting stages function in the natural environment as a mechanism of surviving periods of desiccation or absence of substrate and may explain the presence of viable cells in the anoxic bottom waters of lakes where aerobic methane-oxidizing activity was absent (Cappenberg, 1972; Degens *et al.*, 1973).

In one case another means of survival in the absence of substrate has been to diversify the kinds of utilizable substrates. *Methylobacterium organophilum*, isolated from Lake Mendota (Patt *et al.*, 1974, 1976a) has done this very successfully. In fact, it prefers more complicated organic substrates such as sugars, tricarboxylic acid cycle intermediates, and complex media. It possesses a type II internal membrane system but only when methane is the sole carbon and energy source. The array of internal membranes is most prolific at very low oxygen concentrations, suggesting an adaptive response to this stress (Patt *et al.*, 1976b).

Although the importance of *M. organophilum* in the natural environment has not been established, it has been shown to be present in a number of environments including soil, water and sewage (G. C. Cole, cited in Patt *et al.*, 1976b). It apparently was not important in the eutrophic Lake 227 which receives a constant methane supply from the sediments. The principle methane oxidizer in this lake used only methane as a carbon and energy source (Rudd *et al.*, 1976).

Co-oxidation of non-growth substrates is another means by which many methane oxidizers have diversified their kinds of utilizable substrates. Leadbetter and Foster (1958, 1960) reported the oxidation of a number of hydrocarbons and non-hydrocarbons by *Pseudomonas methanica* (Söhngen) when growing on methane. Although these compounds would not support growth, they were used in the synthesis of a number of amino acids. Patel *et al.* (1975) and Eccleston and Kelly (1972, 1973) have supported and expanded these observations for another type I oxidizer, *Methylococcus capsulatus*. If *Methanomonas methanooxidans* is representative of type II methane oxidizers, their ability to co-oxidize non-growth compounds is more extensive. Wadzinski and Ribbons (1975) reported that *M. methanooxidans* incorporated ^{14}C-acetate into several amino acids, particularly glutamate, and also respired some of the carbon as $^{14}CO_2$. Growth was enhanced in the presence of acetate by up to 15 per cent. Hazeu (1975) has also reported that the addition of 0·1 per cent glucose stimulated growth of certain methane utilizers. At least eighteen co-oxidizable substrates have been identified by Whittenbury *et al.* (1976) and this list is undoubtedly not complete.

Whittenbury *et al.* (1975) have found that ethanol co-oxidation produced both ATP and NADH which was subsequently used by the methane oxidizer to reduce acetylene to ethylene. The capability of methane-oxidizing bacteria to derive part of their carbon and energy requirements from co-oxidizable substrates would tend to stabilize their precarious dependence on a single substrate. For this to be true, however, the methane oxidizers must successfully compete with heterotrophic bacteria for these additional carbon sources at very low *in situ* concentrations (probably $< 1 \cdot 0$ μM). Their ability to do this *in situ* and the contribution this makes to overall production of energy and cell material is unknown.

There have been two reports of methane oxidizers utilizing $C_1 - C_{22}$ *n*-alkanes as sole energy and carbon sources (Ooyama and Foster, 1965; Perry, 1968). However, since alkanes above C_1 are plentiful only in the vicinity of petroleum deposits or spills, this capability is not of general importance in overcoming the dependence on a single substrate.

3.3 AEROBIC METHANE OXIDATION AND THE NITROGEN CYCLE

3.3.1 *Dissolved inorganic nitrogen assimilation*

Methane-oxidizing bacteria participate in the aquatic nitrogen cycle at several points. They all assimilate nitrogen as ammonia and/or nitrate (Whittenbury *et al.*, 1970b). A few also assimilate urea and amino acids (Whittenbury *et al.*, 1970b) although the ability of their competing for organic nitrogen sources at *in situ* concentrations has not been demonstrated.

A rough estimate of their contribution to the uptake of dissolved inorganic nitrogen ($DIN = NH_4^+ + NO_2^- + NO_3^-$) can be made for Lake 227 if it is assumed that DIN is the sole nitrogen source of methane oxidizers and if their C/N ratio is about 6·6 (Redfield, 1958). This estimate can be made for Lake 227 since the production of cellular carbon by methane-oxidizing bacteria during one year is known to have been 6·0 g C m^{-2} $year^{-1}$ (Rudd and Hamilton, 1978). Therefore, approximately 1·0 g N m^{-2} were assimilated by the oxidizers during that year. This amount is equal to 13 per cent of the total nitrogen input into Lake 227 during the same year (D. W. Schindler, unpublished data). If the same C/N ratio is assumed for Lake 227 phytoplankton they would have assimilated 24 g N m^{-2} $year^{-1}$ (based on the annual primary productivity of 138 g C m^{-2} $year^{-1}$ (E. J. Fee, personal communication). This figure is 24 times greater than the nitrogen assimilation by methane oxidizers and is about three times greater than the total nitrogen input into the lake during that year.

This apparent anomaly may be explained by the difference spatial and temporal activities of the phytoplankton and methane oxidizers. Primary productivity occurred almost exclusively during the summer in the epilimnion using nitrogen which was evidently recycled a number of times within the epilimnion during that time. Methane oxidation, on the other hand, peaked during fall overturn (Rudd and Hamilton, 1975a, 1978) during which time the oxidizers principally assimilated DIN which had accumulated over a period of months in the anoxic hypolimnion (Rudd *et al.*, 1976). Therefore on a short term (summer) basis DIN uptake by methane oxidizers was not important. On a yearly basis (with respect to total nitrogen input data) they made a more significant contribution. While these estimates are rough they may be of use in setting the methane oxidizer's DIN assimilation activities in a eutrophic lake in proper perspective.

During spring, fall and throughout the winter when methane oxidation rates are very high and primary productivity is low in Lake 227 (see section 4.5 and Rudd and Hamilton, 1978) methane oxidizers were probably important contributors to DIN assimilation.

3.3.2 *Nitrification*

Methane oxidizers also participate in the first step of the nitrification process. Hutton and ZoBell (1953) found that 72 per cent of soil and marine and samples contained methane oxidizers which oxidized ammonia to nitrite. This was confirmed by Whittenbury *et al.* (1975) who found that every methane oxidizer tested was capable of this activity. It has since been hypothesized that this ammonia oxidation may actually be a co-oxidation by a methane-oxidizing mechanism involving free radicals (Hutchinson *et al.*, 1976).

In freshwater lakes, where aerobic methane oxidation is prevalent (see section 3.6), ammonia oxidation by methane oxidizers may be important since the highest rates of both nitrification and methane oxidation always occur at the same location, namely at the oxic–anoxic interfaces of sediments and stratified water columns or throughout the water column during overturn periods (e.g. Rudd *et al.*, 1974; Patt *et al.*, 1974; Rudd and Hamilton, 1975a; Jannasch, 1975). *In situ* measurement of rates of ammonia oxidation by methane oxidizers have not been attempted. Thus their contribution to nitrification in aquatic environments is unknown.

3.3.3 *Denitrification*

In has been reported (Davies, 1973) that methane oxidizers isolated from activated sludge utilized nitrate as an electron acceptor in the denitrification process. Again the most rapid denitrification rates in lakes occur at oxic–anoxic interfaces (e.g. Chan, 1977) where methane oxidation is also most rapid. However, attempts to demonstrate this in several water samples taken across the oxic–anoxic interface in Lake 227 failed. Enhancement of ^{14}C-CH_4 oxidation by addition of nitrate was not observed (J. Rudd, unpublished data). Several other workers have similarly failed to demonstrate denitrification in pure cultures of CH_4-oxidizing bacteria (Whittenbury *et al.*, 1970b). Though this reaction is thermodynamically favourable (Kosaric and Zajic, 1974) its occurrence has not been substantiated.

3.3.4 *Nitrogen fixation*

It is a well established fact that several methane-oxidizing bacteria fix dinitrogen (e.g. Davis *et al.*, 1964; Coty, 1967; deBont and Mulder, 1974; Whittenbury *et al.*, 1975). A very interesting problem exists with respect to assaying this activity by the acetylene reduction technique both in the laboratory and *in situ*. Methane oxidizers have been shown to co-oxidize ethylene (Fig. 2a and deBont and Mulder, 1974; Flett *et al.*, 1975; Whittenbury *et al.*, 1975). Also, acetylene has been found to inhibit methane oxidation which presumably in turn stops acetylene reduction by preventing the production of reducing equivalents (Fig. 2a and deBont, 1976; deBont and Mulder, 1976). It has been suggested (Whittenbury *et al.*, 1975; deBont and Mulder, 1976) that this problem with the acetylene reduction technique could be circumvented by giving the oxidizers methanol as a carbon and energy source during the assay since its utilization is not inhibited by acetylene so that reduction of acetylene to ethylene can continue (Fig. 2b). However, under these circumstances ethylene also continues to be co-oxidized (deBont and Mulder, 1976) so that nitrogen fixation rates may be underestimated as suggested by Flett *et al.* (1975). In any case this is not a suitable solution for *in situ* estimates of nitrogen fixation rates by methane oxidizers since a change in the type and most likely the concentration of the energy-producing substrate would severely effect *in situ* acetylene reduction estimates by altering the supply of reducing power to the nitrogenase enzyme. Flett (1977) emphasized that $^{15}N_2$ should be used to assay for *in situ* nitrogen fixation by methane oxidizers.

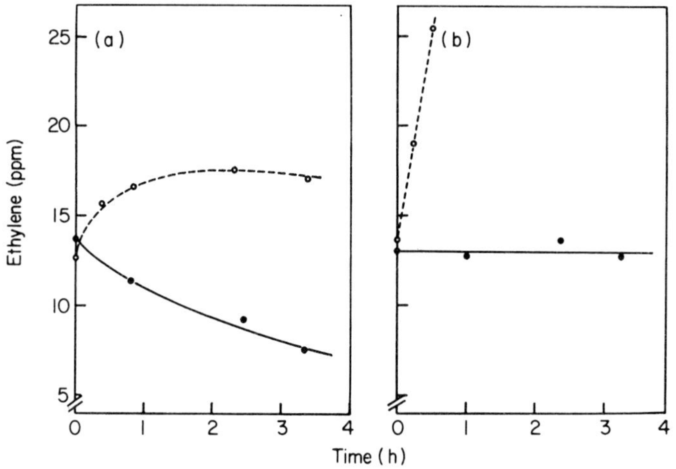

Fig. 2(a). In the absence of acetylene (●) a methane-oxidizing culture grown on a nitrogen-free medium with 10 per cent CH_4, 5 per cent O_2 and 85 per cent nitrogen co-oxidized ethylene (initial concentration 13 ppm). When acetylene was added to the culture (○) ethylene oxidation ceased.

(b). When the same culture was grown on 0·1 per cent CH_3OH instead of CH_4 in the presence (○) or absence (●) of acetylene, ethylene production proceeded and ethylene co-oxidation did not occur.

(From deBont and Mulder, 1974.)

The possible importance of nitrogen fixation by methane oxidizers in aquatic environments was suggested by Whittenbury *et al.* (1975 and 1976). They found that up to 10^8 ml^{-1} of these bacteria were present in sediment–water slurries taken from a eutrophic pond and that nitrogen-fixing methane oxidizers always predominated when DIN was undetectable. Flett (1977) attempted to quantify this nitrogen-fixing activity in the water column of Lake 302, a Canadian Precambrian Shield lake, by observing ^{15}N uptake across the oxic–anoxic interface where methane oxidizers were known to be active (J. Rudd, unpublished data). A peak of nitrogen-fixing activity was observed at

TABLE 1

Estimates of nitrogen fixation rates across the oxic–anoxic interface of Lake 302. The peak in particulate nitrogen at 7·6 meters corresponded with the biomass of non-nitrogen fixing algae. (From Flett, 1977.)

Depth m	O_2 mg l^{-1}	Particulate nitrogen µg N l^{-1}	Nitrogen fixation µg N l^{-1} h^{-1}
7·2	0·2	295	0·011
7·4	0·1	371	0·019
7·6	0·0	475	0·019
8·0	0·0	340	0·006

the interface (Table 1) which was probably due to methane oxidizers since no algal fixers were present in the samples. Although this fixation rate was undoubtedly important to the oxidizers (see section 3.4), it was not an important source of fixed nitrogen to this lake which was also receiving a large amount of nitrogen fertilizer as part of a whole lake eutrophication experiment (Schindler and Fee, 1974). Under different circumstances, for example in lakes receiving a relatively high phosphorus input, e.g. Lake 261 (Schindler, 1977), nitrogen fixation by methane oxidizers may be of some importance.

3.4 QUANTIFICATION OF METHANE OXIDATION RATES IN THE NATURAL ENVIRONMENT

Three general approaches of widely varying precision have been used to estimate *in situ* methane oxidation rates.

1. Rough anaerobic methane oxidation rates have been estimated for Cariaco Trench sediments and overlying deoxygenated water by mass balance calculations based on known methane concentration gradients and estimates of flux rates (Reeburgh, 1976). This approach probably gives only a general impression of the contribution of anaerobic methane oxidation to methane cycling (i.e. within an order of magnitude) because of the difficulties in accurately estimating diffusion coefficients in sediment pore waters and the water column.
2. Jannasch (1975) and Howard *et al.* (1971) have measured methane oxidation rates by monitoring decreases in methane con-

centration with time in water samples incubated in glass bottles. This method is at its best only when oxidation rates are rapid enough to detect significant methane concentration change within a few hours. Prolonged incubation (75 hours) may result in unpredictable bottle effects. Also, if representative rates are to be obtained, amounts of dissolved oxygen and methane must be sufficient so that neither is depleted or changed drastically in concentration during the incubation period. Since *in situ* methane-oxidizing activity very often occurs at oxic–anoxic interfaces where both methane and oxygen concentrations are low these three requirements for successful rate analysis are often not present in the water samples. Howard *et al.* (1971) incubated water samples for 5 days after adding a 30 per cent gas phase (O_2, CH_4 and N_2) to the incubation bottles. These changes in *in situ* substrate concentration as well as the long incubation periods limit the utility of these estimates as accurate indicators of natural activity. Jannasch (1975) encountered similar difficulties during his incubations of Lake Kivu samples. He diluted samples with surface water to increase oxygen concentrations in the sample bottles. As a result he states that these samples are useful only as "potential" rate estimates. Consequently he discussed his estimate of the total oxidation rate for the entire water column in terms of order of magnitude accuracy.

3. Rudd and Hamilton (1975a), Rudd *et al.* (1974) and Belyaev *et al.* (1975b) have used radiotracer methods to monitor *in situ* methane oxidation rates. In both cases small quantities of ^{14}C-methane were introduced into the water samples and equilibrated with natural ^{12}C-methane. After incubation the quantity of ^{14}C-methane converted to ^{14}C-carbon dioxide and ^{14}C-bacterial cell material was assayed and oxidation rates were computed. This approach offers several advantages. Because of its relative sensitivity short incubation periods could be used (maximum 4 hours). This minimized bottle effects as well as the amounts of methane and oxygen consumed during incubation. The addition of unnatural quantities of methane and oxygen were unnecessary. In addition the ratio of methane–carbon conversion either to carbon dioxide or bacterial cell material could be assayed. Thus the significance of carbon dioxide and particulate carbon production by methane oxidizers in terms of whole lake carbon cycling could be accurately assessed (Rudd and Hamilton, 1978).

3.5 FACTORS CONTROLLING AEROBIC METHANE OXIDATION *IN SITU*

Factors controlling the rates and distribution of methane-oxidizing bacteria have been studied in both fresh water and marine ecosystems with similar results.

Rudd and Hamilton (1975a) observed that during summer stratification the oxidizers in Lake 227 were confined to a very narrow zone at the oxic–anoxic interface within the thermocline (Fig. 3). There

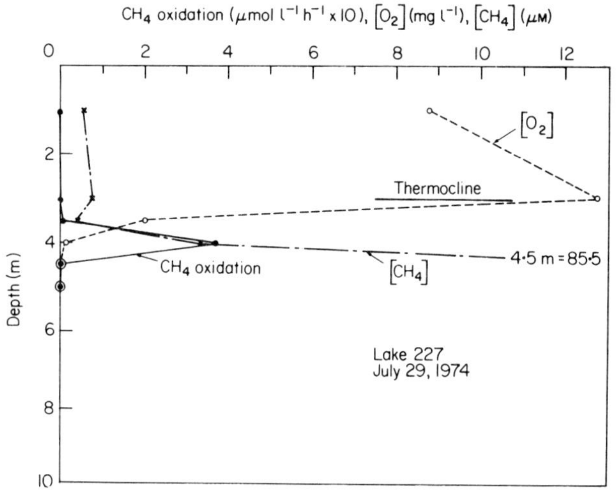

Fig. 3. A profile of methane oxidation in Lake 227 during summer stratification showing the narrow zone of methane-oxidizing activity of the oxic–anoxic interface (from Rudd and Hamilton, 1975a).

was no oxidation below this zone of activity because of anaerobiosis. For reasons which were not immediately apparent there was no epilimnetic methane oxidation even though concentrations of both methane and oxygen were adequate (Fig. 3). They also found that whole lake methane oxidation rates were very low during summer even though the amount of dissolved methane in the hypolimnion was increasing progressively (Rudd and Hamilton, 1975b, 1978). During periods of overturn and throughout the winter methane oxidation rates increased tremendously (see section 4.5) and the narrow zone of methane-oxidizing activity dispersed throughout the oxygenated portion of the water column (Fig. 4). These observations suggested that during summer stratification the distribution and rates of activity of the methane oxidizers were regulated by a sophisticated control mechanism. During periods of overturn and throughout the winter this control mechanism seemed to be switched off, enabling the oxidizers to disperse and become very active (Rudd and Hamilton, 1975a).

Several possible controlling factors were examined (Rudd and Hamilton, 1975a). *In situ* methane oxidation rates were never regulated by pH. Temperature was a controlling factor on a few occasions during fall overturn when methane and oxygen concentrations were within their optima. Methane concentration was a controlling factor only

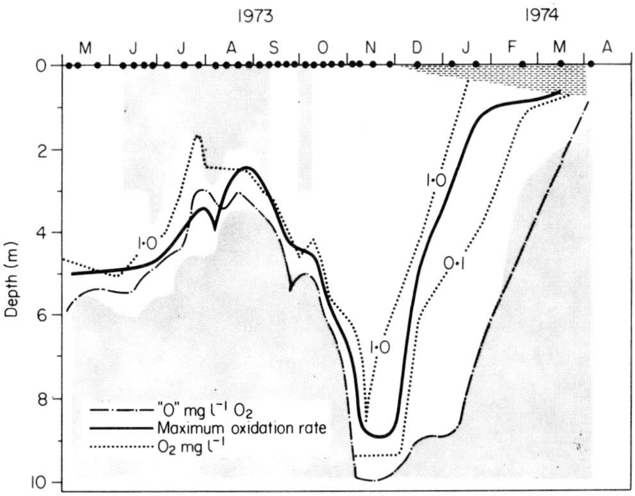

Fig. 4. Distribution of methane-oxidizing activity in Lake 227 during different seasons. The lower shaded area was anoxic. The methane oxidizers always preferred oxygen concentrations of less than 1 mg l^{-1}. (From Rudd and Hamilton, 1975a.)

within the narrow zone of activity during summer stratification. Within this zone methane concentrations were often almost undetectable (Fig. 3). Oxygen concentration was found to be very important and it was concluded that the methane oxidizers were microaerophiles preferring less than 1 mg l^{-1} O_2. Because of this they were forced to exist during summer stratification in the narrow zone of activity within the thermocline where oxygen concentrations were low (Figs 3 and 4).

Their microaerophilic nature was not consistent, however, with the observations that they rapidly oxidized methane at high oxygen concentration throughout the water column during other times of the year (Fig. 4). One explanation that was offered was that the oxidizers somehow adapted to the high oxygen concentration and so became oxygen tolerant during overturn and throughout the winter. This explanation was verified and clarified by Rudd *et al.* (1976) using a mixed culture which was thought to be dominated by the principle methane oxidizer active in Lake 227. They found that the culture's oxygen sensitivity was linked to an oxygen-sensitive nitrogen fixation process which has also been observed in pure cultures of methane-oxidizing bacteria (deBont and Mulder, 1974; Whittenbury *et al.*, 1975; Dalton and Whittenbury, 1976). When the mixed culture was dependent on nitrogen fixation as a sole nitrogen source (0–288 h of Fig. 5) it

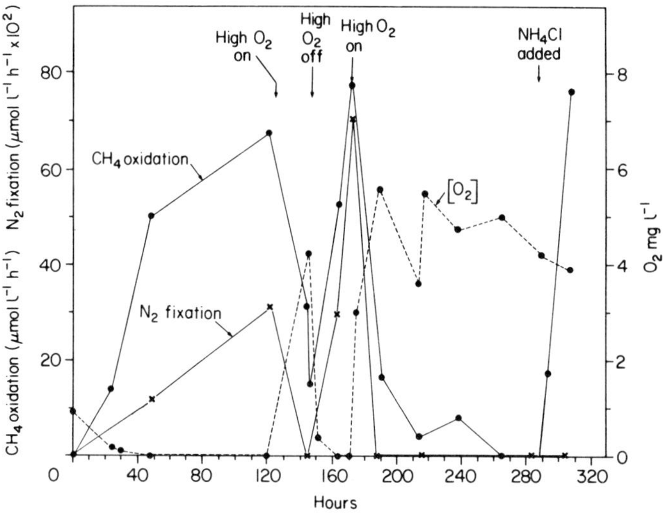

Fig. 5. Effect of oxygen concentration on rates of methane oxidation and nitrogen fixation in the presence and absence of NH_4Cl. The methane-oxidizing culture was obtained from Lake 227 and grown on a 50 per cent lake water medium. (From Rudd *et al.*, 1976.)

was inhibited by oxygen concentrations of above 1 mg l^{-1}. When either ammonium chloride or sodium nitrate was added to the culture as an alternative nitrogen source the oxygen sensitivity was mitigated (Fig. 5). From these data it was concluded that the methane oxidation process itself was not sensitive to high oxygen concentration. Instead it was the nitrogen fixation process which was oxygen sensitive. Thus when the methane oxidizer was dependent on nitrogen fixation as a sole nitrogen source at high concentration it was inactivated because of nitrogen starvation.

Using this information in conjunction with known DIN concentrations in Lake 227 Rudd *et al.* (1976) were able to explain which factors were controlling the rates and distribution of the oxidizers on a yearly basis. Absence of summer epilimnetic methane oxidation resulted from nitrogen starvation due to low DIN concentrations (< 3 μM) and high oxygen concentrations (> 1 mg l^{-1}) which inhibited nitrogen fixation. Consequently during summer the oxidizers were forced to exist in a narrow zone near the depth of zero oxygen where they were able to obtain nitrogen both from nitrogen fixation and from the upward diffusion of hypolimnetic ammonium (Figs 3 and 4). During spring and fall overturn large quantities of methane and ammonia were swept up from the hypolimnion and DIN concentrations often increased to 20 μM throughout the oxygenated portion of the water column. This nullified the inhibitory effect of high oxygen concentration and enabled the oxidizers to consume methane rapidly throughout the oxygenated portion of the water column (Fig. 4). During winter under ice and snow cover primary productivity was severely limited by low light levels. Consequently DIN concentrations remained high (> 40 μM) and the oxidizers remained active throughout the aerated portion of the water column (Fig. 4).

It was also recognized by Rudd and Hamilton (1975a) and Rudd *et al.* (1976) that the overriding controlling factors of whole lake methane oxidation rate were the physical factors controlling rates of diffusion and mixing in Lake 227. These physical factors profoundly influenced the concentration and rate of supply of DIN, O_2 and CH_4 to the methane oxidizers during all seasons of the year.

The oxygen concentration/nitrogen source control mechanism is of general importance at least in Precambrian shield lakes since, in addition to the eutrophic Lake 227, the same annual pattern of methane-oxidizing activity was noted for the mesotrophic Lake 383 and the meromictic Lake 120 (Rudd, unpublished data). The phenomenon may also be important in other types of freshwater lakes since similar narrow zones of methane-oxidizing activity have been observed near the oxic–anoxic interface of Lake Mendota (Patt *et al.*, 1974) and two African rift lakes, Lake Kivu (Jannasch, 1975) and Lake Tanganyika (Rudd, in preparation).

Marine aerobic methane oxidation is apparently also controlled in a similar fashion. Sansone and Martens (1978) have found that aerobic methane oxidation rates in Cape Lookout Bight water samples incubated in the laboratory were reduced by more than a factor of ten (0·01 ± 0·008 versus 0·21 ± 0·026 μmol l^{-1} day^{-1} when oxygen concentrations were greater than 90 μM (2·9 mg l^{-1}) and DIN concentrations were below 14 μM. When DIN was above 14–16 μM sensitivity to oxygen was not observed. An *in situ* experiment suggested that the DIN-controlled sensitivity to oxygen was operative in Cape Lookout Bight. They concluded that the methane oxidizers active in the marine environment were facultative microaerophiles similar to those observed by Rudd and co-workers in freshwater lakes.

Various types of suspended particulate materials may also have a controlling influence on methane oxidation rates. Harwood and Pirt (1972) showed that amberlite CG-120, an ion exchange resin increased growth in batch cultures by > 100 per cent. They speculated that the resin may have been adsorbing inhibitory materials present in the medium. Coal and glass particles were also found to increase methane oxidation rates and this effect was inversely proportional to particle size (Nesterov and Nazarenko, 1975). A similar particle size effect was previously noted by Weaver and Dugan (1972a) for several types of clay particles which were found to decrease lag, and increase the amount of methane consumed. Possible explanations included adsorption of inhibitor(s), increased substrate availability and/or provision of some unknown stimulating factor (Weaver and Dugan, 1972b). Several algal species decreased methane oxidation (Weaver and Dugan, 1972a). This effect was almost always a light-dependent phenomenon and so may possibly have been related to the oxygen concentration/nitrogen source control mechanism discussed above. Weaver and Dugan (1972a) have discussed some of the possible eutrophication implications resulting in increased carbon recycling by enhanced methane oxidation in the presence of particulates, but the *in situ* significance of this has never been tested.

3.6 ANAEROBIC METHANE OXIDATION

3.6.1 *Physical and chemical evidence*

In most marine sediments there is a zone of low methane concentration often extending approximately 20–100 cm below the sediment–water interface (e.g. Fig. 6 and Emery and Hogan, 1958; Barnes and Goldberg, 1976; Martens and Berner, 1974, 1977). Below this zone the methane concentration increased rapidly. The shape of the methane concentration profile is usually concave upwards (e.g. Fig. 6) and may reach saturating levels resulting in the release of bubbles (e.g. Reeburgh, 1969; Martens, 1976). Until recently, the zone of low methane concentration was discussed in terms of aerobic methane oxidation resulting from the injection of oxygen by macroinfaunal bioturbation (Reeburgh, 1969) or by the inhibition of methane production in the presence of sulphate, which occurs down to the base of the low methane zone (see section 2; also Nissenbaum *et al.*, 1972; Martens and Berner, 1974, 1977; Claypool and Kaplan, 1974). A third explanation, inhibition of methane production by high sulphide concentrations, now seems to be unlikely (see section 2).

Recently circumstantial evidence has been reported which suggests

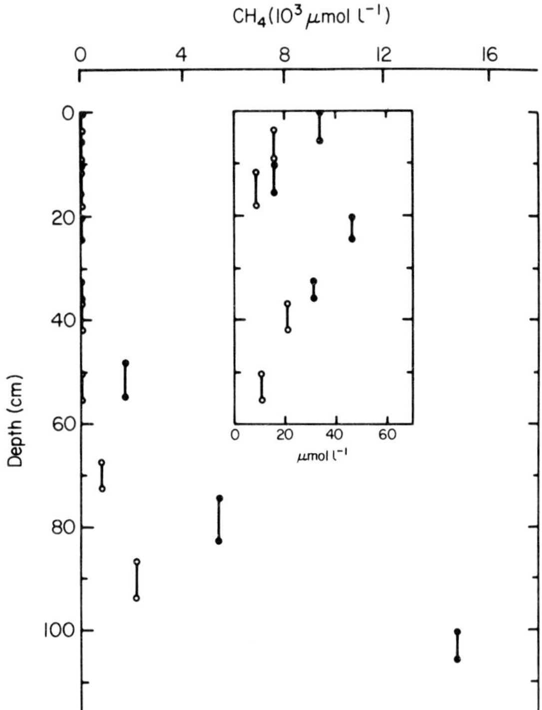

Fig. 6. Methane concentration versus depth in the eastern basin (●), and western basin (○), of Cariaco Trench sediments. The line connecting the points is the depth of the sediment core sampled. The inset shows the methane concentrations in the surface zone replotted on an expanded concentration scale. Inset depth scale is unchanged. (From Reeburgh, 1976.)

that this zone may also be the result of anaerobic methane oxidation possibly by sulphate-reducing bacteria. Then anaerobic methane oxidation could be a major methane sink and an important controlling factor in the distribution and concentration of dissolved methane, particularly in marine environments where sulphate concentrations are high. A summary of the indirect *in situ* evidence is as follows:

1. Based on estimations of eddy diffusion coefficients and natural advective velocities, methane concentration profiles between 300 and 1200 meters in the east and west basins of the Cariaco Trench should be concave downwards if methane is conservative in the absence of oxygen. The observed profiles are linear with depth, suggesting anaerobic methane oxidation in the water column (Reeburgh, 1976; M. I. Scranton, personal communication). A similar phenomenon has also been observed in the anoxic zone of the Black Sea (M. I. Scranton, personal communication).
2. In the Cariaco Trench sediments methane profiles are concave upwards (Fig.6). A low methane concentration zone extends between 0 and 35–55 cm depth. This suggests methane consumption in the surface sediments since aerobic oxidation and bioturbation are precluded by the anoxic environment and because methane removal by bubbling or hydrate formation can be excluded on physical grounds (Reeburgh, 1976). Similar concave upward methane profiles have been observed in other types of marine sediments (e.g. Barnes and Goldberg, 1976; Emery and Hogan, 1958; Martens and Berner, 1974, 1977; Reeburgh, 1969). According to Reeburgh (1978) these concave upward profiles could only be produced by anaerobic methane oxidation (there is some disagreement on this point which, however, does not change the overall conclusions—see below).
3. The sediment depth at which sulphate is completely reduced in the Cariaco Trench and other sediments usually occurs near the base of the zone of low methane concentration. This suggests that sulphate which is diffusing downwards is being used at this point as a terminal electron acceptor of anaerobic methane oxidation and is converted to hydrogen sulphide (Reeburgh, 1976, 1978; Barnes and Goldberg, 1976; Martens and Berner, 1977). This is an energy yielding reaction (~ -6K cal mol^{-1} CH_4) at *in situ* methane and sulphate concentrations (Martens and Berner, 1977). The contribution to sulphate reduction by other types of sulphate-reducing bac-

teria (Le Gall and Postgate, 1973) should also be included in the consideration of this point.

4. The depth profile of total carbon dioxide concentration often changes in slope at the base of the low methane concentration, suggesting the additon of carbon dioxide originating from methane (Reeburgh, 1976).

5. A $\delta^{13}CO_2$ minimum has also been observed at the base of the zone of low methane concentration. Reeburgh (1978) interprets this to mean that there is an addition of light carbon dioxide at this point as a result of methane-oxidizing activity. Kaplan and co-workers had suggested alternative explanations for this prior to the apparent discovery of anaerobic methane oxidation. Both points of view are discussed in detail in Reeburgh (1978).

A schematic summary of this evidence is shown in Fig. 7. Unfortunately no single data set exists in which all of the parameters discussed above were assayed. Therefore pairs or trios of these parameters are intercompared using several data sets to produce the overall picture.

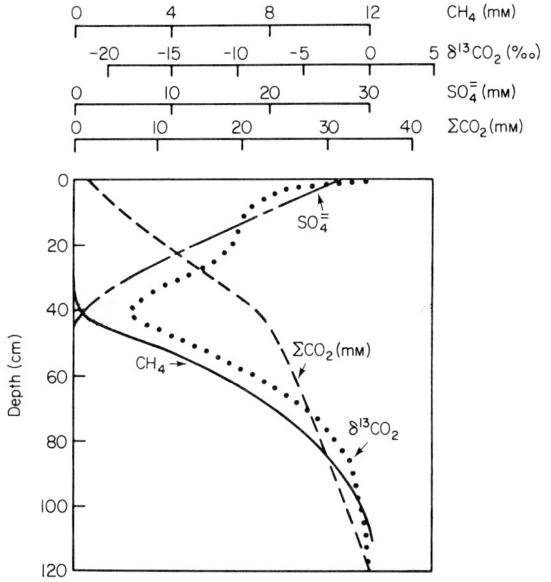

Fig. 7. Schematic diagram showing depth distribution of methane, sulphate, total carbon dioxide, and carbon isotope ratio of carbon dioxide in interstitial waters of marine sediments. All the distributions show breaks or slope changes in the stippled area, which represents the zone of anaerobic methane oxidation. (From Reeburgh, 1976.)

The sediment sulphate and methane concentration profiles shown in Fig. 8 are examples of some of the data used to make these conclusions. Martens and Berner (1977) estimated by extrapolation that methane saturation would have occurred in this core at 175 cm. They are of the opinion that concave upward methane profiles can also be produced by the combined effects of methane production at depth, upward diffusion, and sediment deposition which raises the sediment–water interface. Since deposition and diffusion rates were known for these Long Island Sound sediments, they were able to produce a theoretical

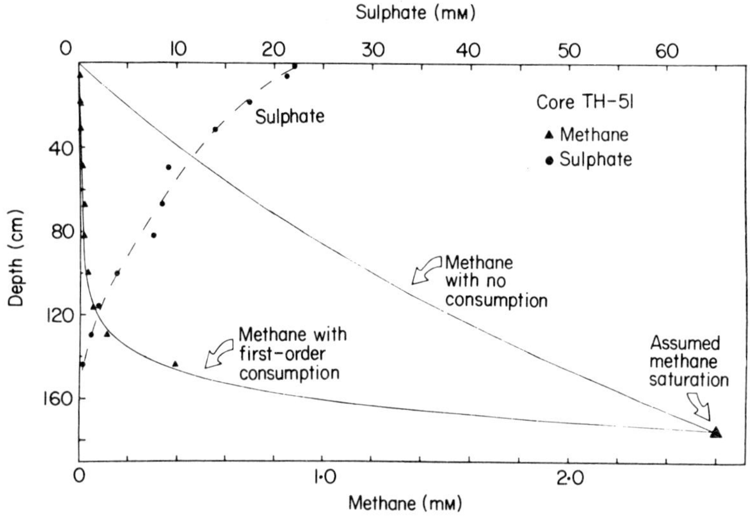

Fig. 8. Methane and sulphate concentration versus depth for core TH-51 taken from Long Island Sound. Solid lines represent plots of theoretical curves for no consumption and for consumption via first-order kinetics with $k_1 = 8 \times 10^{-9}\ s^{-1}$. Dashed line is an exponential fit to sulphate data. (From Martens and Berner, 1977.)

methane concentration profile assuming no methane production or oxidation between the surface and 175 cm (Fig. 8). This curve was only slightly concave upward and did not fit the observed data. Therefore they concluded that the large upward concavity observed could only be indicative of anaerobic oxidation. They also suggested that anaerobic oxidation was a first-order reaction since a theoretical profile based on consumption via first-order kinetics fitted the observed data and estimated depth of methane saturation (Fig. 8). Reeburgh (1976) justifiably objects to this since first-order kinetics would imply that the fastest anaerobic methane oxidation should occur at the highest methane concentrations where sulphate would be absent. Instead he proposed a higher-order reaction involving the *in situ* concentrations of both sulphate and methane.

There are two other indications of anaerobic oxidation in Fig. 8. Sulphate was almost completely depleted at the base of the surface zone of low methane concentration and oxygen injection by bioturbation occurred only to a depth of 20 cm. Both of these observations suggest anaerobic and not aerobic oxidation was determining the distribution of methane in these sediments.

There is one complication which prevents an unambiguous interpretation of much of these observations. Sulphate has also been found to inhibit methane production (section 2). It is possible that inhibition of methanogenesis (and thus also CO_2 reduction) could contribute to the formation of the low methane concentration zones shown in Figs 6 and 8 and affect the conclusions that could otherwise be drawn from points 2, 3, 4 and possibly 5 of the summary of indirect evidence. (pp. 46–47). For this reason Martens and Berner (1977) favour a compromise hypothesis to explain the surface sediment zone of low methane concentration. They feel that a combination of inhibition of methanogenesis and anaerobic methane oxidation may be responsible for the observed data. To resolve this issue finally it will be necessary to obtain microbiological evidence from both the laboratory and *in situ* demonstrating that anaerobic methane oxidation is occurring at rates which will account for the observed data.

3.6.2 *Microbiological evidence*

The general consensus regarding anaerobic methane oxidation has been that it does not occur (e.g. Quayle, 1972; Whittenbury *et al.*, 1970b; Winfrey and Zeikus, 1977; Cappenberg and Jongejan, 1977). This conclusion is reinforced by the inability of Postgate (1969) to detect anaerobic methane oxidation by sulphate reducers (although they did produce small amounts of methane) and by the inability of Sorokin (1957) to grow *Desulfovibrio desulfuricans* on methane using sulphate as an electron acceptor. The use of nitrate as an electron acceptor by known aerobic methane oxidizers is also unlikely (section 3.5).

The co-oxidation of methane by bacteria primarily dependent upon other compounds as carbon and energy sources is another possible mechanism of anaerobic methane oxidation. Two well-documented examples of this involving anaerobic oxidation of ^{14}C-methane have been reported. Wertlieb and Vishniac (1967) demonstrated that the photosynthetic purple non-sulphur bacterium *Rhodopseudomonas gelatinosa*, isolated from river mud, could both oxidize methane to carbon dioxide and incorporate small amounts of methane carbon into cell materials. Methane would not act as the sole electron source and the oxidizing activity was not light dependent. For this to be a practicable mechanism of anaerobic methane oxidation, purple non-sulphur bacteria would have to be capable of growth in the dark in anaerobic sediments. Formerly it was thought that these bacteria could not grow under these conditions (Pfennig, 1967). However, Uffen and Wolfe (1970) demonstrated that pure cultures can grow fermentatively without oxygen and in the dark when strict anaerobic techniques are used; so this may be one of the mechanisms of anaerobic methane oxidation in aquatic environments.

Desulfovibrio desulfuricans has also been shown to co-oxidize methane anaerobically (Davis and Yarbrough, 1966). In this case the primary substrate was lactate and sulphate was used as an electron acceptor. Almost all of the ^{14}C-methane oxidized appeared as ^{14}C-carbon dioxide. Although the amount of methane oxidized in this experiment was small, it would be adequate to account for *in situ* anaerobic methane oxidation according to Reeburgh (1976). It was not occurring at a rate detectable by gas chromatographic techniques during a 25 day sediments jar experiment in which sulphate reduction was occurring (Martens and Berner, 1977). They ascribed this to the preference of sulphate reducers for more easily metabolizable substrates present in the organic rich sediments used in this experiment. If this was so it would not be in agreement with the usual understanding of co-oxidation in which it has been observed that the more rapidly an organisms oxidizes its growth substrate the faster it also oxidizes its co-substrate (e.g. Flett *et al.*, 1975; Patel *et al.*, 1975). Other possible explanations could be that microbiological processes were severely disrupted at the beginning of the experiment by exposure of the sediments to atmospheric oxygen, or anaerobic oxidation may have been proceeding but at rates undetectable to their gas chromatographic techniques. The use of more sensitive ^{14}C-tracers may have been more successful.

Oremland (1976) and Oremland and Taylor (in preparation) added lactate to sediments from *Thalassia* beds and incubated anaerobically in the laboratory. They found that methane levels were higher in flasks containing inhibitors of sulphate reduction (β-fluorolactate or

Na_2MoO_2) or in the absence of sulphate as compared to flasks containing sediment alone in the presence of sulphate. They interpreted this as an indication that measurable anaerobic methane oxidation may have been occurring under these circumstances. However, an alternative explanation for these observations could be that the inhibition of sulphate reduction diverted the electron flow in the sediments away from sulphate reduction and towards methane production.

Recently Panganiban and Hanson (1976) have reported the isolation of a bacterium (strain AP-1) from Lake Mendota, which in the absence of oxygen requires methane for growth. In pure culture it apparently utilizes methane only as an energy source since only ^{14}C-carbon dioxide was produced during ^{14}C-methane oxidation and because the amount of ^{14}C-methane oxidized was approximately proportional to the amount of sulphate supplied as an electron acceptor. Anaerobic methane-oxidizing activity has also been detected in Lake Mendota using $^{14}CH_4$ tracer techniques (R. S. Hanson, personal communication, A. Zehnder, personal communication).

The isolation of this type of bacterium has not been repeated yet by other investigators. However, if it is active in aquatic environments, particularly in marine environments where sulphate concentrations are higher, it may play an important role in *in situ* anaerobic methane oxidation.

3.7 INTERRELATIONSHIPS BETWEEN AEROBIC AND ANAEROBIC METHANE OXIDATION

Although aerobic and anaerobic methane oxidation are believed to occur in both freshwater and marine ecosystems, aerobic oxidation predominates in freshwater while anaerobic oxidation seems to be prevalent in salt water (Reeburgh and Heggie, 1977). Consequently methane distribution and concentrations in freshwater and marine environment differ in several respects.

In salt water sediment methane profiles have the surface zone of low methane concentration with a slightly concave upward methane gradient extending downward (Figs 6, 7 and 8). In freshwater sediments this zone is absent and methane concentrations usually increase linearly beginning near the sediment water interface (Fig. 9). Water column methane concentrations are relatively low in marine anoxic basins reaching approximately 7 μM in the Cariaco Trench and the Black Sea

(Reeburgh and Heggie, 1977) and 80 μM in Lake Nitinat (Atkinson and Richards, 1967). In anoxic freshwater environments maximum concentrations are usually much higher. Concentrations of 500 μM (Cappenberg, 1972) and 1280 μM (Rudd and Hamilton, 1975b) have

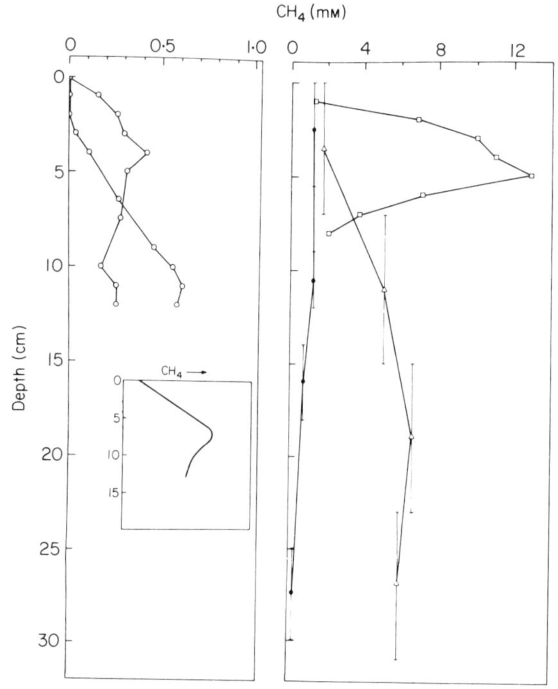

Fig. 9. Selected methane profiles in freshwater sediments. Δ, Koyama (1963), Table 6, Lake Nakatsuna; □, Cappenberg (1974b); ●, Reeburgh and Heggie (1974); ○, Rudd and Hamilton (1975b). Note differences in concentration scales. Vertical bars show reported depth interval of samples. Inset shows a schematic summary. (From Reeburgh and Heggie, 1977.)

been found in the hypolimnia of eutrophic lakes. The methane concentrations in the monimolimnia of meromictic lakes have reached 1200 μM (Weimer and Lee, 1973), 5000 μM (Lake 120 Rudd, unpublished data), or even 22 000 μM in Lake Kivu (Deuser *et al.*, 1973). In fact, methane concentrations are so high in Lake Kivu that a pipeline has been constructed (Anon., 1976) from a depth of 300 meters to produce methane to be used in a brewery!

Reeburgh and Heggie (1977) have suggested that two factors may contribute to the differences in methane concentrations and distributions in freshwater and marine environments. One is the higher marine sulphate concentrations as compared to freshwater (> 10 mM versus ~ 0·1 mM) which would tend to accentuate anaerobic methane oxidation in marine sediments if sulphate is proven to be a principle electron acceptor. The other is the relatively faster flux rate of organic materials to the sediments in freshwater systems, which are usually more productive. These two factors could work together to produce much higher methane to sulphate ratios in freshwater sediment surfaces. This occurs not only because sulphate concentrations are much lower in fresh water but also because the higher organic load causes more rapid sulphate reduction rates in the sediments. The end result is thought to be that in freshwater systems methane production overwhelms sulphate reduction (and thus anaerobic oxidation) resulting in the straight sedimental methane concentration profiles shown in Fig. 9. In freshwater systems if the water over the sediments is anoxic most of the methane diffuses up to and across the sediment–water interface and accumulates in the anoxic bottom waters. Significant methane oxidation does not occur until the methane reaches the oxic–anoxic interface where it is consumed aerobically (sections 3.4 and 4.5).

In marine systems where sulphate reduction is relatively more important and methane production rates are lower most of the methane is thought to be consumed anaerobically, in many cases, before it crosses the sediment–water interface (Figs 6, 7, 8). This results in lower concentrations in anoxic bottom waters (Reeburgh and Heggie, 1977). In considering this last point the reader should also bear in mind the inhibition of methanogenesis by sulphate (section 3.5).

Certain very productive marine and brackish environments resemble freshwater systems because high respiratory rates deplete sulphate near the sediment–water interface. This may have the effects of decreasing the relative importance of anaerobic methane oxidation by increasing methane production, methane ebullition and aerobic oxidation. For example, Martens (1976) reported that in the interior of Cape Lookout Bight during summer and early fall water near the sediment–water interface was periodically anoxic. This suffocated the macro-infauna, which were present at other locations in the Bight, and thus eliminated biological irrigation that was thought to be transporting significant quantities of sulphate into the sediments. As a result methane ebullition was believed to be a major source of dissolved methane to the overlying waters.

Bubbling apparently also occurs in other productive marine and brackish environments (Hammond *et al.*, 1975; Reeburgh, 1969; Martens and Berner, 1974). In these situations ebullition may predominate over methane oxidation as a means of removal of methane from marine sediments.

Aerobic methane oxidation may be stimulated in these productive marine environments by three processes.

1. Partial dissolution of methane containing bubbles was a major methane source to oxygenated water overlying organic rich sediments (Hammond *et al.*, 1975; Sansone and Martens, 1978). This promoted aerobic oxidation especially when oxygen concentrations were low or DIN was high (Sansone and Martens, 1978).
2. Macro-infaunal irrigation may enhance aerobic as well as anaerobic oxidation by transporting oxygen as well as sulphate into the sediments (Sansone and Martens, 1978). The result should be an excellent environment for aerobic oxidation since low oxygen and high DIN would be probable.
3. *Thalassia* (seagrass) beds may provide a similar environment. Oremland and Taylor (1977) monitored diurnal fluctuations of O_2, N_2 and CH_4 in the rhizomes and rhizosphere of these plants in Bimini Harbour. They concluded that there was probably a net transfer of photosynthetically produced oxygen into the sediments during the day. They suggested that an extensive aerobic interface, suitable for aerobic oxidation (i.e. low oxygen concentration promoting N_2 fixation) would be present since the *Thalassia* rhizomes formed an intricate network to a depth of one meter in the sediments. The contribution of aerobic methane oxidation to methane cycling in these highly productive marine environments has not been determined.

3.8 ATMOSPHERIC METHANE OXIDATION

Approximately 80 per cent of the methane entering the earth's atmosphere is of biological origin (Enhalt, 1976). The remainder is abiogenic and escapes to the atmosphere mostly from natural gas deposits and volcanoes. About 20 per cent of the biogenic methane is produced in animal intestines (Enhalt, 1976). Most of the remainder enters the atmosphere from aquatic environments either by ebullition (e.g. Reeburgh, 1969; Martens, 1976) or by evasion from the air–water interface (e.g. Whelan, 1974; Atkinson and Hall, 1976; Rudd and Hamilton, 1978).

Despite the continual flux of methane to the atmosphere it is present only in trace quantities. This is probably as a result of photochemical methane oxidation reactions involving the hydroxyl radical (Levy, 1973; Wofsy, 1976). Although the exact reaction sequence has not been clarified a stepwise oxidation of methane through carbon monoxide and finally to carbon dioxide is believed to occur (Levy, 1973; Wofsy, 1976). These reactions may be increasingly disrupted in the future by the anthropogenic additions of large quantities of carbon monoxide to the atmosphere. Since carbon monoxide effectively competes with methane for the hydroxyl radical a gradual increase in atmospheric methane concentrations has been suggested (Wofsy, 1976; Sze, 1977).

Methane flux to the atmosphere would undoubtedly be greater in the absence of a microbial methane oxidation. Under many circumstances the majority of methane produced is consumed by methane-oxidizing bacteria before it leaves the sediments or water column (e.g. Reeburgh, 1976; Reeburgh and Heggie, 1977; Martens and Berner, 1977; Rudd and Hamilton, 1978, 1975a; Rudd *et al.*, 1976). If this was not the case the additional methane flux to the atmosphere and consequent consumption of hydroxyl radicals could result in many of the same important changes in atmospheric chemistry that have been suggested should carbon monoxide flux to the atmosphere continue at a fast pace. Sze (1977) stated that an increase in methane and decrease in hydroxyl concentration could: (i) increase stratospheric odd hydrogen and water abundances, which could effect atmospheric chemistry; (ii) consume the chloride radical ($Cl + CH_4 \rightarrow HCl + CH_3$), which may increase ozone concentrations; and (iii) increase the greenhouse effect because both methane and water have infrared absorption bands. Thus methane cycling in aquatic environments may make a very significant contribution to both atmospheric chemistry and climate.

4 Examples of methane cycling in aquatic environments

4.1 QUANTIFICATION OF METHANE CYCLING *IN SITU*

Knowledge of the characteristics and factors controlling methane production and oxidation is only part of the information required fully to describe methane cycling in aquatic environments. Other necessary

TABLE 2

Areal methane production rates from various aquatic environments. Accuracy of the estimates will probably vary considerably according to the methods used

	Production rate $\mu mol\ m^{-2}\ h^{-1}$	Location	Methods	Source
1.	0·15–1·86	Seagrass beds	Accumulation of CH_4 in chambers on sediment surface. Higher rates are from chambers that contained less O_2	Oremland (1975)
2.	0·016–0·47	Coral reefs	Same as (1)	Oremland (1975)
3.	18·2	Eastern Basin Cariaco Trench	Calculated from estimates of rates of diffusion in the sediments	Reeburgh (1976)
4.	12·2	Santa Barbara Basin	Same as (3)	Barnes and Goldberg (1976)
5.	~ 7	Salt marsh	Average rate based on several methods with large variability	Atkinson and Hall (1976)
6.	570	Paddy soils	Laboratory incubation followed by extra-polation to the natural environment	Koyama (1963)
7.	33	Lake 227 surface oxygenated sediments	Mass balance calculation including estimates of methane oxidation, accumulation and evasion to the atmosphere	Rudd and Hamilton (1978)

8.	450	Lake 227 anoxic sediments	Same as (7)	Rudd and and Hamilton (1978)
9.	230	Lake Kivu	Calculated on the assumption that 20 per cent of primary production was converted to CH_4	Jannasch (1975)
10.	1010	Lake Kivu	Based on estimated increase in amount of dissolved CH_4 of ~ 1 per cent per year	Deuser *et al.* (1973)
11.	206	Lake Tanganyika	Minimum estimate based on assumption that the rate of methane oxidation at the oxic–anoxic interface equals methane production rate	Rudd (in preparation)
12.	4580	Lake Erie	Analysis of gas bubbles collected in funnels	Howard *et al.* (1971)
13.	230	Third Sister Lake	Mass balance calculation based on estimates of methane accumulation, flux to the oxic–anoxic interface and ebullition	Robertson (1979)
14.	800	Frain's	Same as (13)	Robertson (1979)

components are quantitative estimates of *in situ* methane oxidation and production rates and estimates of movement and storage of dissolved methane within the ecosystem.

It is equally important to be able to relate the methane cycle to the entire carbon cycle of an ecosystem and to know if the activities of the methane oxidizers and producers are having any other important effects. To accomplish these last two objectives a large body of information in addition to methane data is necessary (including rates of carbon input and output, primary production, sedimentation, organic carbon degradation, oxygen concentration, etc.). Probably because of the quantity of ancillary data required there have been few attempts or opportunities to produce this kind of information.

Numerous investigators have monitored methane concentrations in a variety of marine and freshwater environments. (e.g. Atkinson and Richards, 1967; Cappenberg, 1972; Lamontagne *et al.*, 1973). However, these data alone only provide information on the instantaneous quantity of methane and so provide no insight into the rates and effects of methane cycling in aquatic environments. Several other workers (Table 2) have estimated methane production rates from a variety of sediments. Although the accuracy of these estimations probably varies considerably, it is obvious that methane production rates are much faster in freshwater than in salt water environments, probably for the reasons discussed in sections 2.2 and 3.6.

To complete the picture of methane cycling *in situ* estimates of rates of methane oxidation are also necessary. Knowledge of methane oxidation rates are particularly instructive for the following reasons:

1. Methane oxidation rather than production very often determines the distribution of methane in aquatic environments (Reeburgh and Heggie, 1977).
2. Estimates of methane oxidation rates provide information on the quantity of methane–carbon recycled within an ecosystem. The impact of this on the entire carbon cycle can then be assessed.
3. In stratified systems the sum of the rates of methane oxidation within the narrow zone of activity, plus the rate of accumulation of dissolved methane below the zone, and the ebullition rate if it is occurring, is an estimate of methane flux from sediments. This is an accurate method of measurement because the methane production processes in the sediments can function naturally during the rate measurement (Rudd and Hamilton, 1975b).

4. If the production rate of carbon dioxide and cell material by methane oxidizers is known the importance of the methane cycle as a carbon source for primary producers and secondary grazers can be established by comparing these rates to primary production rates.
5. If methane oxidation rates are known consumption rates of O_2 by aerobic methane oxidizers and possibly sulphate by anaerobic oxidizers can be computed. If the supply of O_2 and $SO_4^=$ is also known the importance of consumption via methane cycling and the effects this has on the ecosystem can be assessed.

For these reasons, investigators attempting to quantify methane cycling and look at it in terms of carbon cycling and the effects it has on entire aquatic ecosystems, have almost always concentrated on measurement of *in situ* methane oxidation rates. These efforts are outlined below.

4.2 CARIACO TRENCH

Reeburgh (1976) has studied methane cycling in the Cariaco Trench which has a maximum depth of 1400 m and is permanently anoxic below about 400 meters depth (Presley, 1974). He measured methane

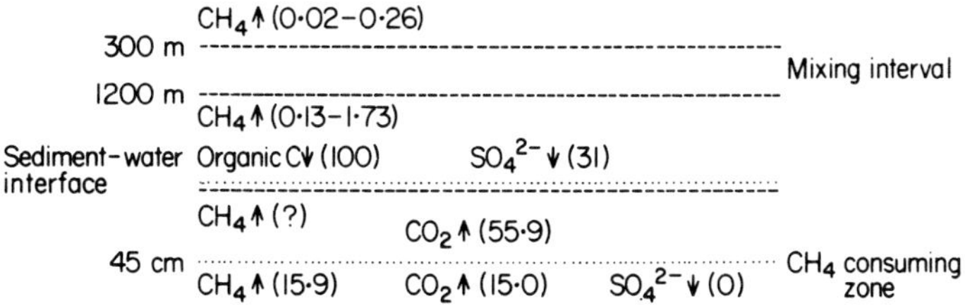

Fig. 10. Summary of methane fluxes in the Cariaco Trench water column and methane, carbon dioxide, sulphate and organic carbon fluxes in the sediments (in μmol cm^{-2} year^{-1}) (from Reeburgh, 1976).

concentration in the anoxic zone of the water column and methane and carbon dioxide concentrations in the sediments and calculated flux rates on the basis of the concentration gradients and diffusion coefficients (corrected for tortuosity in the sediments). He combined these data with estimates of flux of organic carbon to the sediments and sulphate reduction rates within the sediments (Presley, 1974) to produce the methane cycling data shown in Fig. 10. The upward fluxes of methane and carbon dioxide to the methane-consuming zone were approximately equal. Mass balance calculations indicated that anaerobic methane oxidation consumed about 50 per cent of the downward sulphate flux and contributed 25 per cent of the upward carbon dioxide flux from this zone. According to Presley (1974) approximately 40 per cent of the organic carbon flux to the sediments remains there as recalcitrant material. The methane flux across the sediment–water interface could not be quantified but the methane concentration in the anoxic zone of the water column could have been maintained if 1–10 per cent of the methane flux to the methane-consuming zone escaped oxidation and diffused across the sediment–water interface into the water column. Reeburgh was unable to estimate oxidation rates between 1200 m and the bottom but he did estimate that anaerobic methane oxidation consumed 85 per cent of the methane flux into a mixing zone between 300 and 1200 meters. Fifteen per cent of the methane entering this mixing zone (probably less than 1 per cent of the methane production in the sediments) escaped anaerobic methane oxidation and diffused upwards to the oxic–anoxic interface where it may have been largely consumed by aerobic methane-oxidizing bacteria.

About 50–75 per cent of the methane carbon oxidized by aerobic oxidizers is converted to carbon dioxide (Rudd and Hamilton, 1978, Rudd, in preparation) the remainder appears as bacterial cell material. Anaerobic methane oxidation is thought to convert 100 per cent of methane carbon to carbon dioxide (R. S. Hanson, personal communication; Reeburgh, 1976). Therefore in this case since anaerobic methane oxidation was evidently of overwhelming importance virtually all of the methane carbon recycled would have appeared as carbon dioxide. If methane-oxidizing sulphate reducers were responsible for this activity or if sulphate reducers can obtain energy from the co-metabolism of methane, the energy derived from anaerobic oxidation may be indirectly important to the production of organic carbon (bacterial cells) in this deep anoxic environment. This possibility has not been investigated.

The data in Fig. 10 are a first attempt in the marine environment to quantify methane cycling. Although the flux estimates are approximate they provide an interesting insight into the impact methane cycling (particularly anaerobic methane oxidation) may be having on carbon cycling and sulphate reduction.

4.3 LAKE KIVU

Methane cycling has also been studied in Lake Kivu, an African rift lake which has a surface area of about 2370 km^2, a maximum depth of 460 meters and is permanently anoxic below a depth of 50 meters (Jannasch, 1975). It contains an enormous amount of dissolved methane (50 km^3 at STP, Deuser *et al.*, 1973) mostly in the anoxic bottom waters below a depth of 200 meters. This methane has apparently accumulated because of a series of stepwise increases in salinity which severely restrict vertical mixing (Deuser *et al.*, 1973). Methane concentration never exceeds maximum saturation at depth; therefore there is no loss by ebullition.

There have been two estimates of methane production rates from Lake Kivu sediments (Table 2). Jannasch (1975) and Degens *et al.* (1973) assumed that 20 per cent of the photosynthetically fixed carbon ($\sim$ 60 g C m^{-2} $year^{-1}$) reached the sediments; they estimated that about one-third of this was then converted to methane resulting in a methane production rate of about 230 μmol m^{-2} h^{-1}. Deuser *et al.* (1973) have hypothesized that in addition to the methane produced from this biogenic carbon source there was also bacterial methane production from abiogenic hydrogen and carbon dioxide originating from volcanic sources beneath the lake. They site the high ^{13}C content and low ^{14}C content of the dissolved methane, the apparent 20 per cent increase in methane concentration in about 20 years, and the relative scarcity of phosphorous in the bottom waters as evidence that most of the methane did not originate from decomposing organic matter but instead was derived from "old" carbon dioxide of volcanic origin. Based on the apparent rate of increase in the concentration of methane of about 1 per cent per year the methane production rate (calculated from data presented in Deuser *et al.*, 1973) was about 1010 μmol m^{-2} h^{-1}. Thus approximately 80 per cent of methane production may be from volcanic hydrogen and carbon dioxide.

Both of these estimates are rough but do not seem unreasonable when compared to estimates of methane production rates in other productive freshwater lakes (e.g. Lake 227, Third Sister Lake, Frain's Lake and Lake Tanganyika, Table 2). With the small amount of data presently available it is not possible to decide which of the production estimates is more accurate. In fact, there is a possibility that both of the estimates may be too low since bottom water sulphate concentrations are sufficiently high (2·3 mM $SO_4^=$) (Degens *et al.*, 1973) to raise the possibility of significant anaerobic methane oxidation within the sediments.

Whether or not the amount of dissolved methane in the bottom waters is stable or increasing is of little consequence to the aerobic zone above. What is of importance to the oxygenated zone is the amount of methane reaching the oxic–anoxic interface, what happens to it, and what impact this may have on the oxygenated zone. Jannasch (1975) has studied methane oxidation at the interface by monitoring the decrease in methane concentration with time in incubated water samples. Virtually all of the methane that diffused up to the interface was consumed by aerobic oxidizers at an estimated rate of 300 μmol m^{-2} h^{-1} (calculated from data presented by Jannasch, 1975). This is a rough estimate because of the methods used but agrees (within an order of magnitude) with the rate of CH_4 diffusion towards the interface. It is also within 30 per cent of the estimated rate of methane oxidation in the neighbouring Lake Tanganyika (Rudd, in preparation).

If it is assumed that 25 per cent of the methane oxidized was converted to organic matter and the remainder was converted to carbon dioxide, which was the case in Lake Tanganyika (Rudd, in preparation), then the amount of methane carbon appearing as organic matter and carbon dioxide amounted to only about 3 per cent and 8 per cent respectively of the average estimate of annual primary production (calculated from Jannasch, 1975). Thus methane cycling was probably not an important carbon source for either photosynthesizers or secondary grazers. It may, however, have been much more important in longer-term carbon recycling in the lake as was the case for Lake 227 (see section 4.5).

4.4 LAKE TANGANYIKA

Lake Tanganyika is situated south of Lake Kivu in the same rift valley which runs 3000 km across Africa from the Zambesi River in the South to the Red Sea in the North (Hecky and Degens, 1973). It is the second largest lake in the world with a surface area of 34 000 km^3 and a maximum depth of 1400 meters. It is permanently anoxic below a depth of 100–200 meters and has a seasonal thermocline between approximately 25 and 75 meters depth.

During a north to south cruise of the lake, Rudd (in preparation) measured methane concentrations and methane oxidation rates at the oxic–anoxic interface. Methane concentrations in the oxygenated zone of the water column were very low (<0·5 μM) but increased to approximately 80 μM at 350 meters depth. Methane oxidation always occurred in about a 10 meter thick zone at the oxic–anoxic interface consuming virtually all of the methane diffusing upward to the interface. A rough estimate of the whole lake methane oxidation rate was 1·8 mol m^{-2} $year^{-1}$ or 206 μmol m^{-2} h^{-2}. This was considered to be a conservative estimate because it did not account for methane transported to the oxygenated zone by periodic upwelling. Upwelling could simultaneously increase methane and DIN concentration in the oxygenated zone and thus could result in increased methane oxidation rate over a broader depth interval in a manner analogous to periods of overturn in smaller lakes (sections 3.4 and 4.5).

If it is assumed that the total amount of methane was not increasing in the anoxic bottom waters (i.e. the rate of CH_4 diffusion to the oxic–anoxic interface equalled the rate of production) the production rate would equal the oxidation rate (206 μmol m^{-2} h^{-1}) since ebullition did not occur.

Anaerobic methane oxidation would probably have an insignificant effect on estimates of rates of either methane production or oxidation since sulphate concentrations in the bottom waters were very low, approximately 20 μM (Degens *et al.*, 1971).

Hecky *et al.* (1978) have estimated rates of primary production and respiration in Lake Tanganyika during April–May and October–November, 1975. They concluded that the carbon dioxide respired by the entire epilimnetic planktonic community greatly exceeded the amount of carbon dioxide fixed by photosynthesizers. Thus there was a large epilimnetic fixed carbon debt if primary production was considered to be the sole source of fixed carbon production. The methane oxidizers active in Lake Tanganyika assimilate about 26 per cent of methane carbon into cell material (Rudd, in preparation). Thus, some of the carbon deficit (15·4 mg C m^{-2} day^{-1}) could be accounted for by methane oxidation at the oxic–anoxic interface. This contribution by the methane oxidizers may have been even more significant if the additional amount of methane transported to the oxygenated zone by upwelling was known.

Estimates of methane oxidation rates at the oxic–anoxic interface in

Lakes Kivu and Tanganyika agreed to within 30 per cent. Since these oxidation rates must be ultimately controlled by the flux rate of methane from below and since the methane gradient below the depth of zero oxygen concentration was much steeper in Lake Kivu (Degens *et al.*, 1973; Rudd, in preparation) the rate of upward diffusion in Lake Tanganyika must be much faster. This can probably be accounted for by the much smaller lake size and the larger density gradient due to dissolved salts in Lake Kivu. Thus even though the "standing crop" of methane is much lower in Tanganyika (bottom water concentrations were not supersaturated at atmospheric pressure—R. Weiss, personal communication), the effects of methane cycling in terms of carbon recycling, particulate carbon production and oxygen consumption are probably somewhat similar in these two lakes.

4.5 LAKE 227

Lake 227 is located in the Canadian Precambrian Shield. It has a surface area of 5×10^4 m^2 and a maximum depth of 10 meters. Complete circulation usually occurs for only a few days prior to freeze-up. During most of the remainder of the year the bottom waters are anoxic and contain a large quantity of methane (Rudd and Hamilton, 1975b) which is a result of the artificial eutrophication of the lake by additions of phosphorous and nitrogen fertilizers (Schindler *et al.*, 1973).

A comprehensive study of the annual methane cycle of Lake 227 and its effects on whole lake metabolism was conducted over a 26-month period by Rudd and Hamilton (1978). They monitored methane concentrations in the oxygenated and anoxic portions of the lake and evasion rates from the lake surface. They also estimated whole lake methane oxidation rates by a ^{14}C-tracer technique (Rudd *et al.*, 1974; Rudd and Hamilton, 1975a). From these data they were able to estimate methane production rates from sediments whose surfaces were either oxygenated or deoxygenated. Seasonal and annual whole lake methane budgets could then be computed.

4.5.1 *Seasonal methane budgets*

a. *Summer*. During summer stratification vertical diffusion from the hypolimnion was very slow. As a result during the summer of 1974 only 11 per cent of the methane produced by the hypolimnetic sediments (at an areal rate of 450 μmol m^{-2} hypolimnetic sediment per hour) reached the oxic–anoxic interface within the thermocline (Fig. 11). The

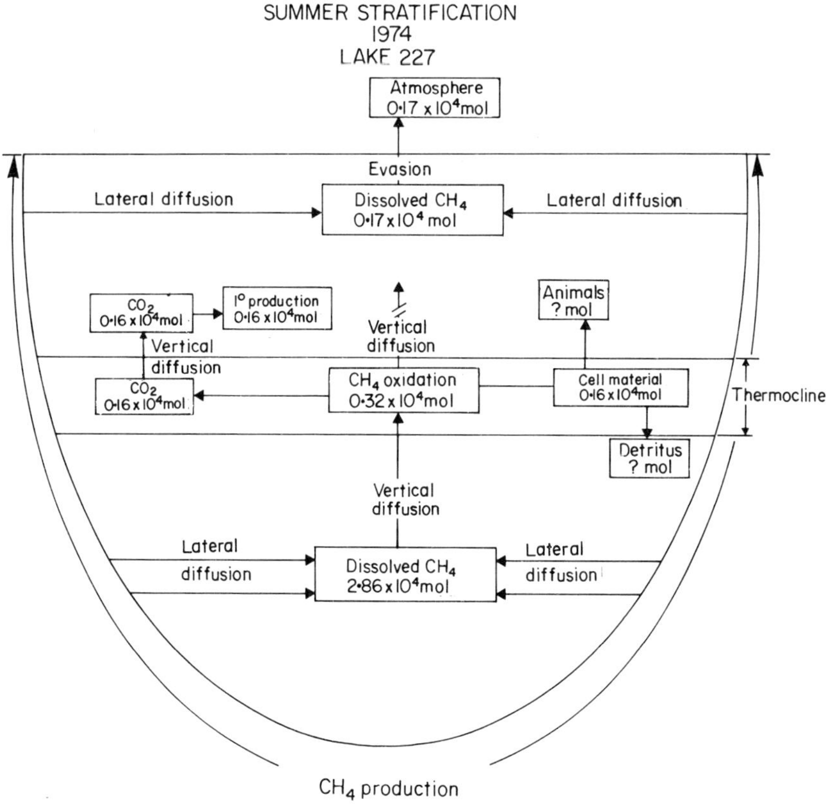

Fig. 11. Production, oxidation and evasion of dissolved methane during summer stratification, May 21 to August 30, 1974. Surface areal rates can be calculated by dividing by 5×10^4 m^2. (From Rudd and Hamilton, 1978).

remainder progressively accumulated in the hypolimnion (Fig. 12a). The small amount of methane carbon that diffused up to the interface was converted (Figs 13a and 13b) to bacterial cell material and carbon dioxide in a 50:50 ratio (Fig. 11). The cell material was either consumed by animal grazers or returned to the sediments as detritus. The carbon dioxide produced via methane oxidation diffused up a carbon dioxide concentration gradient and into the epilimnion was fixed by primary producers since a large carbon dioxide deficit was typical of Lake 227 during summer (Schindler *et al.*, 1972).

There was almost no upward movement of methane into the epilimn-

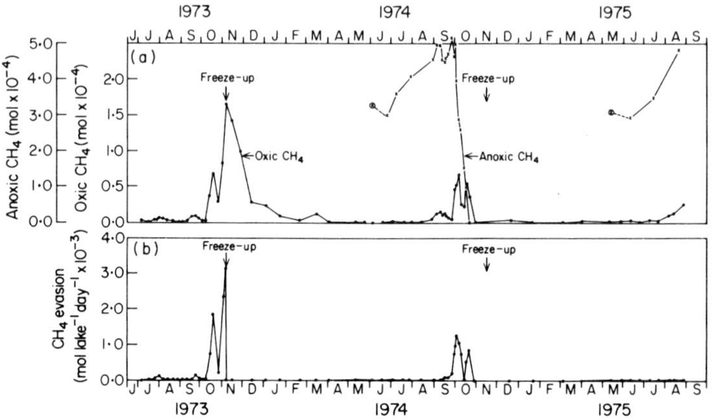

Fig. 12(a). Dissolved methane in oxygenated water (oxic methane) and de-oxygenated hypolimnetic water (anoxic methane) during a 26 month period in Lake 227.

(b). Methane evasion rates from Lake 227 during the same period.

(From Rudd and Hamilton, 1978.)

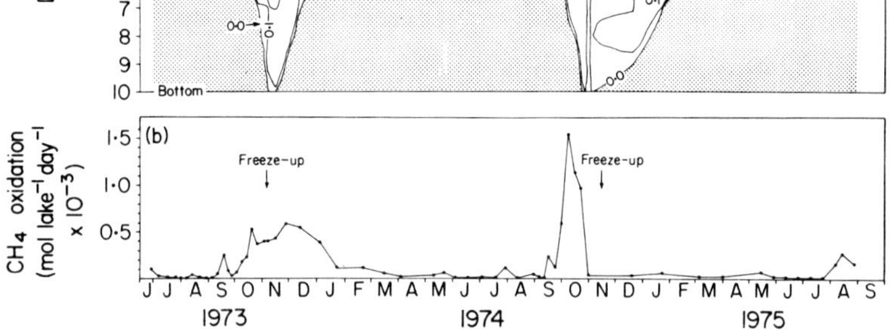

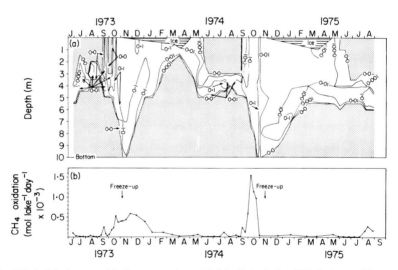

Fig. 13(a). Methane oxidation rates (μmol l^{-1} h^{-1}) in Lake 227 during a 26 month period. The stippled areas are zones of zero methane oxidation.

(b). Whole lake rates of methane oxidation during the same period. Surface areal rates can be calculated by dividing by 5×10^4 m^2.

(From Rudd and Hamilton, 1978.)

ion during summer stratification because of consumption at the oxic–anoxic interface (Fig. 13a) and a lack of ebullition from the hypolimnetic sediments. Thus the relatively small amounts of epilimnetic methane (Figs 3 and 12a) must have originated from the epilimnetic sediments. Most of the methane diffusing from these sediments was probably consumed by aerobic oxidizers known to be active in the surface of sediments. The methane which escaped oxidation diffused horizontally into the epilimnion at an estimated areal rate of 33 μmol m^{-2} epilimnetic sediments per hour. Since epilimnetic oxidation was absent during summer because of the combined inhibitory effects of high oxygen and low DIN concentration (section 3.4 and Fig. 17a) this methane slowly evaded to the atmosphere (Figs 11 and 12b).

b. *Fall.* During fall overturn (e.g. August 31 to November 15, 1974) a total of $4{\cdot}55 \times 10^4$ mol of methane was released from the hypolimnion

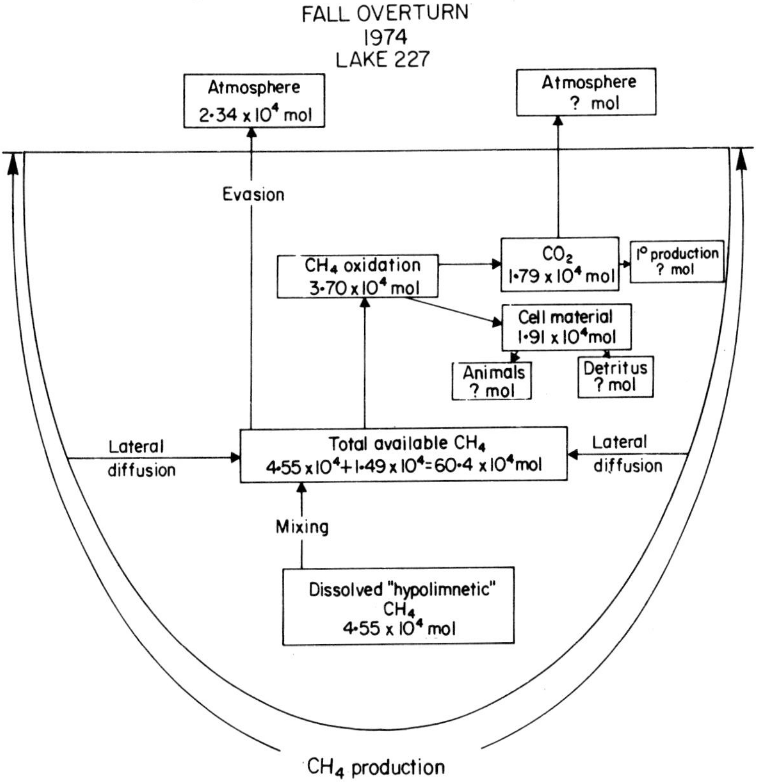

Fig. 14. Production, oxidation and evasion of methane during fall overturn, August 31 to November 15, 1974 (from Rudd and Hamilton, 1978).

(Figs 14 and 12a). This amount included the $2 \cdot 86 \times 10^4$ mol produced during the summer plus the amount of methane accumulated in the hypolimnion since the previous fall overturn. Another $1 \cdot 5 \times 10^4$ mol diffused from the sediments during overturn so that about $6 \cdot 0 \times 10^4$ mol of methane were available for methane oxidation during fall overturn (Fig. 14).

As a result of mixing of oxygenated water with anoxic water containing high concentrations of methane and DIN, methane oxidation rates increased dramatically throughout the oxygenated portion of the water column (Figs 13a and 13b). Since methane concentrations in the surface water increased, evasion rates also increased (Fig. 12b). During the fall of 1974 about 40 per cent of the methane in the lake escaped to the atmosphere (Fig. 14). The remainder was recycled by methane oxidizers in a 50:50 ratio as cell material and carbon dioxide (Fig. 14). In this case most of the carbon dioxide produced would have probably

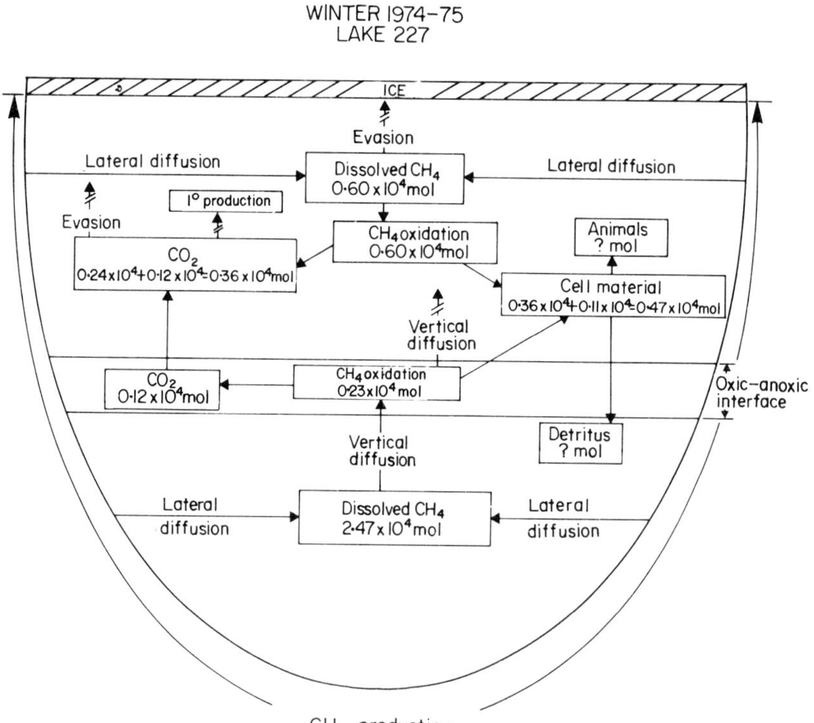

Fig. 15. Production and oxidation of methane under winter ice, November 16, 1974 to May 7, 1975 (from Rudd and Hamilton, 1978).

escaped to the atmosphere since Lake 227 is typically supersaturated with carbon dioxide at this time of year (Schindler *et al.*, 1972) and primary production rates are reduced.

c. *Winter.* During ice-cover the lake restratified and an anoxic hypolimnion developed. The hypolimnetic sediments produced methane at about the same rate as during the summer (533 μmol m^{-2} hypolimnetic sediments per hour), suggesting that primary production and methane production were remotely linked (Rudd and Hamilton, 1978, 1979). During the winter of 1974–75 less than 10 per cent of the hypolimnetic methane diffused to the oxic-anoxic interface and was consumed (Fig. 15). Most of the remainder was still present in the hypolimnion at the beginning of the 1975 fall overturn.

Most of the methane diffusing from the epilimnetic sediments during the winter of 1974–75 was consumed by methane oxidizers (Fig. 15) since they remained active at high oxygen concentration through the winter (Fig. 13 and section 3.4). Consequently the amount of methane in the oxygenated zone of the lake did not increase during the winter (Fig. 12a). The ice and snow cover prevented gas exchange with the atmosphere (Fig. 12b) and prevented penetration of light. Primary production was thus terminated and the CO_2 produced from the oxidation of methane accumulated in the oxygenated portion of the water column (Fig. 15).

4.5.2 *Participation of methane in the carbon cycle of Lake 227*

Rudd and Hamilton (1978, 1979) also attempted to put the annual methane cycle of Lake 227 into proper perspective with the overall carbon cycle of the lake by comparing the annual methane production, oxidation and evasion to annual primary production and to the total carbon input. Total carbon input includes all of the carbon entering the lake from the atmosphere and runoff. In comparison to primary production the amounts of carbon dioxide and bacterial cell material produced by methane oxidizers was insignificant (Table 3). Thus methane oxidation was not an important carbon source to either secondary grazers or primary producers. Methane cycling was much more important when compared to the total carbon input into the lake. The amount of carbon regenerated as methane was equal to about 55 per cent of the total carbon input into the lake during 1974 (Table 3).

TABLE 3

A comparison of annual rates of methane production and oxidation to annual rates of primary production and total carbon input in Lake 227. Primary production and total carbon input occurred mostly during the summer of 1974. The methane production and oxidation year included the succeeding winter since regeneration of methane carbon fixed during the previous summer would be occurring. (From Rudd and Hamilton, 1978.)

	$g\ C\ m^{-2}$ lake surface
CH_4 production (ice-out 1974 to ice-out 1975)	18
Total CH_4 oxidation (ice-out 1974 to ice-out 1975)	12
CH_4 oxidizing bacterial production (ice-out 1974 to ice-out 1975)	6
CO_2 production by CH_4 oxidizers (ice-out 1974 to ice-out 1975)	6
Primary production (1974)[a]	138
Total carbon input (1974)[b]	33

[a]E. J. Fee (personal communication).
[b]D. W. Schindler (personal communication).

Methane oxidation recycled two-thirds of this carbon (an amount equal to 36 per cent of total carbon input during 1974).

The large differences between the amount of carbon entering the lake and the amount fixed by primary producers (Table 3) probably resulted from rapid aerobic recycling of carbon a number of times in the epilimnion during the summer months when most of the primary production occurred. Rudd and Hamilton (1979) estimated that only about 25 per cent of the particulate carbon produced in the epilimnion was deposited on the anaerobic hypolimnetic sediments. The rapid aerobic carbon recycling explains why methane was not an important carbon source for primary or secondary producers but was an important contributor to longer-term (at least annual) whole lake carbon cycling.

4.5.3 *Effects of methane cycling on dissolved oxygen concentrations*

The most dramatic effect of methane cycling was on dissolved oxygen concentration under winter ice-cover (Table 4). If the lake mixed completely for a few weeks prior to freeze-up (e.g. October, 1974, of Figs 12 and 13) almost all of the methane either was oxidized or escaped to the atmosphere before freeze-up. In this case, the amount of methane

TABLE 4

Contribution of methane-oxidizing activity under winter ice-cover to the disappearance of dissolved oxygen in Lake 227 (from Rudd and Hamilton, 1978).

Total Lake 227 values (mol)	Winter of 1973–74	Winter of 1974–75
1. Dissolved CH_4 at freeze-up	$1 \cdot 7 \times 10^4$	11·0
2. Dissolved O_2 at freeze-up	$5 \cdot 5 \times 10^4$	$5 \cdot 9 \times 10^4$
3. Per cent O_2 consumed by (1)	45%[a]	0·03%
4. CH_4 oxidation under ice	$4 \cdot 0 \times 10^4$	$0 \cdot 8 \times 10^4$
5. Per cent of O_2 consumed by (4)	110%	21%
6. Dissolved O_2 one month before ice-out	0·0	$1 \cdot 4 \times 10^4$

[a]An O_2 CH_4 ratio of 1·5:1 was used to calculate the amount of oxygen consumed during methane oxidation.

oxidized during the next winter was small (Fig. 12) and a relatively large amount of dissolved oxygen was present in the lake just prior to spring break-up (Table 4). If, on the other hand, a large quantity of methane was trapped under the ice by a quick freeze-up (e.g. October to November, 1973, Fig. 12), methane oxidation continued rapidly under the ice during the next winter (Fig. 13). Rudd and Hamilton (1978) calculated that during the winter of 1973–74 methane oxidizers consumed 110 per cent of the oxygen present in the lake at freeze-up 1974 (Table 4). Since this oxygen could not be replaced by either atmospheric invasion or photosynthesis, the lake became entirely anoxic before the ice thawed in the spring (Table 4). As a result, widespread suffocation of fish, zooplankton and zoobenthos was observed.

4.6 FRAIN'S AND THIRD SISTER LAKES

Frain's and Third Sister Lakes are small well-stratified hardwater lakes whose bottom waters become anoxic during summer stratification. Robertson (1979) has related rates of particulate carbon deposition on to the anoxic sediments of these lakes to *in situ* methane production rates from the sediments. As in Lake 227 (section 4.5) she found that methane production contributed very significantly to carbon regeneration from the anoxic sediments. In the highly eutrophic Frain's Lake methane production regenerated 59 per cent of the carbon deposited. In the less productive Third Sister Lake methane carbon regeneration amounted to 36 per cent of carbon deposition. Most of this production occurred in an organic rich layer of recently deposited material within 3 cm of the sediment–water interface. Robertson's results were also very similar to Lake 227 in that less than 10 per cent of the methane produced during summer stratification diffused vertically out of the axonic bottom waters into the oxygenated layer. Unlike Lake 227 methane was released directly to the atmosphere via ebullition from the sediments. This ebullition amounted to 6 per cent of carbon input to the sediments in Third Sister Lake and 28 per cent in the more productive Frain's Lake. Thus in certain productive lakes the loss of methane carbon from the lake surface may have a significant impact on the whole lake carbon cycle since this carbon can not be recycled by methane-oxidizing bacteria.

In Fig. 16 the rate of particulate carbon sedimentation is plotted against rate of methane production from sediments of four different lakes. Although the range of ratios of methane carbon release to carbon input varied from 0·36 to 0·60 it appears that there is at least a gross relationship between sedimentation rates and methane production rates. Robertson (1979) has concluded from these data that methane production rates are primarily controlled by rate of carbon input to the sediments and not primarily by temperature as

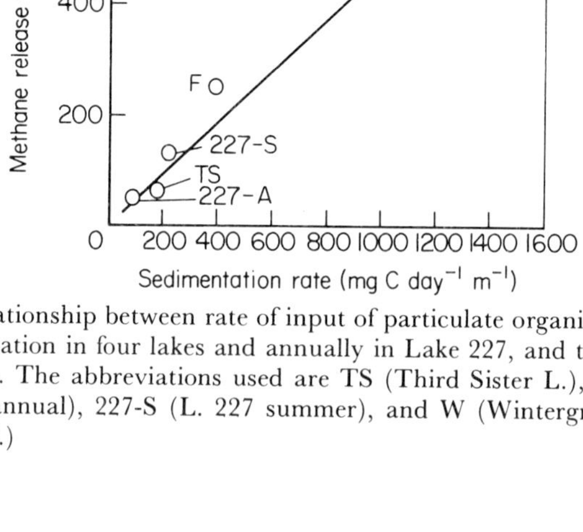

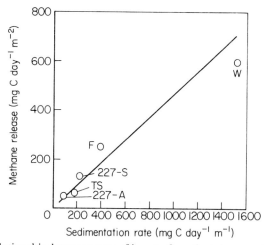

Fig. 16. The relationship between rate of input of particulate organic carbon during summer stratification in four lakes and annually in Lake 227, and the rate of *in situ* methane release. The abbreviations used are TS (Third Sister L.), F (Frain's L.), 227-A (L. 227 annual), 227-S (L. 227 summer), and W (Wintergreen L.). (From Robertson, 1979.)

suggested by Zeikus and Winfrey (1976). This conclusion is supported by *in vitro* incubation of Frain's Lake and Third Sister Lake sediments. In these experiments (Robertson, unpublished data) it was found that although the Q_{10} value for methane production in both sediments was approximately 2, the richer Frain's Lake sediments produced methane at an average rate of 2·2 times faster than the Third Sister Lake sediments. Rudd and Hamilton (1979) have concluded that although methane production rates within sediments are probably influenced by temperature fluctuations the rate of methane diffusion from both aerobic and anaerobic Lake 227 sediments was independent of temperature. This may be explained by the constant year around temperatures of anoxic hypolimnetic sediments and because both methane production and methane oxidation at the epilimnetic sediment–water interface varied with temperature resulting in no net change in the rate of methane diffusion into the epilimnium of Lake 227.

Robertson (1979) also suggested that because of the relationship shown in Fig. 16 the rate of methane release to the water columns may be a good indicator of the rate of carbon entry and turnover in anoxic sediments. This is an attractive possibility because in certain lakes (e.g. Frain's, Third Sister and 227) methane production rates can be accurately monitored by the *in situ* techniques described by Robertson (1979) and Rudd and Hamilton (1975b, 1978).

5 Conclusions

Although methane cycling has always contributed to overall aquatic metabolism it has recently become more important as a result of the cultural eutrophication of many aquatic ecosystems. One of the effects of this eutrophication is increased organic loading of sediments which promotes the development of anoxic hypolimnia in lakes and reducing conditions in freshwater and marine sediments. As a result there is a shift from carbon dioxide production towards methane production in sediments since anaerobically produced carbon dioxide is used as a terminal electron acceptor in place of oxygen or sulphate.

Because of this increased contribution of methane cycling in aquatic environments an understanding of the *in situ* activites of methane producers and oxidizers is important now and will probably be even more important in the future in predicting the effects of progressive cultural eutrophication. Our knowledge of these processes is just beginning to coalesce. Some of the chemical and physical factors controlling *in situ* rates of methane production and oxidation have been identified. The fundamental differences between methane cycling in freshwater and marine environments have been recognized. In a few cases the impact of the methane cycle on the overall carbon cycle of an ecosystem has been assessed more or less completely and has been found to be significant, especially in highly productive ecosystems. Also, when large quantities of methane are cycling through a system there appears to be a significant impact upon cycling of various nutrients and the consumption of dissolved oxygen.

Acknowledgements

The useful advice and constructive criticisms of H. W. Jannasch and C. A. Kelly are gratefully acknowledged.

References

Anonymous (1976). Fire in the lake. *Scientific American*, October, **235**, 65A–65B.
Atkinson, L. P. and Hall, J. R. (1976). Methane distribution and production in the Georgia Salt Marsh. *Estuarine and Coastal Marine Science*, **4**, 677–686.
Atkinson, L. P. and Richards, F. A. (1967). The occurrence and distribution of methane in the marine environment. *Deep-Sea Research*, **14**, 673–684.
Balch, W. E. and Wolfe, R. S. (1976). New approach to the cultivation of methanogenic bacteria: 2-mercaptoethane-sulfuric acid (HS-CoM)-dependent growth of *Methanobacterium ruminantium* in a pressurized atmosphere. *Applied and Environmental Microbiology*, **32**, 781–791.
Barker, H. A. (1956). Biological formation of methane. *In* "Bacterial Fermentation", p. 1. John Wiley, New York.
Barnes, R. O. and Goldberg, E. D. (1976). Methane production and consumption in anoxic marine sediments. *Geology*, **4**, 297–300.
Belyaev, S. S., Finkel'shtein, Z. I. and Ivanov, M. V. (1975a). Intensity of bacterial methane formation in ooze deposits of certain lakes. *Microbiology*, **44**, 272–275.
Belyaev, S. S., Laurinavichus, K. S. and Ivanov, M. V. (1975b). Determination of the intensity of the process of microbiological oxidation of methane using $^{14}CH_4$, *Microbiology*, **44**, 478–481.
Bishop, J. K. B. (1977). The chemistry, biology and vertical flux of oceanic particulate matter. Ph.D. thesis. Massachusetts Institute of Technology/Woods Hole Oceanographic Institution Joint Program in Oceanography. 292 pp.

Brooks, J. M. and Sackett, W. M. (1973). Sources, sinks, and concentrations of light hydrocarbons in the Gulf of Mexico. *Journal of Geophysical Research*, **78**, 5248–5258.

Brooks, J. M., Fredericks, A. D., Sackett, W. M. and Swinnerton, J. W. (1973). Baseline concentrations of light hydrocarbons in Gulf of Mexico. *Environmental Science and Technology*, **7**, 639–642.

Brown, L. R., Strawinski, R. J. and McCleskey, C. S. (1964). The isolation and characterization of *Methanomonas methanooxidans* Brown and Strawinski. *Canadian Journal of Microbiology*, **10**, 791–799.

Bryant, M. P., Wolin, E. A., Wolin, M. J. and Wolfe, R. S. (1967). *Methanobacillus omelianskii*, a symbiotic association of two species of bacteria. *Archiv für Mikrobiologie*, **59**, 20–31.

Bryant, M. P., Tzeng, S. F., Robinson, I. M. and Joyner, A. E. (1971). Nutrient requirements of methanogenic bacteria. *In* "Anaerobic Biological Treatment Processes Advances in Chemistry Series 105" (Ed. F. G. Pohland), pp. 23–40. American Chemical Society, Washington, D.C.

Bryant, M. P., Campbell, L. L., Reddy, C. A. and Crabill, M. R. (1977). Growth of *Desulfovibrio* in lactate or ethanol media low in sulfate in association with H_2-utilizing methanogenic bacteria. *Applied and Environmental Microbiology*, **33**, 1162–1169.

Buswell, A. M. and Sollo, F. W. (1948). The mechanism of the methane fermentation. *Journal of the American Chemical Society*, **70**, 1778–1780.

Cappenberg, Th. E. (1972). Ecological observations on heterotrophic methane oxidizing and sulfate reducing bacteria in a pond. *Hydrobiologia*, **40**, 471–485.

Cappenberg, Th. E. (1974a). Interrelations between sulfate-reducing and methane-producing bacteria in bottom deposits of a freshwater lake. I. Field observations. *Antonie van Leeuwenhoek*, **40**, 285–295.

Cappenberg. Th. E. (1974b). Interrelations between sulfate-reducing and methane-producing bacteria in bottom deposits of a freshwater lake. II. Inhibition experiments. *Antonie van Leeuwenhoek,* **40**, 297–306.

Cappenberg, Th. E. (1975). A study of mixed continuous cultures of sulfate-reducing and methane-producing bacteria. *Microbial Ecology*, **2**, 60–72.

Cappenberg, Th. E. (1976). Methanogenesis in the bottom deposits of a small stratifying lake. *In* "Microbial Production and Utilization of Gases" (Eds H. G. Schlegel, G. Gottschalk and N. Pfennig), pp. 125–134. E. Goltze KG, Göttingen.

Cappenberg, Th. E. and Jongejan, E. (1978). Microenvironments for sulfate reduction and methane production in freshwater sediments. *In* "Biogeochemistry of Defined Microenvironments in Aquatic and Terrestrial Systems" (Ed. W. E. Krumbein). Proceedings of the Third International Symposium of Environmental Biogeochemistry.

Cappenberg, Th. E. and Prins, R. A. (1974). Interrelations between sulfate-reducing and methane-producing bacteria in bottom deposits of freshwater lake. III. Experiments with ^{14}C-labelled substrates. *Antonie van Leeuwenhoek*, **40**, 457–469.

Chan, Y. K. (1977). Denitrification and phytoplankton assimilation of nitrate in Lake 227 during summer stratification. Ph.D. dissertation. University of Manitoba.

Claypool, G. E. and Kaplan, I. R. (1974). The origin and distribution of methane in marine sediments. *In* "Natural Gases in Marine Sediments" (Ed. I. R. Kaplan), pp. 99–139. Plenum Press, New York and London.

Cosper, T. C. and Reeve, M. R. (1975). Digestive efficiency of the chaetognath *Sagitta hispida* (Conant). *Journal of Experimental Marine Biology and Ecology*, **17**, 33–38.

Coty, V. (1967). Atmospheric nitrogen fixation by hydrogen-oxidizing bacteria. *Biotechnology and Bioengineering*, **9**, 25–32.

Craig, H. and Weiss, R. F. (1968). Argon concentrations in the ocean: a discussion. *Earth Planetary Science Letters*, **5**, 175–183.

Dalton H. and Whittenbury, R. (1976). Nitrogen metabolism in *Methylococcus capsulatus* (strain Bath). *In* "Microbial Production and Utilization of Gases" (Eds H. G. Schlegel, G. Gottschalk, and N. Pfennig), pp. 379–388. E. Goltze KG, Göttingen.

Davey, J. F., Whittenbury, R. and Wilkinson, J. F. (1972). The distribution in the methylobacteria of some key enzymes concerned with intermediary metabolism. *Archiv für Mikrobiologie*, **87**, 359–366.

Davies, S. L. and Whittenbury, R. (1970). Fine structure of methane and other hydrocarbon-utilizing bacteria. *Journal of General Microbiology*, **61**, 227–232.

Davies, T. R. (1973). Isolation of bacteria capable of utilizing methane as a hydrogen donor in the process of denitrification. *Water Research*, **7**, 575–579.

Davis, J. B. and Yarbrough, H. F. (1966). Anaerobic oxidation of hydrocarbons by *Desulfovibrio desulfuricans*. *Chemical Geology*, **1**, 137–144.

Davis, J. B., Coty, V. F. and Stanley, J. P. (1964). Atmospheric nitrogen fixation by a methane-oxidizing bacteria. *Journal of Bacteriology*, **88**, 468–472.

de Boer, W. E. and Hazeu, W. (1972). Observations on the fine structure of a methane-oxidizing bacterium. *Antonie van Leeuwenhoek*, **38**, 33–47.

de Bont, J. A. M. (1976). Bacterial degradation of ethylene and the acetylene reduction test. *Canadian Journal of Microbiology*, **22**, 1060–1062.

deBont, J. A. M. and Mulder, E. G. (1974). Nitrogen fixation and co-oxidation of ethylene by a methane-utilizing bacterium. *Journal of General Microbiology*, **83**, 113–121.

deBont, J. A. M. and Mulder, E. G. (1976). Invalidity of the acetylene reduction assay in alkane-utilizing, nitrogen-fixing bacteria. *Applied and Environmental Microbiology*, **31**, 640–647.

Degens, E. T., von Herzen, R. P. and Wong, H. K. (1971). Lake Tanganyika: Water chemistry, sediments, geological structure. *Die Naturwissenschaften*, **58**, 229–241.

Degens, E. T., von Herzen, R. P., Wong, H. K., Deuser, W. G. and Jannasch, H. W. (1973). Lake Kivu: Structure, chemistry and biology of an East African rift lake. *Sonderdruck aus der Geologischen Rundschau*, **62**, 245–277.

Deuser, W. G., Degens, E. T., Harvey, G. R. and Rubin, M. (1973). Methane in Lake Kivu: new data bearing on its origin. *Science*, **181**, 51–54.

Dworkin, M. and Foster, J. W. (1956). Studies on *Pseudomonas methanica* (Söhngen) nov. comb. *Journal of Bacteriology*, **72**, 646–659.

Eccleston, M. and Kelly, D. P. (1972). Competition among amino acids for incorporation into *Methylococcus capsulatus*. *Journal of General Microbiology*, **73**, 303–314.

Eccleston, M. and Kelly, D. P. (1973). Assimilation and toxicity of some exogeneous C_1 compounds, alcohols, sugars, and acetate in the methane-oxidizing bacterium *Methylococcus capsulatus*. *Journal of General Microbiology*, **75**, 211–221.

Emery, K. O. and Hoggan, D. (1958). Gases in marine sediments. *Bulletin of the American Association of Petroleum Geologists*, **42**, 2174–2188.

Enhalt, D. H. (1976). The atmospheric cycle of methane. *In* "Microbial Production and Utilization of Gases" (Eds H. G. Schlegel, O. Gottschalk, and N. Pfennig), pp. 13–22. E. Goltze KG, Göttingen.

Ferry, J. G. and Peck, H. D. (1977). Relationship between sulfate reduction and methane production in salt marsh sediment. Abstracts of the Annual Meeting of the American Society for Microbiology, N49, p. 236.

Ferry, J. G. and Wolfe, R. S. (1976). Anaerobic degradation of benzoate to methane by a microbial consortium. *Archives of Microbiology*, **107**, 33–40.

Flett, R. J. (1977). Nitrogen fixation in Canadian precambrian shield lakes. Ph.D. dissertation. University of Manitoba.

Flett, R. J., Rudd, J. W. M. and Hamilton, R. D. (1975). Acetylene reduction assays for nitrogen fixation in freshwaters: a note of caution. *Applied Microbiology*, **29**, 580–583.

Foster, J. W. and Davis, R. H. (1966). A methane-dependent coccus, with notes on classification and nomenclature of obligate, methane-utilizing bacteria. *Journal of Bacteriology*, **91**, 1924–1931.

Fuhs, W. G. (1961). Der mikrobielle Abbay von Kohlenwasserstoffen. *Archiv für Mikrobiologie*, **39**, 374–422.

Hammond, D. E., Simpson, H. J. and Mathieu, G. (1975). Methane and Radon-222 as tracers of mechanisms of exchange across the sediment–water interface in the Hudson River Estuary. *In* "Marine Chemistry in the Coastal Environment" (Ed. I. Church), pp. 119–132. American Chemical Society, Washington, D.C.

Harwood, J. H. and Pirt, S. J. (1972). Quantitative aspects of growth of the methane oxidizing bacterium *Methylococcus capsulatus* on methane in shake flask and continuous chemostat culture. *Journal of Applied Bacteriology*, **35**, 597–607.

Hazeu, W. (1975). Some cultural and physiological aspects of methane-utilizing bacteria. *Antonie van Leeuwenhoek*, **41**, 121–134.

Hazeu, W. and Steenis, P. J. (1970). Isolation and characterization of two vibrio-shaped methane-oxidizing bacteria. *Antonie van Leeuwenhoek*, **36**, 67–72.

Hecky, R. E., and Degens, E. T. (1973). Late pleistocene-holocene chemical stratigraphy and paleolimnology of the rift lakes of central Africa. Woods Hole Oceanographic Institution Technical Report 73–28.

Hecky, R. E., Fee, E. J., Kling, H. J. and J. W. M. Rudd (1978). Studies of the planktonic ecology of Lake Tanganyika. Fisheries and Marine Services Technical Report. In press.

Hesslein, R. H. (1976a). An *in situ* sampler for close interval pore water studies. *Limnology and Oceanography*, **21**, 912–914.

Hesslein, R. H. (1976b). The fluxes of CH_4, ΣCO_2, and NH_3-N from sediments and their consequent distribution in a small lake. Ph.D. dissertation. Columbia University, Columbia, Ohio.

Hobson, P. N., Bousfield, S. and Summers, R. (1974). Anaerobic digestion of organic matter. *CRC Critical Reviews in Environmental Control*, **4**, 131–191.

Howard, D. L., Frea, J. I. and Pfister, R. M. (1971). The potential for methane carbon cycling in Lake Erie. *Proceedings of the 14th Conference on Great Lakes Research*. pp. 236–240.

Hungate, R. E. (1966). "The Rumen and its Microbes." Academic Press, New York and London.

Hungate, R. E. (1967). Hydrogen as an intermediate in the rumen fermentation. *Archiv für Mikrobiologie*, **59**, 158–164.

Hungate, R. E. (1975). The rumen microbial ecosystem. *Annual Review of Ecology and Systematics*, **6**, 39–66.

Hungate, R. E., Bryant, M. P. and Mah, R. A. (1964). The rumen bacteria and protozoa. *Annual Review of Microbiology*, **18**, 131–166.

Hutchinson, D. W., Whittenbury, R. and Dalton, H. (1976). A possible role of free radicals in the oxidation of methane by *Methylococcus capsulatus*. *Journal of Theoretical Biology*, **58**, 325–335.

Hutchinson, G. E. (1957). "A Treatise on Limnology," vol. 1. John Wiley, New York.

Hutton, W. E. (1948). Studies on bacteria which oxidize methane. Ph.D. Dissertation, University of California at Los Angeles.

Hutton, W. E. and Zobell, C. E. (1949). The occurrence and characteristics of methane-oxidizing bacteria in marine sediments. *Journal of Bacteriology*, **58**, 463–473.

Hutton, W. E. and Zobell, C. E. (1953). Production of nitrite from ammonia by methane oxidizing bacteria. *Journal of Bacteriology*, **65**, 216–219.

Iannotti, E. L., Kafkewitz, D., Wolin, M. J. and Bryant, M. P. (1973). Glucose fermentation products of *Ruminococcus albus* grown in continuous culture with *Vibrio succinogenes*; changes caused by interspecies transfer of H_2. *Journal of Bacteriology*, **114**, 1231–1240.

Jannasch, H. W. (1960). Versuche über Denitrifikation und die Verfügbarkeit des Sauerstoffes in Wasser und Schlamm. *Archiv für Hydrobiologie*, **56**, 355–369.

Jannasch, H. W. (1975). Methane oxidation in Lake Kivu (Central Africa). *Limnology and Oceanography*, **20**, 860–864.

Jeris, J. S. and McCarty, P. L. (1965). The biochemistry of methane fermentation using ^{14}C-tracers. *Journal of the Water Pollution Control Federation*, **37**, 181–197.

Jørgensen, B. B. (1977a). Bacterial sulfate reduction within reduced microniches of oxidized marine sediments. *Marine Biology*, **41**, 7–17.

Jørgensen, B. B. (1977b). The sulfur cycle of a coastal marine sediment (Limfjorden, Denmark). *Limnology and Oceanography*, **22**, 814–832.

Jørgensen, B. B. (1978a). A comparison of methods for the quantification of bacteria sulfate reduction in coastal marine sediments. I. Measurement with radiotracer techniques. *Journal of Microbial Geochemistry.* In press.

Jørgensen, B. B. (1978b). A comparison of methods for the quantification of bacterial sulfate reduction in coastal marine sediments. II. Calculation from radiotracer techniques. *Journal of Microbial Geochemistry.* In press.

Jørgensen, B. B. and Fenchel, T. (1974). The sulfur cycle of a marine sediment model system. *Marine Biology*, **24**, 189–201.

Kaserer, H. (1906). Über die Oxydation des Wasserstoffes und des Methans durch Mikroorganismen. *Zentralblatt für Bakteriologie Parasitenkenkunde Infektionskrankheiten.* Abt. 2, **15**, 573–576.

Kosaric, N. and Zajic, J. E. (1974). Microbial oxidation of methane and methanol. *Advances in Biochemical Engineering*, **3**, 89–125.

Koyama, T. (1963). Gaseous metabolism in lake sediments, paddy soils, and the production of atmospheric methane and hydrogen. *Journal of Geophysical Research*, **68**, 3971–3973.

Lamontange, R. A., Swinnerton, J. W., Linnenbom, V. J. and Smith, W. D. (1973). Methane concentrations in various marine environments. *Journal of Geophysical Research*, **78**, 5317–5324.

Lasker, R. (1970). Utilization of zooplankton energy by a Pacific sardine population in the California current. *In* "Marine Food Chains" (Ed. Steele, J. H.), pp. 265–284. University of California Press, Berkeley, California.

Latham, M. J. and Wolin, M. J. (1977). Fermentation of cellulose by *Ruminococcus flavefaciens* in the presence and absence of *Methanobacterium ruminantium. Applied and Environmental Microbiology*, **34**, 297–301.

Lawrence, A. J. and Quayle, J. R. (1970). Alternative carbon assimilation pathways in methane-utilizing bacteria. *Journal of General Microbiology*, **63**, 371–374.

Leadbetter, E. R. and Foster, J. W. (1958). Studies on some methane-utilizing bacteria. *Archiv für Mikrobiologie*, **30**, 91–118.

Leadbetter, E. R. and Foster, J. W. (1960). Bacterial oxidation of gaseous alkanes. *Archiv für Mikrobiologie*, **35**, 92–104.

LeGall, J. and Postgate, J. R. (1973). The physiology of sulfate-reducing bacteria. *Advances in Microbial Physiology*, **10**, 81–133.

Levy, H. (1973). Tropospheric budgets for methane, carbon monoxide and related species. *Journal of Geophysical Research*, **78**, 5325–5332.

Macgregor, A. N. and Keeney, D. R. (1973). Methane formation by lake sediments during *in vitro* incubation. *Water Resources Bulletin*, **9**, 1153–1158.

Mah, R. A., Ward, D. M., Baresi, L. and Glass, T. L. (1977). Biogenesis of methane. *Annual Review of Microbiology*, **31**, 309–341.

Marshall, S. M. and Orr, A. P. (1955). The biology of a marine copepod *Calanus finmarchicus* (Gunnerus). Oliver and Boyd, London.

Martens, C. S. (1976). Control of methane sediment–water bubble transport by macroinfaunal irrigation in Cape Lookout Bight, North Carolina. *Science*, **192**, 998–1000.

Martens, C. S., and Berner, R. A. (1974). Methane production in the interstitial waters of sulfate-depleted marine sediments. *Science*, **185**, 1167–1169.

Martens, C. S. and Berner, R. A. (1977). Interstitial water chemistry of anoxic Long Island Sound sediments. I. Dissolved gases. *Limnology and Oceanography*, **22**, 10–25.

Miller, T. L. and Wolin, M. J. (1973). Formation of hydrogen and formate by *Ruminococcus albus*. *Journal of Bacteriology*, **116**, 836–846.

Naguib, M. (1971). On methane-oxidizing bacteria in fresh water. III. The capacity of methane utilization by methane-oxidizing enrichment cultures as revealed by gas chromatographic analyses. *Zeitschrift für Allgemeine Mikrobiologie*, **11**, 39–47.

Nelson, D. R. and Zeikus, J. G. (1974). Rapid method for the radioisotopic analysis of gaseous end products of anaerobic metabolism. *Applied Microbiology*, **28**, 258–261.

Nesterov, A. I. and Nazarenko, A. B. (1975). Activity of methane-oxidizing bacteria in the adsorbed state. *Microbiology*, **44**, 769–772.

Nissenbaum, A., Presley, B. J. and Kaplan, I. R. (1972). Early diagenesis in a reducing fjord, Saanich Inlet, British Columbia—I. Chemical and isotopic changes in major components of interstitial water. *Geochemica et Cosmochimica Acta*, **36**, 1007–1027.

Ooyama, J. and Foster, J. W. (1965). Bacterial oxidation of cyclo-paraffinic hydrocarbons. *Antonie van Leeuwenhoek*, **31**, 45–65.

Oremland, R. S. (1975). Methane production in shallow-water tropical marine sediments. *Applied Microbiology*, **30**, 602–608.

Oremland, R. S. (1976). Studies on the methane cycle in tropical marine sediments. Ph.D. Dissertation. University of Miami, Florida.

Oremland, R. S. and Taylor, B. F. (1977). Diurnal fluctuations of O_2, N_2 and CH_4 in the rhizosphere of *Thalassia testudinum*. *Limnology and Oceanography*, **22**, 566–570.

Oremland, R. S. and Taylor, B. F. Sulfate reduction and methanogenesis in tropical marine sediments. In preparation.

Ortner, P. B., Hulburt, E. M. and Wiebe, P. H. (1978). Phytohydrography and herbivore habitat contrasts in the western North Atlantic. *Deep-Sea Research*. In press.

Patel, R., Hoare, S. L., Hoare, D. S. and Taylor, B. F. (1975). Incomplete tricarboxylic acid cycle in a type I methylotroph, *Methylococcus capsulatus*. *Journal of Bacteriology*, **123**, 382–384.

Panganiban, A. and Hanson, R. S. (1976). Isolation of a bacterium that oxidizes methane in the absence of oxygen. Abstract of Annual Meeting of the American Society for Microbiology I, **59**, 121.

Patt, T. E., Cole, G. C., Bland, J. and Hanson, R. S. (1974). Isolation and characterization of bacteria that grow on methane and organic compounds as sole sources of carbon and energy. *Journal of Bacteriology*, **120**, 955–964.

Patt, T. E., Cole, G. C. and Hanson, R. S. (1976a). *Methylobacterium*, a new genus of

facultatively methylotrophic bacteria. *International Journal of Systematic Bacteriology*, **26**, 226–229.
Patt, T. E., O'Connor, M., Cole, G. L., Day, R. and Hanson, R. S. (1976b). Characteristics of a facultative methylotrophic bacterium. *In* "Microbial Production and Utilization of Gases" (Eds H. G. Schlegel, G. Gottschalk and N. Pfennig), pp. 317–327. E. Goltze KG, Göttingen.
Perry, J. J. (1968). Substrate specificity in hydrocarbon utilizing micro-organisms. *Antonie van Leeuwenhoek*, **34**, 27–36.
Pfennig, N. (1967). Photosynthetic bacteria. *Annual Review of Microbiology*, **21**, 285–324.
Pfennig, N. and Biebl, H. (1976). *Desulfuromonas acetoxidans* gen. nov. and sp. nov., a new anaerobic, sulfur reducing, acetate-oxidizing bacterium. *Archives for Microbiology*, **110**, 3–12.
Pine, M. J. (1971). The methane fermentations. *In* "Advances in Chemistry Series 105" (Ed. R. F. Gould), pp. 1–10. American Chemical Society, Washington, D.C.
Pine, M. J. and Barker, H. A. (1956). Studies on the methane fermentation XII. The pathway of hydrogen in acetate fermentation. *Journal of Bacteriology*, **71**, 644–648.
Pine, M. J. and Vishniac, W. (1957). The methane fermentations of acetate and methanol. *Journal of Bacteriology*, **73**, 736–742.
Postgate, J. R. (1969). Methane as a minor product of pyruvate metabolism by sulphate-reducing and other bacteria. *Journal of General Microbiology*, **57**, 293–302.
Presley, B. J. (1974). Rates of sulfate reduction and organic carbon oxidation in the Cariaco Trench. (Abstract). *EOS Transaction American Geophysical Union*, **55**, 319–320.
Proctor, H. M., Norris, J. R. and Ribbons, D. W. (1969). Fine structure of methane-utilizing bacteria. *Journal of Applied Bacteriology*, **32**, 118–121.
Quayle, J. R. (1972). The metabolism of one-carbon compounds by micro-organisms. *Advances in Microbial Physiology*, **7**, 119–203.
Quayle, J. R. (1976). Mechanisms of C_1-oxidation by methane utilizers and their correlation with growth yields. *In* "Microbial Production and Utilization of Gases" (Eds H. G. Schlegel, G. Gottschalk and N. Pfennig), pp. 353–357. E. Goltze KG, Göttingen.
Reddy, C. A., Bryant, M. P. and Wolin, M. J. (1972). Characteristics of S organism isolated from *Methanobacillus omelianskii*. *Journal of Bacteriology*, **109**, 539–545.
Redfield, A. C. (1958). The biological control of chemical factors in the environment. *American Scientist*, **46**, 205–221.
Reeburgh, W. S. (1969). Observations of gases in Chesapeake Bay sediments. *Limnology and Oceanography*, **14**, 368–375.
Reeburgh, W. S. (1972). Processes affecting gas distributions in estuarine sediments. *In* "Environmental Framework of Coastal Plain Estuaries" (Ed. B. W. Nelson), pp. 383–389. Geol. Soc. America Memoir 133.
Reeburgh, W. S. (1976). Methane consumption in Cariaco Trench waters and sediments. *Earth and Planetary Science Letters*, **28**, 337–344.
Reeburgh, W. S. (1978). A major sink and flux control for methane in marine sediments. Anaerobic oxidation. *In* "Benthic Processes and the Geochemistry of Marine Interstitial Waters" (Eds K. Fanning and F. Manheim). In press.
Reeburgh, W. S. and Heggie, D. T. (1977). Microbial methane consumption reactions and their effect on methane distributions in freshwater and marine environments. *Limnology and Oceanography*, **22**, 1–9.
Reeve, M. R., Cosper, T. C. and Walter, M. A. (1975). Visual observations on the process of digestion and the production of faecal pellets in the chaetognath *Sagitta hispida* (Conant). *Journal of Experimental Marine Biology and Ecology*, **17**, 39–46.

Ribbons, D. W., Harrison, J. E. and Wadzinski, A. M. (1970). Metabolism of single carbon compounds. *Annual Review of Microbiology,* **24**, 135–158.

Robertson, C. K. (1979). Quantitative comparison of the significance of methanogensis in the carbon cycle of two lakes. *Archiv Für Hydrobiologie ergbenis der Limnologie,* **12**, 123–135.

Roman, M. R. (1977). Feeding of the copepod *Acartia tonsa* on the diatom *Nitzschia closterium* and brown algae (*Fucus vesiculosus*) detritus. *Marine Biology,* **42**, 149–155.

Rosenthal, H. and Hempel, G. (1970). Experimental studies in feeding and food requirements of herring larvae (*Clupea harengus* L.). *In* "Marine Food Chains" (Ed. J. H. Steele), pp. 344–364. University of California Press, Berkeley, California.

Rudd, J. W. M. Methane oxidation in Lake Tanganyika (East Africa). Submitted.

Rudd, J. W. M. and Hamilton, R. D. (1975a). Factors controlling rates of methane oxidation and the distribution of the methane oxidizers in a small stratified lake. *Archiv für Hydrobiologie,* **75**, 522–538.

Rudd, J. W. M. and Hamilton, R. D. (1975b). Two samplers for monitoring dissolved gases in lake water and sediments. *Limnology and Oceanography,* **20**, 902–906.

Rudd, J. W. M. and Hamilton, R. D. (1978). Methane cycling in a eutrophic shield lake and its effects on whole lake metabolism. *Limnology and Oceanography,* **23**, 337–348.

Rudd, J. W. M. and Hamilton, R. D. (1979). Methane cycling in Lake 227 in perspective with some components of the carbon and oxygen cycles. *Archiv für Hydrobiologie ergbenis der Limnologie,* **12**, 115–122.

Rudd, J. W. M., Hamilton, R. D. and Campbell, N. E. R. (1974). Measurement of microbial oxidation of methane in lake water. *Limnology and Oceanography,* **19**, 519–524.

Rudd, J. W. M., Furutani, A. F., Flett, R. J. and Hamilton, R. D. (1976). Factors controlling rates of methane oxidation in shield lakes: the role of nitrogen fixation and oxygen concentration. *Limnology and Oceanography,* **21**, 357–364.

Sansone, F. J. and Martens, C. S. (1978). Methane oxidation in Cape Lookout Bight, North Carolina. *Limnology and Oceanography,* **23**, 349–355.

Schindler, D. W. (1977). Evolution of phosphorous limitation in lakes. *Science,* **195**, 260–262.

Schindler, D. W., Kling, H., Schmidt, R. V., Prokopowich, J., Frost, V. E., Reid, R. A. and Capel, M. (1973). Eutrophication of Lake 227 by addition of phosphate and nitrate: the second, third and fourth years of enrichment, 1970, 1971, and 1972. *Journal of the Fisheries Research Board of Canada,* **30**, 1415–1440.

Schindler, D. W. and Fee, E. J. (1974). Experimental lakes area: whole-lake experiments in eutrophication. *Journal of the Fisheries Research Board of Canada,* **31**, 937–953.

Schindler, D. W., Brunskill, G. J., Emerson, S., Broecker, W. S. and Peng, T. H. (1972). Atmospheric carbon dioxide: its role in maintaining phytoplankton standing crops. *Science,* **177**, 1192–1194.

Scranton, M. I. (1977). The marine geochemistry of methane. Ph.D. dissertation. Massachusetts Institute of Technology and Woods Hole Oceanographic Institution, Massachusetts.

Scranton, M. I. and Brewer, P. G. (1977). Occurrence of methane in the near-surface waters of the western subtropical North Atlantic. *Deep-Sea Research,* **24**, 127–138.

Scranton, M. I. and Farrington, J. W. (1977). Methane production in waters off Walvis Bay. *Journal of Geophysical Research,* **82**, 4947–4953.

Seiler, W. and Schmidt, U. (1974). Dissolved nonconservative gases in sea-water. *In* "The Sea" (Ed. E. D. Goldberg), vol. 5, pp. 219–243. John Wiley, New York.

Sheehan, B. T. and Johnson, M. J. (1971). Production of bacterial cells from methane. *Applied Microbiology*, **21**, 511–515.
Silverman, M. P. (1964). Methane-oxidizing bacteria—a review of the literature. United States Bureau of Mines Information Circular 8246.
Smith, P. H. (1966). The microbial ecology of sludge methanogenesis. *Developments in Industrial Microbiology*, **7**, 156–161.
Smith, P. H. and Hungate, R. E. (1958). Isolation and characterization of *Methanobacterium ruminantium*, n. sp. *Journal of Bacteriology*, **75**, 713–718.
Smith, P. H. and Mah, R. A. (1966). Kinetics of acetate metabolism during sludge digestion. *Applied Microbiology*, **14**, 368–371.
Söhngen, N. L. (1906). Über Baktevien, welche Methan als Kohlenstoffnahrung und Energiequelle gebrauchen. *Zentralblatt für Bakteriologie Parasitenkenkunde Infektions Krankheiten*, Abt. 2, **15**, 513–517.
Sorokin, Y. I. (1957). The question of the ability of sulfate-reducing bacteria to utilize methane for reduction of sulfates to hydrogen sulfide. *Doklady Akademiia Nauk SSSR*, **115**, 816–818.
Stadtman, T. C. (1967). Methane Fermentation. *Annual Review of Microbiology*, **21**, 121–142.
Stadtman, T. C. and Barker, H. A. (1949). Studies on the methane fermentation. VII. Tracer experiments on the mechanism of methane formation. *Archives of Biochemistry*, **21**, 256–264.
Stadtman, T. C. and Barker, H. A. (1951). Studies on the methane fermentations. IX. The origin of methane in the acetate and methanol fermentation by Methanosarcina. *Journal of Bacteriology*, **61**, 81–86.
Swinnerton, J. W. and Linnenbom, V. J. (1967). Determination of the C_1 to C_4 hydrocarbons in seawater by gas chromatography. *Journal of Chromatographic Science (J. ges. Chromat.)*, **5**, 570–573.
Swinnerton, J. W., Linnenbom, V. J. and Cheek, C. H. (1962a). Determination of dissolved gases in aqueous solutions by gas chromatography. *Analytical Chemistry*, **34**, 483–485.
Swinnerton, J. W., Linnenbom, V. J. and Cheek, C. H. (1962b). Revised sampling procedure for determination of dissolved gases in solution by gas chromatography. *Analytical Chemistry*, **34**, 1509.
Swinnerton, J. A., Linnenbom, V. J. and Cheek, C. H. (1969). Distribution of CH_4 and CO between the atmosphere and natural waters. *Environmental Science and Technology*, **3**, 836–838.
Sze, N. D. (1977). Anthropogenic CO emissions: Implications for the atmospheric CO-OH-CH_4 cycle. *Science*, **195**, 673–675.
Taylor, C. D. and Jannasch, H. W. (1976). Subsampling technique for measuring growth of bacterial cultures under high hydrostatic pressure. *Applied and Environmental Microbiology*, **32**, 355–359.
Trotsenko, Y. A. (1976). Isolation and characterization of obligate methanotrophic bacteria. *In* "Microbial Production and Utilization of gases" (Eds H. G. Schlegel, G. Gottschalk and N. Pfennig), pp. 329–336. E. Goltze KG, Göttingen.
Turner, J. T. (1977). Sinking rates of fecal pellets from the marine copepod *Pontella meadii*. *Marine Biology*, **40**, 249–259.
Uffen, R. L. and Wolfe, R. S. (1970). Anaerobic growth of purple nonsulfur bacteria under dark conditions. *Journal of Bacteriology*, **104**, 462–472.
Vary, P. S. and Johnson, M. J. (1967). Cell yields of bacteria grown on methane. *Applied Microbiology*, **15**, 1473–1478.

Wadzinski, A. M. and Ribbons, D. W. (1975). Utilization of acetate by *Methanomonas methanooxidans*. *Journal of Bacteriology*, **123**, 380–381.
Wake, L. V., Rickard, P. and Ralph, B. J. (1973). Isolation of methane utilizing micro-organisms: a review. *Journal of Applied Bacteriology*, **36**, 93–99.
Weaver, T. L. and Dugan, P. R. (1972a). The eutrophication implications of interactions between naturally occurring particulates and methane oxidizing bacteria. *Water Research*, **6**, 817–828.
Weaver, T. L. and Dugan, P. R. (1972b). Enhancement of bacterial methane oxidation by clay minerals. *Nature*, **237**, 518.
Weimer, W. C. and Lee, G. F. (1973). Some considerations of the chemical limnology of meromictic Lake Mary. *Limnology and Oceanography*, **18**, 414–425.
Weimer, P. J. and Zeikus, J. G. (1977). Fermentation of cellulose and cellobiose by *Clostridium thermocellum* in the absence and presence of *Methanobacterium thermoautotrophicum*. *Applied and Environmental Microbiology*, **33**, 289–297.
Wertlieb, D. and Vishniac, W. (1967). Methane utilization by a strain of *Rhodopseudomonas gelatinosa*. *Journal of Bacteriology*, **93**, 1722–1724.
Whelan, T. (1974). Methane, carbon dioxide and dissolved sulfate from interstitial water of coastal marsh sediments. *Estuarine and Coastal Marine Science*, **2**, 407–415.
Whittenbury, R., Davies, S. L. and Davey, J. F. (1970a). Exospores and cysts formed by methane-utilizing bacteria. *Journal of General Microbiology*, **61**, 219–226.
Whittenbury, R., Phillips, K. C. and Wilkinson, J. F. (1970b). Enrichment, isolation and some properties of methane-utilizing bacteria. *Journal of General Microbiology*, **61**, 205–218.
Whittenbury, R., Dalton, H., Eccleston, M. and Reed, H. L. (1975). The different types of methane oxidizing bacteria and some of their more unusual properties. *In* "Microbial Growth on C-1 Compounds". Proceedings of the International Symposium on Microbial Growth on C_1-Compounds (Eds The Organizing Committee), September 5, 1974, Tokyo.
Whittenbury, R., Colby, J., Dalton, H. and Reed, H. L. (1976). Biology and ecology of methane oxidation. *In* "Microbial Production and Utilization of Gases" (Eds H. G. Schlegel, G. Gottschalk, and N. Pfennig), pp. 281–292. E. Goltze KG, Göttingen.
Widdel, F. and Pfennig, N. (1977). A new anaerobic, sporing, acetate-oxidizing, sulfate-reducing bacterium: *Desulfotomaculum* (emend.) *acetoxidans*. *Archives of Microbiology*, **112**, 119–122.
Wilkinson, J. F. (1971). Hydrocarbons as a source of single-cell protein. *In* "Microbes and Biological Productivity, Society for General Microbiology" (Eds Hughes and Rose), pp. 15–47. Cambridge University Press, London.
Williams, R. T. and Bainbridge, A. E. (1973). Dissolved CO, CH_4, and H_2 in the Southern Ocean. *Journal of Geophysical Research*, **78**, 2691–2694.
Winfrey, M. R., Nelson, D. R., Klevickis, S. C. and Zeikus, J. G. (1977). Association of hydrogen metabolism with methanogenesis in Lake Mendota sediments. *Applied and Environmental Microbiology*, **33**, 312–318.
Winfrey, M. R. and Zeikus, J. G. (1977). Effect of sulfate on carbon and electron flow during microbial methanogenesis in freshwater sediments. *Applied and Environmental Microbiology*, **33**, 275–281.
Wofsy, S. C. (1976). Interactions of CH_4 and CO in the earth's atmosphere. *Annual Review of Earth and Planetary Sciences*, **4**, 441–469.
Wolfe, R. S. (1971). Microbial formation of methane. *In* "Advances in Microbiol Physiology" (Eds A. H. Rose and J. F. Wilkinson), vol. 6, pp. 107–146. Academic Press, London and New York.

Wolin, M. J. (1975). Interactions between the bacterial species of the rumen. *In* "The Proceedings of the 4th International Symposium of Ruminant Physiology" (Eds I. W. McDonald and A. C. I. Warner), pp. 134–148. The University of New England Publishing Unit, Armidale.

Wolin, M. J. (1976). Interactions between H_2-producing and methane-producing species. *In* "Microbial Formation and Utilization of Gases" (Eds H. G. Schlegel, G. Gottschalk and N. Pfennig), pp. 141–150. E. Goltze KG, Göttingen.

Zeikus, J. G. (1977). Biology of methanogenic bacteria. *Bacterial Reviews,* **41**, 514–541.

Zeikus, J. G. and Ward, J. C. (1974). Methane formation in living trees: a microbial origin. *Science,* **184**, 1181–1183.

Zeikus, J. G. and Winfrey, M. R. (1976). Temperature limitation of methanogenesis in aquatic sediments. *Applied and Environmental Microbiology,* **31**, 99–107.

Zeikus, J. G. and Wolfe, R. S. (1972). *Methanobacterium thermoautotrophicum* sp. n., an anaerobic, autotrophic, extreme thermophile. *Journal of Bacteriology,* **109**, 707–713.

Zeikus, J. G., Weimer, P. J., Nelson, D. R. and Daniels, L. (1975). Bacterial methanogenesis: acetate as a methane precursor in pure culture. *Archives of Microbiology,* **104**, 129–134.

Continuous culture in phytoplankton ecology

G-YULL RHEE

Environmental Health Center, Division of Laboratories and Research, New York State Department of Health, Albany, NY, USA

1 Introduction

The abundance and distribution of phytoplankton in nature are regulated by a multitude of environmental factors such as nutrients, light, temperature, and grazing. Understanding these complex processes in both qualitative and quantitative terms is an ultimate objective of algal ecology for the purpose of predicting and controlling primary productivity.

In experimental phytoplankton ecology, these environmental factors are reduced to a manageable number and investigated under defined conditions. This is usually carried out in a closed system (batch culture) or in an open continuous culture system in which spent medium is regularly or continuously replaced with fresh medium.

The use of continuous culture systems for ecological studies of phytoplankton is relatively recent. This is an attempt to report the current progress in this field through 1977 and to indicate where information is needed and to suggest directions for future research. It will deal exclusively with the ecological studies of both marine and freshwater phytoplankton. Some bacteriological studies are discussed to elucidate the work with algae, since applications of continuous culture in algal ecology are basically not much different from those in the ecology of aquatic bacteria. There are several reviews on aquatic bacterial ecology (Jannasch, 1965; Veldkamp and Jannasch, 1972; Jannasch and Mateles, 1974; Meers, 1974; Veldkamp, 1976; 1977) and many extensive theoretical treatments of continuous culture as applied to bacteriology (Herbert *et al.*, 1956; Herbert, 1958; Tempest, 1969, 1970; Kubitschek, 1970; Pirt, 1975). Readers are referred to these for a better understanding of continuous culture applications and theory.

2 Batch and continuous culture

The main advantage of batch culture is its simplicity. Its usefulness, however, is severely limited by some of its inherent properties; cell growth in a confined enclosure creates an everchanging environment and there is no input or output of materials. Also, in most instances artificially high nutrient concentrations are employed to ensure measurable growth, in contrast to natural conditions, in which the ambient levels of limiting nutrients are low. Moreover, in nature the rate of input of nutrients rather than their levels at a given time determines growth limitations. The growth cycle in a closed system is characterized by artificial phases of growth during which morphological, physiological, and biochemical changes take place; the lag phase is a period of metabolic adjustment, and the stationary phase is induced by nutrient depletion and/or accumulation of inhibitory products released during growth. During exponential growth, the average cell properties remain more or less unchanged, but this phase is short-lived and overlaps the other phases to a great extent.

To avoid these difficulties associated with use of batch culture, some investigators (Droop, 1961; Guillard *et al.*, 1973) used a very dilute cell suspension, containing only a few hundred cells per ml, with low nutrient levels. Although such a system may be satisfactory for a study of growth only, to have a sizeable population for other studies the nutrient levels must be increased to unnaturally high concentrations or the culture volume has to be inordinately large.

In continuous culture, on the other hand, very low nutrient concentrations can be used, and at the same time a dynamic equilibrium between nutrient input and growth is obtained, similar in principle to that existing in natural waters. A desired cell density can be maintained by controlling the levels of limiting nutrients in the reservoir or the rate of their introduction. Growth rate can be easily selected and maintained in a well-defined chemical environment for a long period of time, during which the mean cell composition is uniform. The continuous overflow of spent medium keeps the culture from accumulating excretory products.

Continuous culture systems are, however, neither capable of reproducing nor intended to replicate natural environments (see Jannasch, 1974). The chemostat reaches a steady state solely by virtue of a constant rate of nutrient input; the turbidostat, by density-dependent flow rate. In nature, on the other hand, if a steady state ever existed, it would be the result of the combined effects of a number of environmental factors. Continuous culture is designed principally to achieve a steady state because this state is amenable to simple kinetic treatments of growth. Thus, the use of continuous culture is, as also pointed out by Veldkamp and Jannasch (1972), to elucidate ecological principles and physiological mechanisms of organisms as related to natural conditions.

3 Types of continuous culture

There are two different types of continuous culture: chemostat and turbidostat. In the chemostat, growth rate is the independent variable, while in the turbidostat the independent variable is cell density. However, steady-state kinetic theory applies equally to both systems.

3.1 CHEMOSTAT

3.1.1 *Theory*

A chemostat consists of a culture system in which a fresh supply of nutrient solution is introduced at a constant rate into a culture vessel of finite volume and at the same time growth medium is withdrawn at the same rate. Consequently, when the growth rate (μ) of an organism becomes equal to the flow rate, a steady state is achieved. The conventional mathematical theory of a steady state is based on the following three *a priori* conditions: (i) one nutrient must be limiting for growth; (ii) the culture is perfectly mixed; and (iii) a constant proportion of cells is viable (see below).

Dilution rate (D) is the ratio of inflow volume (F) per unit time to the volume of the culture vessel (M): $D = F/M$, and has the dimension of time^{-1}. The reciprocal of D is mean residence time. The exponential growth of cells is expressed as

$$\mathrm{d}x/\mathrm{d}t = \mu x \tag{1}$$

where μ is growth rate and x, cell number. The rate of overflow from the vessel is:

$$-\mathrm{d}x/\mathrm{d}t = Dx. \tag{2}$$

The change in cell number in the culture vessel thus depends on the difference between (1) and (2):

$$\mathrm{d}x/\mathrm{d}t = \mu x - Dx = x(\mu - D). \tag{3}$$

Since there is no net change in cell number during a steady state,

$$D = \mu. \tag{4}$$

In a chemostat, therefore, growth rate equals dilution rate and thus can be easily manipulated by simply changing the medium dilution rate.

The relationship between growth rate and substrate concentration is best expressed by the Monod equation (1942):

$$\mu = \mu_m S/(K_s + S) \qquad (5)$$

where S is the substrate concentration; μ_m, a maximum growth rate; and K_s, the half-saturation constant which is equal to S when $\mu = \frac{1}{2}\mu_m$.

A steady-state relationship between substrate concentration and dilution rate is found by combining equations (3) and (5):

$$S = K_s \left(\frac{D}{\mu_m - D} \right). \qquad (6)$$

A quantitative relationship between substrate concentrations and biomass can be derived if we know the amount of biomass produced per unit substrate, S, usually expressed as yield (Y)

$$Y = -dx/dS \qquad (7)$$

If Y is a constant, from equations (3), (5) and (7):

$$ds/dt = D(S_R - S) - \frac{\mu_m x}{Y} \left(\frac{S}{K_s + S} \right)$$

where S_R is the substrate concentration in the medium reservoir.

Solving this equation for $ds/dt = 0$, the steady-state value of biomass becomes

$$x = Y(S_R - S). \qquad (8)$$

In nutrient-limited algal cells, however, Y is rarely constant, but changes with growth rate (see below) in contrast to the bacterial culture on which Monod based his theory (1942). For further discussion of chemostat theory see the review articles cited above.

3.1.2 *Operation*

It cannot be overemphasized that the kinetic theory is applicable only when the three *a priori* conditions are met. It appears, however, that in algal research, the third condition (a constant proportion of viable cells in the total population) has too often been overlooked. At extremely low dilution rates, the proportion of nonviable cells increases; in bacterial studies, nonviable cells accounted for as high as 60–70 per cent of the total cell number (Postgate and Hunter, 1962; Tempest *et al.*, 1967). When dead cells are present in significant numbers, particularly at low dilution rates, the growth rate of viable cells is higher than the dilution rate. Cryptic growth (i.e. growth utilizing the autolysis products of other cells) may also occur at low dilution rates, which can alter the nutritional conditions and affect the metabolism and growth rate of the surviving cells. The excretion of metabolites by living cells also tends to increase at slow growth rates (Hellebust, 1975). These excretory products may inhibit growth by themselves or by combining with a limiting nutrient and rendering it unavailable for absorption (Droop, 1968, 1974). Tempest *et al.* (1967), in fact, suggested that there may be a lower limit of dilution rate below which simple chemostat theory may not apply. The increase in external substrate concentrations at low dilution rates observed by Müller (1972), Paasche (1973a) and Harrison *et al.* (1976) could have resulted from these characteristics of slow-growing cells.

Determination of viable cells is not an easy task with algae. There is no effective method for measuring viable cells of colonial and filamentous forms. Although viable cells of some unicellular forms can be measured on proper nutrient agar plates in a manner similar to bacteria, the growth of aquatic forms on a solid medium sometimes produces self-toxic substances which prevent growth (Van Baalen, 1965).

At higher dilution rates approaching the washout rate, on the other hand, a steady state is hard to maintain because even small fluctuations in the dilution rate result in large changes in external substrate concentrations. The operation of chemostats in this range should therefore be avoided or be executed with extreme care.

Prolonged continuous operation of a chemostat with a single inoculum may also lead to erroneous conclusions because of possible spontaneous mutation. Because a chemostat environment is invariant with time and rigidly controlled, it exerts a tremendous pressure for selection. In fact, Novick and Slizard (1950), soon after the introduction of the chemostat theory, made clear that a chemostat culture should be viewed as a selection process. When a mutant is better adapted to the given environment, it will displace the original type. Such mutation and displacement have been reported for bacteria (Braun *et al.*, 1951; Cocito and Bryson, 1958). Although this has not been documented for algae, it may still occur (Droop, 1974; Sager, 1975). Admittedly, longer generation times reduce the chances for its occurrence in algae, but careful genetic, physiological, and biochemical examination is required to unveil a mutant which may not be morphologically distinct.

To minimize such selection, chemostat cultures should be started afresh frequently from carefully maintained stock cultures.

3.2 TURBIDOSTAT

In a turbidostat, cell concentration is predetermined and dilution rate is allowed to find its own level. Cell concentration in a turbidostat is usually sensed by a photocell; when the cell density exceeds the predetermined value, an electric impulse from the photocell to the solinoid valve lets fresh medium from a reservoir enter the culture vessel and simultaneously allows overflow of an equal volume of culture.

The steady-state kinetics of chemostats are equally applicable to turbidostats; in a steady state, the dilution rate equals growth rate and the substrate/growth rate relationship of equation (5) holds. In fact, Fuhs (1969) used both types of continuous culture very successfully for phosphate limitation studies of two marine diatoms.

Chemostats and turbidostats are complementary in their practical uses; while chemostats are most sensitive well below the critical dilution rate (washout), turbidostats operate most effectively near it. Indeed, with a turbidostat a steady state can be achieved in a nutritionally unrestricted medium (see below). For further discussion of turbidostats, see the review by Munson (1970) and Watson (1972).

4 Nutrient-limited growth

4.1 GENERAL

The survival and distribution of a species in nature ultimately depends on its ability to grow at a rate sufficient to replace continual loss due to sinking, grazing, parasitism, etc.

Nutrient-limited growth rate is described in general by equation (5). In batch culture experiments, growth rate at a given nutrient concentration can be determined by the equation

$$\mu = \ln (N_1/N_0)/(t_1 - t_0), \quad (9)$$

following kinetics of unrestricted growth in nutrient-sufficient conditions. N_1 and N_0 are the cell numbers or biomass at the time t_1 and t_0. Growth rate when expressed in doublings per day is given by

$$\mu = \log_2 (N_1/N_0)/(t_1 - t_0) = 3{\cdot}32 \log_{10} (N_1/N_0)/(t_1 - t_0). \quad (10)$$

Therefore, when the time course of growth is plotted on a semi-log scale, μ is the slope of a straight line. In such experiments, μ is usually expressed as a function of initial nutrient concentrations in the medium (Droop, 1961; Thomas and Dodson, 1968; Golterman *et al.*, 1969; Eppley and Thomas, 1969; Rhee, 1972; Swift and Taylor, 1974). It is difficult to apply this correlation between growth rate and initial concentrations of nutrients to predict growth rate or the nutritional status of natural populations, because in natural waters the continual loss, replacement, and interconversion of available and unavailable nutrients makes it difficult adequately to determine the relevant nutrient fluxes. It is difficult, if not impossible, to define the concentration equivalent to the initial external concentration of batch culture.

To avoid such problems, Droop (1961) and Guillard *et al.* (1973) measured growth rate in a culture of very low cell density before there were any appreciable changes in the initial concentrations. Growth under such conditions approaches a steady state. In this type of experiment, it is essential to characterize the inoculum precisely (see below).

4.2 GROWTH RATE AND INTERNAL NUTRIENT CONCENTRATIONS

In zinc-limited batch cultures of *Euglena gracilis*, Price and Quigley (1966) found that growth rate declined progressively with time instead of forming a straight line despite the zinc concentration in the medium remaining relatively constant at a low value. This decline was a consequence of the dilution of intracellular zinc concentrations which had been stored, in part, as a result of its luxury consumption (Knauss and Porter, 1954). When the instantaneous growth rate was related to the internal zinc concentrations, therefore, a linear relationship was obtained. The authors interpreted the linear relationship as indicating zinc complexing with a ligand to form an essential enzyme and predicted a hyperbolic relationship between growth rate and internal zinc concentrations. Eppley and Strickland (1968) also indicated independently the dependence of growth rate to intracellular nutrient levels by analysing batch culture data.

Such growth kinetics can best be studied in a chemostat, where various dynamic equilibria between the nutrient supply and population growth can be maintained and the effect of the past history of the cells on growth can also be effectively eliminated.

Droop (1968), in his work on the growth kinetics of *Monochrysis lutheri* under vitamin B_{12} limitation in continuous culture, also found that growth rate is not a function of external substrate concentrations but is a hyperbolic function of intracellular vitamin B_{12} levels. The intracellular levels of vitamin B_{12}, or cell quota (q), had a linear relationship with μq, with the line intercepting the q axis at a point greater than 0 (Fig. 1). This is described by an empirical formula:

$$\mu q = \mu_m' (q - q_0) \text{ and thus,}$$
$$\mu/\mu_m' = 1 - q_0/q, \tag{11}$$

where q_0 is the minimal cell quota, or subsistence quota, the intercept on the q axis. This threshold equation produces a rectangular hyperbola. The maximum growth rate, μ_m', is represented by the slope of μq plotted against q. This μ_m' is differentiated from the μ_m in equation (5) because it refers to infinite internal substrate concentrations. μ_m is smaller than μ_m' by the factor of $(1 - q_0/q_m)$ (Droop, 1973a, b; 1974), where q_m is the cell quota at which this nutrient ceases to limit growth.

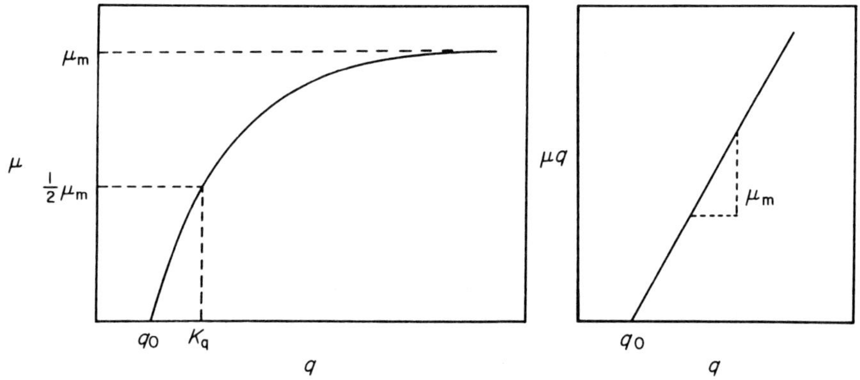

Fig. 1. Relationship between cell quota (q) and growth rate (μ) q_0 is the subsistence cell quota.

Caperon (1968), in his nitrate-limited chemostat study of *Isochrysis galbana*, also found that growth rate showed no relationship to the residual nitrate concentrations in the culture medium but was a direct function of q expressed by

$$\mu/\mu_m' = (q - q_0)/[K_q + (q - q_0)]. \tag{12}$$

For an unknown reason, the value of K_q, the internal nutrient level when $\mu = \frac{1}{2}\mu_m'$ was found to be very close to the value of q_0 in this study and other cases of nutrient limitation (Fuhs, 1969; Fuhs *et al.*, 1972; Rhee, 1973; Gotham and Rhee, manuscript in preparation). Then, the equations takes the same form as equation (11). Silicate-limited *Thalassiosira pseudonana* (Paasche, 1973a) appears to be an exception in which q_0 is significantly larger than K_q, but a recalculation of the data using non-linear regression shows no significant difference between the two values.

Phosphate-limited growth of *Thalassiosira fluviatilis* and *Cyclotella nana* (*Thalassiosira pseudonana*) in a turbidostat and a chemostat respectively also showed a direct relationship between μ and cell phosphorus (Fuhs, 1969) expressed by

$$\mu/\mu_m' = 1 - 2^{-(q-q_0)/q}, \tag{13}$$

which is similar to the Baule–Mitscherlich limiting-factor equation (Baule, 1918). Fuhs' growth data, however, can also be adequately described by equation (11) (Droop, 1973b).

In the internal nutrient model of Droop, the relationship between external substrate levels and growth rate can be found only indirectly. In a steady state,

$$v = \mu q \tag{14}$$

where v represents the nutrient uptake rate at a particular external substrate concentration in a steady state. Nutrient uptake by steady-state cells at various external nutrient levels follows the Michaelis–Menten-type equation:

$$v = \frac{VS}{K_m + S}, \tag{15}$$

where V is the dynamic maximum uptake velocity of a limiting nutrient by the cells with cell nutrient level at q (see below) and K_m, the affinity constant for uptake.

By combining equations (5), (11), (14) and (15), one can find (Droop, 1968):

$$K_s' = \mu_m' q_o \frac{K_m}{V}, \tag{16}$$

where K_s' is the apparent half-saturation constant for growth. V is in general inversely related to q (see section 5.1), and therefore the true maximum uptake rate (V_m) is observed only in a cell in which $q = q_o$.

Thus the value of K_s' derived from equation (16) will vary with different values of V and consequently the true K_s is obtained only when the kinetic constants for uptake are determined using completely starved cells. It was observed in a batch culture of P-rich *Scenedesmus* sp. that even when $S = 0$, the growth rate was about one doubling per day (Rhee, 1972). Because of the capability of growth at the expense of internal phosphorus, the K_s' value found using this culture was much higher (about 0·2 μM) than the value found using equation (16) with uptake data of starved cells (less than 0·06 μM) (Rhee, 1973). This shows clearly that even in dilute batch cultures, the true K_s is found only when the inoculum has been completely starved with respect to the limiting nutrient.

A hyperbolic relationship between growth rate and intracellular nutrient concentrations has also been found in vitamin B_{12} and iron-limited growth of *Skeletonema costatum* (Droop, 1970, 1973a, b); phosphate limitation in *Thalassiosira fluviatilis*, *Thalassiosira pseudonana* (Fuhs, 1969; Fuhs *et al.*, 1972), *Scendedesmus* sp. (Rhee, 1973), *Monochrysis lutheri* (Droop, 1974; Goldman, 1977a), *Asterionella formosa* and *Cyclotella meneghiniana* (Tilman and Kilham, 1976), *Anabaena flos-aquae* and *Microcystis* sp. (Gotham and Rhee, manuscript in preparation); ammonia limitation in *Chaetoceros gracilis* (Thomas and Dodson, 1972); nitrate limitation in *Chlorella pyrenoidosa* (Williams, 1971), *Dunaliella tertiolecta*, *Monochrysis lutheri*, *Cyclotella nana* (*Thalassiosira pseudonana*), *Coccochloris stagnina* (Caperon and Meyer, 1972a), *Scenedesmus* sp. (Rhee, 1974, 1978) and *Chlorella pyrenoidosa* (Pickett, 1975); combined nitrogen limitation in *Dunaliella tertiolecta* (Bienfang, 1975) and silicate-limited growth of *Asterionella formosa* and *Cyclotella meneghiniana* (Tilman and Kilham, 1976). In all cases the yield constant Y decreases with increasing growth rate because it is the reciprocal of cell quota, q. There are exceptions, however, which do not follow Droop's model of growth. Ammonium-limited growth of *Monochrysis lutheri* and *Dunaliella ter-*

tiolecta (Caperon and Meyer, 1972a; Laws and Caperon, 1976), on the other hand, could not be described by either Droop's or Monod's equations, but was a function of cell nitrogen and carbon ratios which is difficult to explain. This is also in contrast with findings of Thomas and Dodson (1972) in ammonium-limited *Chaetoceros gracilis*, and Goldman and McCarthy (1978) in ammonium-limited *Thalassiosira pseudonana*. Although inorganic carbon-limited growth (Goldman *et al.*, 1974) appears to follow only the Monod model, growth in this case was measured by dry weight, not by cell numbers.

Although Droop's equation for growth is expressed in terms of the total cell level of a limiting nutrient, it is likely that growth rate is more directly related to a certain intracellular fraction or "pool" because various fractions serve different cellular functions. It is necessary to identify this fraction not only to understand the growth mechanism but also for the ecological application of equations (11) and (12). Because these equations require determination of the minimum cell quota, q_0, which is impractical, if not impossible, to determine for all phytoplankters occurring in nature, it would be better if the cell quota term of the equation could be replaced by a term representing only the active intracellular fraction that affects growth rate and approaches 0 when $\mu = 0$. For example, in phosphate- and nitrate-limited *Scenedesmus* sp. (Rhee, 1973, 1978), growth rate is related to polyphosphates (the intracellular pool of phosphorus) and intracellular free amino acids in the form of a saturation curve (see below). In non-growing cells ($\mu = 0$) both fractions are negligible. Therefore, when their levels are used in equation (12) it simplifies to

$$\mu/\mu_m' = q_f/(K_{qf} + q_f), \tag{17}$$

q_f here representing the levels of the intracellular compounds. This equation suggests that differences in K_{qf} and μ_m among various phytoplankters would determine the relative effectiveness in competition. It would be interesting to see if K_{qf} when expressed in proper biomass units is species-specific or species-independent. (Biomass in this case requires definitions in terms of methodology.) If K_{qf} is species-independent, the growth rate of natural populations can be estimated from the values of q and the μ_m of the populations, and also measurements of q_f will give a direct indication of the nutritional conditions of a given habitat.

The dependence of growth rate on the intracellular level of a limiting nutrient is of ecological significance. Many nutrients, including silicon (Darley, 1969; Kilham *et al.*, 1977; Barlow, 1979), can accumulate within a cell in excess of their immediate demand. (However, the mechanism of such silicate accumulation and the form of accumulated silicate are not clear.) Thus, the ambient concentrations of nutrients in natural waters, which may sometimes provide a general indication of their availability, may not be reliable indicators of nutrient limitation because they do not provide information concerning the dynamics of their turnover. Indeed, considerable algal blooms have been observed when phosphate concentrations in lakes were extremely low (Pearsall, 1932; Berman and Gophen, 1972; Serruya, 1974). For the same reason, phytoplankton may grow at different rates even with the same concentrations of external nutrients. Therefore, a measurement of intracellular nutrient levels or growth-rate-related nutrient pools would give better estimations of nutritional conditions of a given water and the growth rate of a natural assemblage.

4.3 GROWTH RATE AND EXTERNAL NUTRIENT CONCENTRATIONS

The Monod model has also been used in many continuous culture studies (Müller, 1970, 1972; Paasche, 1973a, b; Goldman *et al.*, 1974; Swift and Taylor, 1974; and Algren, 1978). Although this model appears to differ from Droop's model, both predict the same results under steady-state conditions.

First, in Droop's model an internal nutrient pool, which is utilized as a result of cell growth and multiplication, is continuously replenished by the concentration-dependent uptake of external nutrients; in steady-state conditions this uptake is in equilibrium with the internal pool. Second, the calculated values of the growth half-saturation concentration of external nutrients from Droop's internal pool model are generally lower than or approaching the limit of analytical detectability. For example, in *Scenedesmus* sp. the K_s value for phosphate-limited growth is below 0·06 μM and, for nitrate-limited growth, less than 1·00 μM (Rhee, 1973, 1978). Thus, the apparent lack of a relationship between external nutrients and growth rate may be due mainly to the inadequacy of analytical methods used to detect residual nutrients. Residual substrate can be easily detected near washout rate, but in this region, true steady state is difficult to maintain. The existence of a finite substrate concentration at which the net uptake is zero further confuses the picture (Müller, 1972; Caperon and Meyer, 1972a; Paasche, 1973a; Droop, 1974). Another factor that masks the relationship is the "binding factor" excreted by cells which ties up the limiting nutrient, making it unavailable (Droop, 1968, 1974). The frequently observed tendency for external substrate levels to increase at low dilution rate (Droop, 1968, 1974; Müller, 1972; Paasche, 1973a, Harrison *et al.*, 1976) has been attributed to various factors other than the binding factor such as experimental artifacts, increased cell mortality, and unknown organic growth factors (Müller, 1972; Paasche, 1973a; Harrison *et al.*, 1976).

In a steady state, therefore, there is no incompatibility nor contradition between the two models. In fact, the silicate-limited study of *Thalassosira pseudonana* by Paasche (1973a) and silicate- and phosphate-limited work using two freshwater diatoms by Tilman and Kilham (1976) show that growth can be described by either model.

The latter study further shows that the half-saturation constant for growth measured directly from residual nutrient levels is almost identical to the constant calculated by equation (16). Goldman (1977a) also reported a similar conclusion.

4.4 GROWTH KINETICS OF MULTINUTRIENT LIMITATION

The concept that a single limiting nutrient controls growth originates from the Liebig law of the "minimum". This concept has subsequently been challenged by many investigators. For example, Baule (1918) and Verduin (1964) suggested that the effect of a limiting nutrient on yield could be varied by the simultaneous effects of other nutrients, and that growth or yield could be predicted by the polynomial form of the equation

$$\mu = \mu_m (1 - e^{-Sx}). \qquad (18)$$

Recently multiplicative models for growth have also been preferred to the presumably too simple model of one limiting nutrient, particularly in natural waters where several potentially limiting nutrients may be at critical concentrations. From theoretical considerations, Droop (1973a) proposed a multiplicative equation as an extension of his growth model of internal nutrients:

$$\mu/\mu_m' = (1 - q_{o_A}/q_A)(1 - q_{o_B}/q_B)(\ldots) \qquad (19)$$

where q_{0_A} and q_{0_B} are subsistence quotas; and q_A and q_B, cell quotas for nutrients A and B. In subsequent experimental studies, however, Droop (1974) found that growth did not follow this pattern but was regulated by the single nutrient in shorter supply. *Monochrysis lutheri* was grown in a chemostat at various dilution rates with different phosphorus and vitamin B_{12} ratios in the reservoir. Using experimentally determined values of q_0 and μ_m, q was calculated for each nutrient at various dilution rates according to equations (11) and (19). These calculated values were then compared to measured values. The results clearly showed that the cell quota values obtained with equation (11), the threshold equation, were not different from the measured cell quotas, while the multiplicative model gave values consistently lower than the measured values. The slope of the regression of calculated μ/μ_m' against μ which should be the value of $1/\mu_m'$ also demonstrated that a better fit was provided by the threshold hypothesis than by the multiplicative model. The limiting nutrient is the one which has the smallest ratio of cell quota to subsistence quota (Droop, 1974). This shows clearly that growth control is exerted by a single limiting nutrient, without any effect of the other nutrients.

Although Droop (1974) showed in his experiments that the changes in the cell quota of a non-limiting nutrient was also a saturation function of growth rate and introduced the concept of the luxury coefficient, this does not appear to be always the case. Sometimes when a non-limiting nutrient is in great excess, its cell quota is independent of growth rate as will be discussed below (section 5.1).

I have also investigated the effects of possible dual nutrient limitations on the growth of *Scenedesmus* sp. (Rhee, 1974, 1978). The alga was grown at a fixed dilution rate in medium with various nitrogen to phosphorus atomic ratios (N/P). N/P were varied by adjusting N concentrations to a constant level of P. When steady-state cell numbers were plotted against N/P, it was found that cell numbers increased linearly with increasing N/P up to a ratio of 30 and thereafter remained constant (Fig. 2). The change in the slope of the curves was rather sharp at N/P = 30, indicating no multiple nutrient limitation. A test of multiplicative and threshold models using cell quotas also confirmed this. It was previously reported (Rhee, 1974) and reaffirmed (Rhee, 1978) that an optimum cell N/P ratio for this organism is 30. Therefore, below N/P of 30, growth is solely determined by nitrate limitation and above 30, by phosphate limitation.

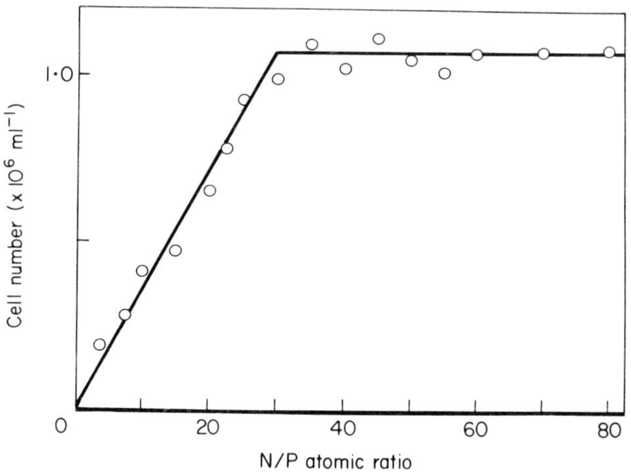

Fig. 2. Steady-state cell number of *Scenedesmus* sp. growing at a fixed growth rate of 0·59 day^{-1} (relative growth rate of 0·437) in a medium with varying N/P atomic ratios. (Cell N/P ratios are the same as those in the inflow medium, since there was no detectable residual nitrogen or phosphorus in the culture medium.) (From Rhee, 1978; with permission.)

The determination of optimal atomic ratios of nutrients may seem to be a tedious process. It can be shown, however, that this value can be obtained with relative ease; given $\mu'_{m_N} = \mu'_{m_P}$, from equation (11) at a given μ when either nutrient is limiting,

$$\frac{q_P}{q_N} = \frac{q_{o_P}}{q_{o_N}},$$

where μ'_{m_P} and μ'_{m_N} are a maximum growth rate under P and N limitation: q_P and q_N are cell quotas of P and N when either nutrient is limiting; and q_{o_P} and q_{o_N} are subsistence quotas of P and N. Therefore, all one needs to know is subsistence cell quotas or cell quotas at one growth rate when either nutrient is limiting (see Rhee, 1978).

5 Nutrient uptake

Nutrient uptake in steady-state chemostat cells is expressed as

$$v = \frac{D(S_R - S)}{x} \tag{20}$$

where S_R is the nutrient concentration of medium in the reservoir and x, cell number or biomass. The uptake velocity calculated represents the rate at S, the substrate concentration in the culture medium, because a steady state must satisfy equations (14) and (15) simultaneously. Using this relationship I investigated the change in the apparent maximum uptake velocity of phosphate in nitrate-limited and of nitrate in phosphate-limited chemostat cultures (Rhee, 1974, 1978). Paasche (1973a) was able to construct a Michaelis–Menten-type hyperbola of silicate uptake from steady-state chemostat cultures using equation (20). The apparent fit of each data to a hyperbola, however, should be interpreted with caution because intracellular nutrient concentrations also affect the uptake rate.

From batch culture studies it is well known that with most nutrients, deficiency increases the uptake rate of the deficient nutrient (Ketchum, 1939; Syrett, 1953; Hattori, 1960; Blum, 1966; Coombs *et al.*, 1967; Busby and Lewin, 1967; Fitzgerald, 1968; Jeanjean, 1969; Harrison, 1976), but it is difficult to relate the increase quantitatively to the degree of deficiency. The chemostat is better suited for these studies because it provides nutritionally well-defined cells.

5.1 PHOSPHATE UPTAKE

Fuhs *et al.* (1972) first investigated nutrient uptake with steady-state cells at various dilution rates that represented different degrees of nutrient limitation. Short-term phosphate uptake experiments with a phosphate-limited chemostat culture of *Thalassiosira fluviatilis* showed that increases in uptake rate were associated with decreases in cell phosphorus concentrations.

Phosphate uptake was examined in detail in steady-state cultures of *Scenedesmus* sp. (Rhee, 1973, 1974), *Anabaena flos-aquae* and *Microcystis* sp. (Gotham and Rhee, manuscript in preparation). It was measured in short-term experiments with minimal disturbance of the steady state by transferring steady-state cells directly into a series of flasks with various phosphate concentrations without harvesting or washing, since residual phosphate was not detectable in the culture medium. Uptake velocity was calculated using measurements made within 10 minutes. Phosphate uptake in these organisms is a function of both internal and external substrate concentrations, described by an equation resembling noncompetitive enzyme inhibition:

$$v = \frac{V_m}{(1 + K_m/S)\,(1 + i/K_i)}, \qquad (21)$$

where V_m is the true maximum uptake velocity which is equivalent to apparent maximum uptake rate (V) in most-starved cells; K_i, an inhibition constant; and i, cellular P concentration or the total inorganic polyphosphate (PP_i) concentrations. A similar negative effect of PP_i on phosphate uptake was also reported for batch culture studies of *Chlorella* (Jeanjean, 1969; Aitchison and Butt, 1973). In *Scenedesmus* sp., *Anabaena flos-aquae* and *Microcystis* sp., the apparent maximum uptake velocity varies with growth rate or PP_i levels, but K_m is constant at all growth rates. Investigation of phosphate uptake in nitrate-limited cultures of *Scenedesmus* sp. (Rhee, 1974) further revealed that i should have the value of the trichloroacetic acid (TCA)-soluble PP_i rather than the total PP_i concentration. Under nitrate limitation, V is lower than the maximum rate under phosphate limitation by a factor of 8 because the pattern of the distribution of the acid-soluble and acid-insoluble fractions of PP_i differs in nitrate- and phosphate-limited cells.

Phosphate- and combined nitrogen-limited chemostat cultures of three different clones of *Thalassiosira pseudonana* (Perry, 1976) also showed that V was much higher in P-limited cultures than in N-limited cultures and increased with P-starvation. K_m also remained constant. However, no definite relationship between V and PP_i levels was found. This may be due to the different method of PP_i determination; Perry used ultraviolet oxidation and subsequent acid hydrolysis, while Jeanjean (1969), Rhee (1973, 1974) and Aitchison and Butt (1973) used chemical extraction methods which can resolve PP_i into various fractions.

Euglena gracilis growing rhythmically in a chemostat also showed variation in V with diel periodicity, but K_m remained invariant (Chisholm and Stross, 1976) and generally V is inversely related to the levels of TCA-soluble PP_i (Gotham, 1977). Such daily oscillation of V but not K_m was also found in *Fragilaria crotonensis* and *Tabellaria fenestrata* in natural phytoplankton assemblages which were measured by autoradiography (Stross and Pemrick, 1974). My recalculation of the data of Fuhs *et al.* (1972), on the other hand, shows increasing K_m with increasing cell phosphorus levels.

5.2 NITROGEN UPTAKE

Nitrate and ammonium uptake also show similar variation in V with cell nitrogen contents in *Thalassiosira pseudonana* (Eppley and Renger,

1974). In this organism K_m also varied but no clear trend was observed. The data of Giddings (1975), on the other hand, appear to show that both K_m and V decrease with dilution rate in *Scenedesmus abundans* and Rhee (1978) also found a similar variation with *Scenedesmus* sp. My data suggested that free amino acids (FAA) in the amino acid pool were related to the variation of V, but the large standard errors of K_m values did not allow accurate evaluation of a relationship between FAA and K_m. Recently, Solomonson and Spehar (1977) proposed a model for feedback control of nitrate uptake by FAA and intracellular ammonia.

There is also a distinct diel rhythm in inorganic nitrogen uptake and activities of the associated enzymes, nitrate and nitrite reductases and glutamic dehydrogenase (Eppley and Renger, 1974). These and a similar variation in phosphate uptake (Stross and Pemrick, 1974; Chisholm and Stross, 1976) prompted investigators to speculate that such temporal differences may allow the coexistence of various competing species in the same habitat. Circadian variation, however, does not appear to be a feature common to all phytoplankters. Perry (1976) found no variation in phosphate uptake in *Thalassiosira pseudonana* throughout the day.

5.3 FEEDBACK AND ECOLOGICAL SIGNIFICANCE

It is clear now that nutrient uptake is not only a function of substrate concentrations but is also regulated by a feedback mechanism. Therefore, the practice of using the simple Michaelis–Menten equation to imply the competitive advantage of uptake systems between species may be in error. If we use the phosphate uptake kinetics of *Scenedesmus* sp. as an example, equation (21) shows that between two competing species which follow the kinetic equation, the one with the lower K_m is not necessarily advantageous, because the disadvantage of a higher K_m can be compensated for by a higher value of K_i.

Kinetic constants for growth may provide useful information concerning the relationship between ambient nutrient levels and species distribution. The measurement of these values is, however, tedious and time-consuming. Uptake, on the other hand, can be determined with relative ease in a short time. Using batch culture Eppley and Thomas (1969) investigated whether nutrient uptake kinetics can provide the same information as growth kinetics and found that the half-saturation constant for uptake (K_m) was the same as the constant for growth (K_s)

within the limits of experimental error. Droop (1968), Fuhs *et al.* (1972), and Rhee (1973) among other investigators, however, found that K_m is much larger than K_s. Droop (1973b) eloquently showed that they cannot be identical; in order for K_m to be equal to K_s

$$S/(K_m + S) = S/(K \quad + S) \text{ and therefore}$$
$$v/V_m = \mu/\mu_m. \tag{22}$$

Because v/μ is the value of the cell quota, q, it is variable, while V_m/μ_m is constant, and therefore K_m cannot be the same as K_s. The reason why Eppley and Thomas were able to obtain identical values for K_m and K_s was, as pointed out by Droop (1973b), that they measured uptake as an increase of cell nitrogen per unit cell nitrogen, which is the specific growth rate in the steady state. It seemed therefore that K_s could be determined by uptake.

Yet this approach is also open to question. In batch uptake experiments using this approach, it is essential to equilibrate inoculum cells in the uptake medium before any measurement to bring the cells to a condition approaching a steady state. This preincubation of nutrient-deficient cells, however, alters the nutritional states of inoculum cells, depending on external nutrient levels in part, which in turn may affect K_m as discussed above.

In the non-steady state, uptake is uncoupled from growth and the relationship between the rate of uptake and the rate of conversion of the absorbed nutrient into organic compounds is obscure. A good example of uncoupling is the luxury consumption of limiting nutrients by starved cells.

It appears, therefore, that nutrient uptake cannot directly provide information on growth.

In recent years, some investigators have used the "perturbation technique" to investigate uptake kinetics (Caperon and Meyer, 1972b; Bienfang, 1975; Harrison *et al.*, 1976; Caperon and Zieman, 1976): experiments are carried out in a "batch mode" steady-state culture by stopping the inflow of medium and adding a limiting nutrient at levels higher than saturation concentrations. Uptake is then followed over time and the time derivation of the curve of nutrient decline is taken as the rate of uptake at a given instantaneous concentration. Such experiments last more than an hour and sometimes as long as 10 hours. Although this approach is mathematically sound and easier to carry out than the conventional short-term method, it suffers a serious flaw in ignoring feedback control of nutrient uptake. The velocity at each substrate concentration represents the uptake velocity of cells with different physiological conditions, which are impossible to define. The resulting data usually seem to fit a Michaelis–Menten-type hyperbola, but the apparent fit includes varying degrees of feedback regulation of uptake. Depending on the nutritional conditions of a cell, K_m can vary as much as fivefold (Fuhs *et al.*, 1972—my recalculation) and V can vary more than eightfold (Rhee, 1974). It is important to recognize that the Michaelis–Menten kinetic equation was originally derived from *in vitro* enzyme reactions in which metabolic regulations are absent. However, enzyme reactions of a whole cell, including nutrient uptake, are subject to metabolic control (see Kornberg, 1973; Tempest and Neijssel, 1976). Therefore, nutrient uptake by a whole cell needs a term for the regulation by intracellular pools.

6 Characteristics of steady-state cells

6.1 NON-LIMITING NUTRIENTS

6.1.1 *Cell carbon, chlorophyll* a *and assimilation rate*

The total carbon per cell appears to generally increase as nutrient limitation becomes more severe. For example, in P-limited *Thalassiosira pseudonana* (Fuhs *et al.*, 1972), severe phosphate limitation, particularly at high temperature and light intensity, resulted in an increase in cell carbon. Similar increases were also reported for *Dunaliella tertiolecta* and *Monochrysis lutheri* under ammonium limitation and *Thalassiosira pseudonana* and *Coccochloris stagnina* (Caperon and Meyer, 1972a) under nitrate limitation. On the other hand, *Thalassiosira pseudonana* under combined nitrogen limitation (Eppley and Renger, 1974) and *Dunaliella tertiolecta* under nitrate limitation (Caperon and Meyer, 1972a) showed increases in cell carbon at higher growth rates and no clear trend was observed in *Chaetoceros gracilis* in combined nitrogen limitation (Thomas and Dodson, 1972). In nitrate-limited *Scenedesmus* sp. (Rhee, 1978) cell carbon remained unchanged at all growth rates examined.

Cell carbon concentration is also affected by the type of limiting nutrients. Thus, when *Scenedesmus* sp. was grown in a medium with differing N/P atomic ratios at a fixed dilution rate, the cell carbon level in the phosphate-limited state was higher than in the nitrate-limited state, but in each state, the level was constant regardless of the excess amount of nitrate or phosphate (Rhee, 1978). The difference in carbon concentrations between phosphate and nitrate limitation was found to be related to cell chlorphyll *a*. Richardson *et al.* (1969) also found a constant cell carbon concentration irrespective of nitrogen concentration in the inflow medium as far as nitrogen was limiting, but it appears that their culture could be light-limited by self-shading at high levels of nitrogen supply.

Chlorophyll *a* contents generally increase with increasing growth rate under phosphate and nitrogen limitations (Caperon and Meyer, 1972a; Thomas and Dodson, 1972; Eppley and Renger, 1974; Healey and Hendzel, 1975; Rhee, 1978). In phosphate-limited *Thalassiosira pseudonana* (Fuhs *et al.*, 1972), however, its concentration was constant independent of growth rate, but its level per unit cell volume increased with growth rate, which was interpreted as a mechanism for the formation of carbohydrate reserve materials at slower growth rates. In a silicate-limited diatom (Harrison *et al.*, 1976), its level appears to decrease with increasing growth rate. Under light limitation chlorophyll *a* content remained unchanged irrespective of growth rate (Eppley and Dyer, 1965).

Assimilation rate (carbon fixed/chlorophyll *a*/time) can be measured with the ^{14}C technique or can be estimated in a steady state by the equation $V_c = \mu q_c chl^{-1}$, where V_c is the assimilation rate; q_c, cell quota of carbon; and *chl*, cellular contents of chlorophyll *a*. Under nitrogen limitation, the rate in *Chaetoceros gracilis* (Thomas and Dodson, 1972) showed a general increase with growth rate. However, in nitrate-limited *Thalassiosira pseudonana* (Caperon and Meyer, 1972a), *Scenedesmus abundans* (Giddings, 1975) and *Scenedesmus* sp. (Rhee, 1978) and nitrogen- and phosphate-limited *Thalassiosira pseudonana* (Fuhs *et al.*, 1972; Eppley and Renger, 1974), the rate appeared to be invariant with growth rate.

Although cell carbon concentrations were different depending on whether phosphate or nitrate was limiting in *Scenedesmus* sp., the cell chlorophyll *a* contents also changed in the same manner and therefore, the assimilation rate remained constant independent of the type of limitation (Rhee, 1978).

6.1.2 *Cell nitrogen*

Under nitrogen limitation, cell nitrogen or N/C ratio or both has a direct relationship with growth rate, with the values increasing at faster growth rates (Thomas and Dodson, 1972; Caperon and Meyer, 1972a; Eppley and Renger, 1974; Giddings, 1975; Laws and Caperon, 1976; Rhee, 1978). This ratio therefore has been suggested as a promising parameter for estimating the growth rate of natural phytoplankton (Eppley and Renger, 1974). In *Scenedesmus* sp. maintained at the same dilution rate with varying N/P atomic ratios in inflow medium, the N/C ratio also remained constant under nitrate limitation but varied widely under phosphate limitation (Rhee, 1978).

There is also luxury accumulation of nitrogen when other nutrients are limiting. In phosphate-limited *Thalassiosira pseudonana* (Fuhs *et al.*, 1972) and silicate-limited *Skeletonema costatum* (Harrison *et al.*, 1976), cell nitrogen remained generally uniform at all growth rates. Under phosphate limitation, *Scenedesmus* sp. (Rhee, 1974), *Anabaena variabilis* and *Scenedesmus quadricauda* (Healey and Hendzel, 1975), also had relatively uniform cell nitrogen contents except at slower growth rates. It appears that the constant level of cell nitrogen is maintained only when nitrogen is in great excess relative to phosphorus. Therefore, when *Scenedesmus* sp. was grown at the same dilution rate while N/P atomic ratios in the reservoir were increased from 30 to 80, cell nitrogen increased in a linear fashion without leaving any detectable amount of residual nitrogen in culture medium. The optimum N/P ratio of this organism is 30, and a maximum cell N/P ratio found in this organism under phosphate limitation was about 140 (Rhee, 1974, 1978).

Most excess nitrogen accumulates in the form of protein, which accounts for more than 70 per cent of the total cell nitrogen; the greater the phosphate limitation, the higher the proportion of protein (Rhee, 1978). It is also interesting to note that under phosphate-limited conditions (Healey and Hendzel, 1975) the proportion of heterocysts in the total biomass of a blue-green alga increased with growth rate.

6.1.3 *Cell phosphorus*

Droop (1974) found that in *Monochrysis lutheri*, cell phosphorus increased with growth rate under vitamin B_{12} limitation whether phosphate was slightly or greatly in excess. (Cell phosphorus was lower under conditions of slight excess.) This led him to conclude that a modification of equation (11) could also apply to non-limiting nutrients. This is not, however, universal, since under nitrogen- and silicate-limited conditions, cell phosphorus contents remain generally the same regardless of growth rate (Rhee, 1974; Tilman and Kilham, 1976; Harrison *et al.*, 1976).

When *Scenedesmus* sp. was grown at a fixed dilution rate while the N/P atomic ratio of the inflow medium was increased, cell phosphorus decreased as nitrate limitation gradually approached the optimum N/P ratio of 30. Under phosphate limitation above the ratio, cell phosphorus level remained constant at the concentration characteristic of the growth rate, but in nitrate limitation, cell phosphorus increased as the N/P ratio decreased (Rhee, 1978). It appears, therefore, that cell phosphorus increases until the N/P atomic ratio reaches the lowest possible cellular N/P ratio of the organism and that only when the ratio in the medium is lower than this value does residual phosphate become detectable. The cellular N/P ratio in this organism ranges from less than 5 to over 140.

6.1.4 *Silicon and vitamin* B_{12}

The silicon level in phosphate-limited *Asterionella formosa* and *Cyclotella meneghiniana* (Tilman and Kilham, 1976) was also constant or independent of growth rate. Of interest is the fact that in these organisms silicon accumulated up to about sixfold in excess of immediate needs. Such excess silicon accumulation was also found in the natural phytoplankton populations of Cayuga Lake when enriched with silicate (Barlow, 1979).

The accumulation of vitamin B_{12} in *Monochrysis lutheri* in phosphate limitation was similar to phosphorus accumulatioin under vitamin B_{12} limitation in this organism (Droop, 1974).

Thus it appears that all nutrients investigated so far are subject to luxury accumulation up to a maximum limit set by species characteristics, probably including cell size. It follows then that cellular elemental ratios vary within a very wide range. It also appears that a significant amount of residual nutrient becomes detectable in the medium only when the nutrient is present in excess of this maximum level. This supports the view that the analysis of water samples gives little information regarding nutrient limitation and that intracellular level is a reliable measure of nutrient supply in a given habitat.

The excess storage of nutrients may be an inevitable evolutionary development for organisms which are subject to rapid and wide fluctuations of immediate environmental conditions in nature. The stored nutrients may protect the cells from environmental changes, helping them to maintain the intracellular homeostasis necessary for their metabolic function and also making them less susceptible to the situations in which nutrients may become limiting.

6.2 CARBOHYDRATE AND LIPIDS

Carbohydrate storage increased with increased nutrient limitation (Fuhs *et al.*, 1972; Healey and Hendzel, 1975). Storage of lipids, however, appears to vary from species to species. While there was no change in lipid contents in phosphate-limited *Thalassiosira pseudonana* (Fuhs, 1969), *Scenedesmus quadricauda* (Healey and Hendzel, 1975) showed an increase with decreasing growth rate.

6.3 NUCLEIC ACIDS

The main interest in nucleic acids in ecological studies are in the possible use of their concentrations in estimating biomass and the growth rate of natural phytoplankton. Since DNA is a structural component basic to cellular function, its use as a biomass parameter was attempted (Holm-Hansen *et al.*, 1968). The change in DNA, however, appears to be an unreliable measure for biomass. In *Thalassiosira pseudonana* (Fuhs, 1969), its concentration seems to decrease with progressive phosphate limitation, whereas in *Scenedesmus* sp. (Rhee, 1973), *Scenedesmus quadricauda* and *Anabaena variabilis* (Healey and Hendzel, 1975) it increased with decreasing growth rate. The apparent inverse relationship between DNA concentrations and growth rate in *Scenedesmus* sp. is due to variation in cell size and therefore, when expressed per unit cell volume, DNA concentration remains unchanged at all growth rates.

The measurement of RNA was suggested as an estimate of growth rate (Sutcliffe, Jr., 1965; Rhee, 1973) because of a close correlation between its variation and growth rates in many procaryotes (see references in Rhee, 1978). RNA has a positive relationship with growth rate in all these cases. In both phosphate- and nitrate-limited *Scenedes-*

mus sp., its concentration is related by a saturation function to growth rate (Rhee, 1973, 1978). Furthermore, RNA concentrations are the same at a given growth rate regardless of the kind of limiting nutrients (Rhee, 1978). It therefore seems promising to use this parameter to assess the growth rate of natural populations with or without prior knowledge of the kind of limiting nutrient.

6.4 OTHER COMPOUNDS

6.4.1 *Inorganic polyphosphate*

Inorganic polyphosphates (PP_i), linear polymers of orthophosphate linked by energy-rich bonds, are well known intracellular reserves for phosphorus. Their metabolism, function and location have been reviewed (Kuhl, 1960; Harold, 1966; Kulaev, 1975), and discussion here will be confined to their relationship to growth. They can be divided into various fractions using different fractionation schemes (See Kuhl, 1960; Miyashi *et al.*, 1964; Kulaev, 1975; Niemeyer, 1976) but in this discussion, PP_i refers to the total amount unless otherwise indicated.

In chemostat-grown phosphate-limited algal cultures, the level of these compounds is related to growth rate by a saturation function in a manner similar to the total cell phosphorus, RNA, and other phosphorus compounds (Rhee, 1973). In batch cultures where growth was measured by cell volume only, the concentration of the total cell phosphorus when the log phase ended was $6 \cdot 0 \times 10^{-12}$ $\mu M/\mu m^3$. At this cell phosphorus concentration, PP_i concentration dropped to a critical value near zero; thereafter the organism increased fourfold in volume (Rhee, 1972). When the total cell phosphorus per volume was converted to the level per cell in chemostat-grown cells where biomass was measured by volume and number, it was found that cell division stopped at that concentration. Thus growth rate under P limitation appears to be closely associated with intracellular PP_i levels, probably via RNA synthesis, and PP_i was suggested as a measure of growth rate in phosphate-limited conditions (Rhee, 1973). It has also been reported that in *Neurospora crassa*, one form of PP_i is directly related to RNA concentrations (Kulaev, 1975).

Under nitrate-limited conditions, PP_i contents increased with increasing nitrogen deficiency at slower growth rates and the acid soluble fraction appears to be involved in the regulation of phosphate uptake (Rhee, 1974). A better-known metabolic function of PP_i is the regulation of the intracellular orthophosphate level (Harold, 1966).

An interesting quantitative relationship also appears to exist between PP_i levels and the activity of alkaline phosphatase, an enzyme associated with phosphorus deficiency. For *Scenedesmus* sp., the reciprocal values of both show an inversely linear slope (Rhee, 1973). Earlier, Fitzgerald and Nelson (1966) used the amount of hot-water extractable inorganic phosphorus, called surplus P, as an estimate of phosphorus limitation in algae. This surplus P is also inversely related to alkaline phosphatase activity. This fraction was later found to be quantitatively equivalent to the total PP_i extracted by the scheme of Miyachi (Rhee, 1972, 1973). The measurement of PP_i or surplus P thus may be an expedient and accurate method to estimate P limitation, and it can be substituted for the determination of particulate phosphorus of living cells in natural waters where the interference from detritus is unacceptably high.

6.4.2 *Free amino acids*

Free amino acids (FAA) in the amino acid pool appear to have an important function in growth. They serve as obligatory intermediates of protein synthesis in many organisms (see Britton and McClure, 1962; Holden, 1962; Cowie, 1962; Dawson, 1965) and are also directly related to RNA synthesis in *Escherichia coli*. A possible mechanism for the control of RNA synthesis is that transfer-RNA (t-RNA) uncharged with amino acids acts as a repressor of RNA synthesis (Stent and Brenner, 1961; Maaløe and Kjeldgaard, 1966). Under nitrogen-rich conditions, therefore, the t-RNA tends to be saturated with amino acids, derepressing RNA synthesis, whereas in nitrogen deficiency the relative unavailability of amino acids leaves a large portion of t-RNA unadenylated.

The total concentration of FAA in nitrate-limited *Scenedesmus* sp. is hyperbolically related to growth rate, as is the total cell nitrogen. When this alga was grown at a fixed growth rate while the N/P atomic ratio in the inflow medium was varied, FAA concentrations were constant at the level characteristic of the growth rate under nitrate limitation, but in the phosphate-limited state the level increased linearly with N/P ratios. The proportion of FAA in total cell nitrogen, however, remained uniform at about 5·7 per cent. Thus, the total FAA concentrations may be used as a sensitive indicator of nitrogen depletion if nitrogen limitation is known to exist. One can also estimate the total cell nitrogen from the amount of FAA. Such substitution would be desirable when detrital nitrogen prevents accurate measurement of the particulate nitrogen of living organisms. The level of FAA also indicates growth rate under nitrogen limitation, as the level of PP_i indicates phosphorus limitation in phosphate-limited conditions.

7 Light and temperature

7.1 LIGHT

In photoautotrophic organisms, light can be considered as a substrate in the sense that it serves as the energy source by providing electrons. The effects of light on growth are particularly difficult to study in batch cultures because of its rapid change in intensity with population growth. A chemostat or turbidostat is therefore an ideal tool for such investigation.

A light-limited chemostat is not amenable to the mathematical treatment used for other substrates because the supply of light is independent of medium flow rate. Effective light intensity can be controlled by the source of incident light, but fine control can be made by taking advantage of the mutual shading of cells at different densities.

Light-limited growth was studied by Azad and Borchardt (1969) in *Scenedesmus* and *Chlorella* and Gons and Mur (1975) and Mur *et al.* (1978) in *Scenedesmus protuberans* and *Oscillatoria agardhii.* As for other limiting substrate, the relationship between effective light intensity and growth rate was hyperbolic. A distinct feature of the curve is the clear existence of a maintenance intensity, indicated by a positive intercept on the abcissa. The existence of a maintenance intensity has been suspected by Myers (1962) who called it the compensation point or compensation intensity (Myers and Graham, 1975). This is similar in concept to the light intensity at the compensation depth in natural waters. It is defined as the light energy required for the maintenance of endogenous metabolism without growth. Maintenance energy in carbon-limited bacterial growth is well known (Pirt, 1975) and is conceptually the same as maintenance intensity.

In light limitation, Eppley and Dyer (1965) showed a linear relationship between effective light intensity and growth rate in a turbidostat; the line intersected the origin, indicating no very low or no compensation intensity. Pipes and Koutsoyannis (1962) made similar observations. The relationship would be hyperbolic if higher intensities were examined and may be described by the equation (Eppley and Dyer, 1965)

$$\mu = \bar{I} \cdot \mu_m/(K_I + \bar{I}), \tag{23}$$

where $\bar{I}$ is mean light intensity and K_I the half-saturation constant.

In light-limited chemostat work a useful ecological concept regarding maintenance intensity and maintenance rate has been developed by Gons and Mur (1975). An energy balance equation can be written as:

$$dE'/dt\ (c) = \mu x - \mu_e x, \tag{24}$$

where E' is energy absorbed; c, an efficiency factor for converting light energy to biomass; and μ_e, specific maintenance rate. When (dE'/dt) $(1/x)$ is plotted against μ, the slope has the value of c (Fig. 3). The value of (dE'/dt) $(1/x)$ when $\mu = 0$ is the energy for maintenance and the value of μ when (dE'/dt) $(1/x) = 0$ is μ_e. The ratio of μ_e to μ_m gives the percentage of biomass to be lost if the light intensity is below the maintenance level. For example, μ_e/μ_m for *Scenedesmus protuberans* is 11 per cent, while *Oscillatoria agardhii* has a value of 6 per cent. This means that in order to maintain themselves without loss, the growth of each population has to be equivalent to the growth at the maximum growth rate for 11 per cent and 6 per cent of a day respectively (Mur *et al.*, 1976). The difference in μ_e between the two species indicates that the compensation depth for each organism is different and that the respective depths in a given habitat can be experimentally determined from μ_e and the extinct coefficient of water. It should be noted, however, that the compensation or maintenance intensity varies with day length and temperature (Holt and Smayda, 1974; Smayda, personal communication).

Maximum productivity below the saturation light intensity occurs when phytoplankters are growing at half-maximal growth rates (Eppley and Dyer, 1965; Gons and Mur, 1975). Eppley and Dyer attributed the decrease in productivity at higher dilution rate to decreases in energy absorption, but in the chemostat system of Gons and Mur the reason was the efficiency of energy utilization.

The effect of light on nutrient uptake was studied by Davis (1976);

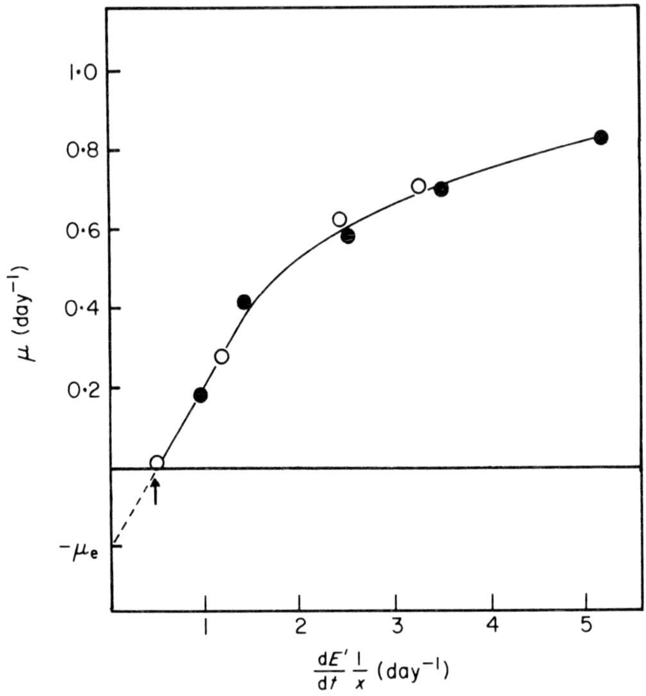

Fig. 3. Growth rate (μ) as a function of energy absorption per unit biomass ($dE'/dt)(1/x)$ measured in cultures receiving 13 000 (●) and 30 000 J l^{-1} day^{-1} (○). The extrapolated value of μ represents the negative maintenance constant μ_e and the intercept on the abscissa (arrow) is the quantity of light required for maintenance. (Redrawn from Gons and Mur, 1975; with permission.)

the uptake rate increased at higher light intensities, probably because at low intensities growth was limited by light while above the saturation intensity silicate limited growth. At suboptimal intensities, the uptake was higher in light-adapted than in non-adapted cells.

7.2 TEMPERATURE

As for light-limited growth, the mathematical theory of the chemostat does not apply to growth limited by temperature. However, the chemostat is useful in investigating the effects of temperature, because the same growth rate can be maintained at different temperatures. This enables us to see the dissociation of mutually dependent metabolic processes which have different temperature optima.

Maximum growth rate of micro-algae increases with temperature to an upper limit of about 40°C (Eppley, 1972). Temperature affects growth by changing the rate of enzyme reactions, the molecular configurations of cellular constituents such as proteins, the fluidity of membrane lipids, and the structure of water, etc.

Chemical reactions at different temperatures are described by the Arrhenius equation

$$k = Ae^{-E/RT} \tag{25}$$

where k is the reaction rate; A, a constant representing the frequency of formation of the activated reaction complex; E, activation energy; R, the gas constant; and T, the absolute temperature. Goldman and Carpenter (1974) suggested that maximum growth rate is numerically related to temperature by the same function, with k representing the maximum growth rate. For this equation to hold true with respect to growth rate, the plot of log μ_m on $1/T$ should be inversely linear, with the slope having the value of $-E/R$. When these investigators plotted values compiled from various continuous culture studies, the relationship was indeed linear with a Q_{10} value of about 2·1 within the temperature range 13–40°C.

Synchronous culture studies of two strains of *Chlorella* (Sorokin, 1960) showed, on the other hand, that the relationship of log μ_m to $1/T$ is not linear but has different slopes in different temperature ranges, suggesting that the limiting step in growth may be different at different temperatures. Therefore, although the temperature–growth rate relationship was clearly definable by one function in the data analysis of Goldman and Carpenter, this does not necessarily mean that the relationship is so simple and clear-cut in individual species.

The importance of the effect of temperature on primary production has been discounted on the ground that in natural environments a maximum growth potential which is set by temperature is seldom achieved because of other limiting factors, notably nutrient levels (Eppley, 1972). Results of investigation of the change in chemical composition appear to support this view. At suboptimal growth temperatures, there are general increases in steady-state cellular constituents such as carbon, nitrogen and phosphorus (Williams, 1971; Goldman, 1977b; Rhee and Gotham, 1979) and the total biomass production in terms of dry weight remains unchanged and the cellular elemental ratios also do not change (Williams, 1971). Cell size generally increases at low temperatures as does chlorophyll content (Williams, 1971; Rhee and Gotham, 1979).

In nitrogen-limited *Scenedesmus* sp. (Rhee, manuscript in preparation) *Phaeodactylum tricornutum, Nitzschia* sp., *Oscillatoria* sp. (Goldman, 1977b), *Chlorella pyrenoidosa* and *Selenastrum gracile* (Williams, 1971), cell numbers were lower at lower temperatures at a given growth rate, but there was no detectable amount of nitrogen left in the culture media except at temperatures lower than 13°C in *Scenedesmus* sp. Nitrogen uptake rate thus seems to increase at lower temperatures according to equation (14).

An interesting observation was made, however, in turbidostat cultures of *Chlorella pyrenoidosa* (Shelef *et al.*, 1970); the residual nitrogen at a given growth rate decreased as the temperature increased up to 35°C, but at 39°C, the optimal temperature for growth, the residual nitrogen level increased. A similar uncoupling of uptake mechanism from growth was also found in nitrate-limited steady-state *Scenedesmus* sp. (Rhee and Gotham, 1979). This alga was grown at a fixed dilution rate under various temperature regimes. Short-term nitrate uptake experiments at corresponding temperatures showed that the optimal temperature for the uptake is 15°C, whereas that for growth was found to be 20–25°C. K_m value appeared unaffected within the temperature range examined (10–25°C). This seeming discrepancy in uptake patterns between the above calculated uptake and the direct measurement probably stems from difficulties in accurate measurement of residual substrate levels in steady state; when residual N levels are different, calculated rates by equation (14) may not reflect such differences because of (i) the insensitivity of q to concentration changes when substrate levels are low and (ii) difficulties in maintaining absolute steadiness of dilution rate. Since residual substrate levels are below the half-saturation level for uptake, changes in uptake rates are most sensitive at these substrate concentrations. The calculated rates, therefore, may actually be values at different substrate concentrations represented by equation (15). Hence, they are not comparable with each other unless steady-state residual concentrations are identical. (A steady-state must satisfy both equations (14) and (15).)

The two different temperature optima for growth and uptake would effectively widen temperature ranges for survival; at suboptimal temperatures for growth, increased efficiency of the uptake system would expeditiously increase cell quota, thereby maintaining growth potential for times when temperature becomes optimal for growth.

An analysis of the short-term phosphate-uptake data by Fuhs *et al.* (1972) shows higher uptake rates at higher temperatures; the relationship of log V to $1/T$ and log K_m to $1/T$ were not single linear functions within the temperature range examined (5–22°C). This apparently contradictory behaviour of uptake to that observed in *Scenedesmus* sp. and turbidostat and chemostat cultures may be explained by the fact that the cells used by Fuhs *et al.* had been grown at an optimal temperature before the uptake experiments while the cells in other cases were temperature-adapted.

Phytoplankters have different optimum growth temperatures, and their upper limits are also different. Temperature effects in five different species of phytoplankters were investigated by growing them at a fixed dilution rate in chemostats (Goldman and Ryther, 1976). As temperature increased from 0 to 30°C, steady-state cell numbers of *Dunaliella tertiolecta* increased progressively, while *Skeletonema costatum* showed an opposite trend. The optimum temperature of *Phaeodactylum tricornutum* and *Monochrysis lutheri*, on the other hand, was 20°C, and their steady-state cell numbers decreased at both higher and lower temperatures. Such differences are of course contributing factors in their competition. Competition experiments among these species at various temperatures thus produced results essentially along the line predicted by temperature characteristics.

7.3 LIGHT, TEMPERATURE AND NUTRIENT INTERACTIONS

Differences in light and temperature requirements among various phytoplankters undoubtedly play an important role in their occurrence in time and space. In natural waters, however, light and temperature conditions change in various combinations: for example, spring is characterized by high light intensity and low temperatures; summer, high light and high temperatures; autumn, low light and high temperatures; winter, low light and low temperatures (Fogg, 1975). It appears, therefore, that the effects of their interactions on growth are probably more important than their individual effects.

Using the turbidostat to maintain low cell density and low nutrient levels, Maddux and Jones (1964) found that at low temperatures the optimum light intensity for growth in *Nitzschia closterium* was higher; at 16°C the optimum intensity was 3230 lux, while at 23°C it was 1883 lux.

They also investigated the effects of nutrient levels on optimum temperature and light levels in *Nitzschia closterium* and *Tetraselemis* sp. In one set of experiments, phosphate and nitrate levels were kept high at typical concentrations employed in batch culture studies and in the other they were similar to the levels occurring in sea water. The optimum temperatures and light intensity were different in those two sets of experiments with their values consistently lower at lower nutrient conditions. Therefore, these investigators correctly warned that the results of batch culture experiments with high nutrient concentrations may not be applicable to natural situations. Although such interaction studies are extremely important and valuable, very few have been done.

8 Diel periodicity

It is well known that in nature many organisms divide at certain times of day. This does not constitute true sychrony, however, because each cell does not undergo division each day. This is termed a phased division (Hastings and Sweeney, 1964). Such circadian division may be due to an endogenous mechanism (biological clock) or to exogenous factors, chiefly the diel changes in illumination, or both. Diel periodicity is not confined only to cell division; it is also observed in photosynthesis, chlorophyll synthesis, nutrient uptake and other biosynthetic processes (Lorenzen, 1970; Sournia, 1974). To elucidate and understand these diel variations, it is essential to grow algal cultures under a light and dark cycle.

Batch systems have been widely used for studies of cell cycles. In these cultures only a few cycles in a preselected population range can be used. The changing light intensity and nutrient concentrations, however, still pose a problem. To avoid such problems, a semicontinuous culture method has been used in which the population level is regularly diluted to the original concentration by adding fresh medium. In this case, however, the timing of the addition and/or the addition itself could have entrainment effects (Terry and Edmunds, 1969). Although more elaborate care of the system is required, continuous culture methods have also been used to provide constant reproducibility of growth at a given rate and for a desired period of time.

Light–dark change is probably the most profound feature of rhythmic cycles in nature that affect the cell cycle. Indeed, this change has been widely used for inducing synthronization (see Lorenzen, 1970). Many laboratory cultures under light–dark cycles also result in phased growth. For example, most species examined in semicontinuous or continuous culture divide in the dark, although not at the same time of the dark period. *Dunaliella tertiolecta* (Eppley and Coatsworth, 1966) and *Euglena gracilis* (Chisholm *et al.*, 1975; Gotham, 1977) divided at the beginning of the dark period; *Coccolithus huxleyi, Skeletonema costatum* (Eppley *et al.*, 1971), and *Thalassiosira pseudonana* (Eppley and Renger, 1974) in the latter part of the dark period; and *Ditylum brightwellii* (Eppley *et al.*, 1967) divides primarily in the light period. Assimilation rate is generally highest at midday and lowest at midnight.

In phased growth, dilution rate is equal to period average growth rate, $1/T\int_0^T \mu(t)\,dt$, where T is period length and $\mu(t)$ is instantaneous growth rate. Gotham (1977) and Frisch and Gotham (1977, 1979), in their extensive mathematical treatment of phased growth, also show that period average growth rate in P-limited *Euglena gracilis* is a function of the initial physiological condition of the population. This initial physiological state is represented by the cell quota at the beginning of photocycle, which corresponds to the trough of cell quota oscillation. This cell quota is related to the growth rate by the equation (12).

Phosphate uptake in *Euglena gracilis* in synchronous batch and phased-chemostat cultures (Chisholm and Stross, 1976; Gotham, 1977) also show diel periodicity in V, with its maximum occurring before the beginning of dark period. Phosphate uptake in synchronized *Scenedesmus*, however, was maximum during the dark (Sundberg and Nilshammer-Holmval, 1975). Diel rhythm in phosphate uptake was also noted in *Chlorella* and *Scenedesmus* in turbidostat culture (Borchardt and Azad, 1968; Azad and Borchardt, 1970).

Nitrite uptake in *Ditylum brightwellii* took place only in light but nitrate assimilation by nitrite-grown cells took place in both light and dark. Because these cells had a low level of NADH-dependent nitrate reductase, the existence of another nitrate reductase was speculated. Most nitrate taken up during the dark was recovered from the cells as free nitrate but only 40 per cent of the nitrate taken up in light period could be recovered in these short-term uptake experiments (Eppley and Coatsworth, 1968). *Skeletonema costatum* assimilated nitrate and ammonia primarily during the day while *Coccolithus huxleyi* utilized them during both day and night (Eppley *et al.*, 1971). In these organisms NADH-dependent nitrate reductase and glutamic dehydrogenase activities were highest after the onset of the light period, while nitrite reductase activity was highest toward the end of the day.

In *Ditylum brightwellii* (Eppley *et al.*, 1967) cell carbon, nitrogen and carbohydrate contents increased during the day and the increase in cell silicon was associated with cell division, but cell phosphorus increased in the dark.

Such temporal differences in phase and amplitudes of physiological mechanisms and rates may play an important role in the survival of species if limiting nutrients also fluctuate over a 24-hour period. It is also clear that rates measured at a particular time of day may not be a sound basis for comparison among different species; integrated rates over a 24-hour period would provide a better basis.

9 Mixed culture studies

9.1 COMPETITIVE EXCLUSION AND COEXISTENCE

Mixed culture studies in continuous culture have received surprisingly little attention, although the distribution of phytoplankters in a given habitat has often been attributed to species differences in growth and nutrient uptake kinetics (Eppley *et al.*, 1969; Kilham, 1971; Guillard *et al.*, 1973). For example, the half-saturation constant and maximum growth rate of the estuarine clone of *Thalassiosira pseudonana* for silicate-limited growth were higher than the respective values of the oceanic clone. These differences between clones were viewed as resulting from selection by the oceanic habitat of low silicate levels and by the estuarine environment of relatively high silicate concentrations.

Selection or competitive elimination in a chemostat is achieved by the difference in substrate affinity or maximum growth rates of competing species. This has been elegantly demonstrated in competition studies of aquatic bacteria (Veldkamp and Jannasch, 1972; Jannasch and Mateles, 1974). In competition between two hypothetical species of algae (Fig. 4), therefore, A will become dominant at a limiting substrate concentration lower than S_1 by virtue of its lower substrate affinity, while B will become dominant at substrate levels higher than S_1 because of its faster growth rate. The two organisms could only coexist at the unstable equilibrium point $S = S_1$.

Any number of environmental factors, such as temperature, light, pH, toxic substances or extracellular products which alter the substrate

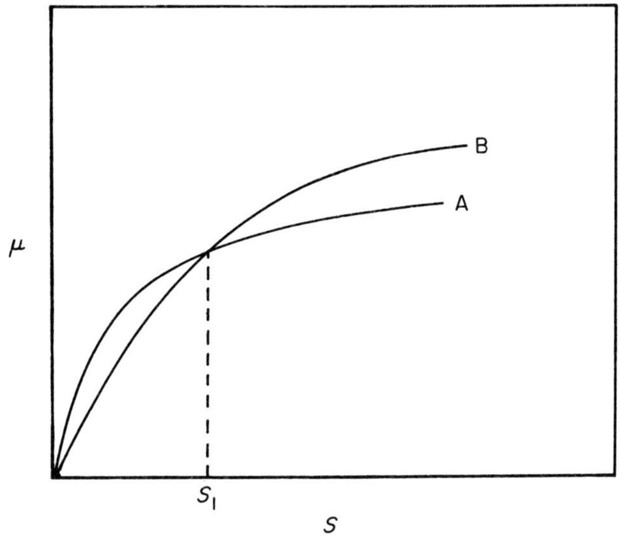

Fig. 4. Competition between two hypothetical organisms A and B which have different half-saturation constants for growth (K_s) and maximum growth rates (μ_m) S is substrate concentration.

affinity or maximum growth rate will influence the outcome of competition.

Under constant environmental conditions, competition for a single limiting nutrient should result in the selection of only one species. In nature, however, closely competing species coexist, violating the principle of competitive exclusion. Hutchinson (1961) termed this as "the paradox of plankton". Such coexistence in competitive equilibrium, however, is possible for example if several nutrients are in relatively short supply and growth of each species is limited by different nutrients. Petersen (1975) and Taylor and Williams (1975) developed theoretical models for such coexistence. Titman (1976), using a competition model based on growth kinetics and a semicontinuous culture, proved that indeed such coexistence took place. The K_s for silicate in *Asterionella formosa* is higher than that for *Cyclotella meneghiniana*. The affinity constant for phosphate, however, is lower in *Asterionella formosa*. Their μ_m values are approximately the same. Since the boundary between growth rates limited by phosphate and silicate should occur when phosphate and silicate cause equal growth rate,

$$S_{Si}/(K_{Si} + S_{Si}) = S_P/(K_P + S_P) \text{ and thus}$$
$$S_{Si}/S_P = K_{Si}/K_P$$

where S_P and S_{Si} are phosphate and silicate concentrations, and K_P and K_{Si} are the half-saturation constants for growth. Then, the boundary for *Asterionella formosa* would occur when Si/P = 97, whereas the boundary for *Cyclotella meneghiniana* would be 5·6. Thus, the two organisms should coexist in the interval between the two ratios, with each organism limited by different nutrients. At ratios larger than 97, *Asterionella formosa* would have a competitive advantage over *Cyclotella meneghiniana* and below the ratio of 5·6, *Cyclotella meneghiniana* should become dominant. In this case such prediction was possible because of the similar μ_m of both organisms. It should also be noted that although diel phases of nutrient uptake in these organisms were not investigated, their diel division cycles appeared to be in phase (Tilman, personal communication).

Tilman (1977) also developed new models which applied irrespective of differences in μ_m. These were derived by solving steady-state growth kinetic equations of either the Monod model or Droop's model. The boundary was calculated as follows at various growth rates:

$$S_{R1} = S_1 + (S_{R2} - S_2)(Y_2 - Y_1),$$
$$S_{R1} = S_1 + (S_{R2} - S_2)(q_{o1}/q_{o2}),$$

where S_{R1} and S_{R2} are concentrations of limiting nutrients 1 and 2 in the inflow medium; S_1 and S_2, steady-state concentrations of each limiting nutrient; Y_1 and Y_2, yield coefficients of each limiting nutrient; and q_{o1} and q_{o2}, the subsistence cell quotas of each nutrient. Both equations predicted almost identical values of the boundaries and the boundary varied with growth rate. The predictions were verified with competition results of semicontinuous culture experiments with *Asterionella formosa* and *Cyclotella meneghiniana*.

As the result of determining that there is no simultaneous limitation

TABLE 1

N/P	0	10		20		30
	Organism A dominant	↑	P-limited A and N-limited B coexist	↑	Organism B dominant	
		Optimum N/P for A		Optimum N/P for B		

of nitrate and phosphate, I (Rhee, 1974, 1978) suggested that possible differences in the optimal cell N/P ratio (or the difference in requirements for these nutrients) would be an important factor determining competitive elimination and coexistence as indicated in Table 1.

This optimal cell N/P ratio might not be affected by temperature because at suboptimal temperatures, cell nitrogen and phosphorus contents appear to increase in proportion to each other (Rhee and Gotham, 1979).

9.2 BIOASSAY OF LIMITING NUTRIENTS AND TOXIC SUBSTANCES

9.2.1 *Limiting nutrients*

Bioassay methods have been widely used to assess the fertility of water, particularly by investigators in the field of water resources management. Bioassay methods include two approaches: one uses the response of a carefully studied single organism in a batch culture as a measure of fertility and the other, the response of confined natural populations in an enclosure, large or small, to the enrichment of various nutrients. The response is usually measured either by a short-term change in the photosynthetic rate or by a long-term change in growth, or both. From the results inferences are made about which nutrient might be limiting in the open system (natural waters). From an ecological point of view, however, there are several serious problems associated with these methods. No single organism can represent the multitudes of species and thus biochemical processes and their inter- and intraspecies associations. In the method using a natural population, enrichment may bring changes not only in photosynthetic rate and growth, but also generally in the composition of the species present and their relative proportions and thus affect final results. In this regard, the level of enrichment and the resulting change in nutrient ratios are factors affecting species composition (Rhee, 1978).

In the light of such problems, an imaginative approach using natural populations in continuous culture by Barlow and his coworkers (Barlow *et al.*, 1973a, b; Peterson *et al.*, 1974; Barlow, 1979) and Jones *et al.* (1978a, b) is of great importance. They employed both the "turbidostat" and "chemostat" at the same time, the former to create nutrient-sufficient conditions and the latter to determine the response of populations to various nutrient enrichments.

Growth vessels were first filled with natural populations from Cayuga Lake or an experimental pond after removing zooplankters. Prior to pumping in enriched medium from a reservoir, phosphate and/or trace metal levels in the culture vessels were raised to the levels in the reservoir by adding a concentrated stock solution. Dilution was kept at a constant rate either that at which the population in the vessel had the same C/P ratio as the natural population or the same rate as the *in situ* turnover rate of carbon.

In chemostat cultures, it was found that up to 7 to 10 days after the beginning of experiments, (i) population could reach and maintain equilibrium for 3 to 4 days with the proportions of major species essentially remaining constant and (ii) initially added phosphate was rapidly utilized to an almost *in situ* level or less in several hours but there was delay in any increase in photosynthetic rate and growth for one or more days. The continuous culture results consistently showed phosphate limitation and trace metals had no effect. These results were significantly different from those obtained from concurrently run batch bioassay in which the change in photosynthetic rate was erratic. Previous bioassay results in Barlow's laboratory indicated silicate or trace metal limitations. Such contrasting outcomes of batch bioassay may be due to the long delay in responses found in enrichment experiments of the continuous culture system.

By comparing C/P ratios of turbidostat and chemostat cultures at two dilution rates with the ratio of lake populations, it could also be determined that Cayuga Lake phytoplankters were limited by phosphorus but rarely phosphorus-deficient.

When silicate and phosphate were enriched in continuous culture, there was the usual delay in responses (Barlow, 1978). One added feature was that the diatom populations increased rapidly after the initial delay, lowering silicate levels and indicating a transient silicate limitation, but eventually they all became phosphorus limited. It was possible to relate cell quotas of phosphorus and silicon to observed growth rates using Droop's growth model (equation 11). When the variation of the particulate Si/P ratio was measured, it could also be seen that diatoms accumulated four times the amount of silicon as required under phosphorus limitation.

A nitrogen and phosphorus-enriched chemostat of experimental pond water where nitrogen was limiting showed that the utilization of both elements was not in phase; for example, for the first three days the utilization rates of both nutrients were similar, but when *Fragilaria* bloomed nitrogen utilization increased rapidly although the phosphorus concentrations remained unchanged. This appears to indicate that different populations have different nitrogen and phosphorus requirements. The succession of natural populations in the culture was similar to that in a similarly fertilized pond.

It can be seen from the above results that properly designed, continuous-culture bioassays can yield information that may be distorted in batch assays and can provide much more valuable insights into nutrient limitation than batch experiments.

9.2.2 *Toxic substances*

Fisher *et al.* (1974) investigated the sublethal effects of polychlorinated biphenyls (PCBs) on *Dunaliella tertiolecta* and *Thalassiosira pseudonana* in batch culture and a chemostat. At a concentration of 0·1 ppb, pure cultures did not show any toxic effects in either batch or chemostat experiments. In a mixed-culture chemostat, however, toxic effects were clearly evident; the PCB-treated chemostat showed a reduction of the *Thalassiosira pseudonana* population by more than 50 per cent of the control. A mixed batch culture did not show any such effect except in the late stationary phase, in which a small decline of *Thalassiosira pseudonana* and a light increase of *Dunaliella tertiolecta* were observed. It appears, therefore, that the competitive ability is very sensitive to toxic chemicals, at least to PCBs, although the exact nature of the impaired competitive mechanism(s) is not clear.

As in the monoculture studies of the above marine algae, the results of batch culture and a turbidostat investigation of the toxicity of mercuric acetate in *Scrippsiella faeroense*, *Prorocentrum micans* and *Gymnodinium splendens* (Kayser, 1976) were not much different. Within the tested concentration range of 1 ppb to 1 ppm, the sensitivity to mercury, measured as growth rate, *in vivo* chlorophyll fluorescence, and morphology, differed from species to species, *Gymnodinium splendens* being most sensitive and *Prorocentrum micans* least.

It seems that long-term sublethal effects of toxic agents can best be studied in continuous culture by investigating competitive abilities or other physiological parameters, rather than measuring growth responses alone. The impairment of competitive ability has important ecological consequences. As indicated by Fisher *et al.* (1974), differential impairments in different species reduce diversity indexes in natural populations, which may initiate or make eutrophication more severe.

10 Transient state

Very little is known about the transient-state dynamics of phytoplankton growth. The transient state represents the dissociation of the well-regulated metabolic processes of the steady state. The mode and site of metabolic regulations, therefore, can be revealed in this state.

Perturbations of the steady-state populations results in oscillations, which eventually damp to another steady state (see Williams, 1971). True oscillations are also present in steady-state conditions, although the amplitude of oscillations is very small. Some of the underlying mechanisms for chemostat oscillations (Harrison and Topiwala, 1974) are (i) poor feedback loop of metabolic processes, (ii) poor control of environmental parameters (e.g. temperature) in the most sensitive range, and (iii) poor coupling between extracellular metabolites and environmental factors such as pH.

The responses of dynamic transient conditions are two types: rapid responses, resulting from processes such as increases in reaction rates, and responses requiring a long lag period generally due to enzyme induction, feedback, selection of mutants (Harrison and Topiwala, 1974) and genome replication (Williams, 1971).

The type of response is determined by the degree of perturbation and the initial physiological conditions of organisms. A slight increase in dilution rate in a steady-state cell brings a rapid adjustment, either because a slight excess of RNA is already present (Harrison and Topiwala, 1974) or because the organism increases its rates of protein synthesis at a constant RNA concentration (Tempest and Hunter, 1965). A large increase in dilution rate, however, results in a considerable lag period, probably because enzyme requirements can be met only with increased synthesis of RNA. On the other hand, when dilution rate is decreased, the adjustment of growth rate to the new rate is very rapid, although metabolic adjustment may be still in progress (Harrison and Topiwala, 1974).

It is possible, then, that steady-state kinetics might apply to a transient state within a certain limit of perturbation, the extent of which is dependent upon the organism's intrinsic physiological capability. Droop (1975) demonstrated that his steady-state kinetic model for chemostat growth (1974) can satisfactorily describe batch growth of *Monochrysis lutheri*. In batch culture, nutrient changes resulting from growth are equivalent to gradual and continuous perturbation, particularly when population density is low, and the perturbation of Droop's batch system might have been within a range at which the organism could adjust very rapidly. The work by Tett *et al.* (1975) and Barlow (1979) on natural populations also suggests that a steady-state model can be extended to relatively complex systems.

To apply a steady-state model to phytoplankton ecology, it is essential to know the upper limit of perturbation for instantaneous adjustment, as related to natural conditions, and the physiological capability of the organism.

The dependence of growth rate on intracellular nutrient "pools" may reflect a physiological adaptation, developed in the long course of evolution, to buffer changes in nutrient supply in the environment. Differences in physiological capability to adjust to perturbations may be a significant factor contributing to succession and the outcome of competition.

11 Concluding remarks

The application of continuous culture techniques in ecology is still in its infancy. Its potential and value have yet to be fully utilized. There are many vital problems to be investigated using the methods for fundamental physiological processes related to ecology and in extrapolating existing findings to natural waters. Among other needs, we must have parameters for estimating various nutrient limitations and a common reliable biomass term, if one can be found, to adapt the kinetics of laboratory growth and nutrient uptake to natural populations. A requirement for these parameters is that they can be measured with relative ease and by simple methods. Regulatory mechanisms of nutrient uptake by a whole cell have to be understood clearly. On the basis of steady-state information, growth and metabolic processes during transient states should be known, because, after all, natural populations are seldom in the steady state.

The finding that there are no additive or multiplicative effects for nutrient limitation has an important ecological ramification; this implies that different organisms may require various nutrients in various proportions. It is possible then that even in the same nutrient environment, the growth of some species may be limited by one nutrient, while others are regulated by another. In nutrient limitations, therefore, it is not only the absolute amounts of available limiting nutrients that are important but also their relative concentrations to other nutrients. Although we have long been using average C/N/P ratios of natural communities to determine a limiting nutrient (Redfield, 1958; Ketchum, 1969; Ryther and Dunstan, 1971), information regarding optimum nutrient ratios and their variability in individual species is almost non-existent.

The importance of interactions between nutrient, light and temperature on phytoplankton growth was clearly demonstrated more than a decade ago (Maddux and Jones, 1964), but not much progress has been made in this area. It has been suggested, for example, that temperature and nutrient limitation interact in a multiplicative fashion to affect growth (Goldman and Carpenter, 1974), but experimental confirmation is lacking.

Temporal variation in diel periodicity in growth, nutrient absorption and physiological processes may be an important ecological strategy for the survival and coexistence of various phytoplankters. This can best be evaluated using continuous cultures.

New potential applications of continuous culture methods in investigating natural populations have been demonstrated by the work of Barlow and his co-workers (Barlow *et al.*, 1973a, b; Peterson *et al.*, 1974; Barlow, 1979 and Jones *et al.* (1978 a, b). These studies may serve as a good example of how one can bridge the gap between laboratory work and field investigations. The field continuous culture apparatus designed by de Noyelles and O'Brien (1974) may also prove to be valuable in studying natural populations *in situ*.

Acknowledgement

This work was supported in part by Environmental Protection Agency grant R-804689-01 and National Science Foundation grant DEB 75-19519.

References

Aitchison, P. A. and Butt, V. S. (1973). The relationship between the synthesis of inorganic polyphosphate and phosphate uptake by *Chlorella vulgaris*. *Journal of Experimental Botany*, **24**, 497–510.

Algren, G. (1978). Growth of *Oscillatoria agardhii.* Gom. in chemostat culture. I. Investigations of nitrogen and phosphorus requirements. *Oikos,* **29**, 209–224.

Azad, H. S. and Borchardt, J. A. (1969). A method for predicting the effect of light intensity on algal growth and phosphorus assimilation. *Journal of Water Pollution Control Federation,* **41**, R392–R404.

Azad, H. S. and Borchardt, J. A. (1970). Variations in phosphorus uptake in algae. *Environmental Science and Technology,* **4**, 737–743.

Barlow, J. P. (1979). Cell quotas of phosphorus and silicon and growth of diatoms in Cayuga Lake. Submitted to *Limnology and Oceanography.*

Barlow, J. P., Peterson, B. J. and Savage, A. E. (1973a). Continuous flow bioassay of phosphorus as a limiting nutrient for Cayuga Lake phytoplankton. *In* "Proceedings of the 16th Conference of Great Lakes Research", pp. 7–14. International Association of Great Lakes Research.

Barlow, J. P., Schaffner, W. R., deNoyelles, F. and Peterson, B. J. (1973b). Continuous flow nutrient bioassays with natural phytoplankton populations. *In* "Bioassay Techniques and Environmental Chemistry" (Ed. G. E. Glass), pp. 299–319. Ann Arbor Science, Ann Arbor, Michigan.

Baule, B. (1918). Zu Mitscherlichs Gesetz der physiologischen Beziehungen. *Landwirtschaftliche Jahrbuecher,* **51**, 363–385.

Berman, T. and Gophen, M. (1972). Lake Kinneret: Planktonic populations during seasons of high and low phosphorus availability. *Verhandlungen Internationale Vereinigung für Theoretische und Angewandte Limnologie,* **18**, 588–598.

Bienfang, P. K. (1975). Steady state analysis of nitrate-ammonium assimilation by phytoplankton. *Limnology and Oceanography,* **20**, 402–411.

Blum, J. J. (1966). Phosphate uptake by phosphate starved *Euglena. Journal of General Physiology,* **49**, 1126–1137.

Borchardt, J. A. and Azad, H. S. (1968). Biological extraction of nutrients. *Journal of Water Pollution Control Federation,* **42**, 1739–1754.

Braun, W., Goodlow, R. J., Kraft, M., Altenbern, R. and Mead, D. (1951). The effects of metabolites upon interactions between variants in mixed *Brucella abortus* populations. *Journal of Bacteriology,* **62**, 45–52.

Britten, R. J. and McClure, F. T. (1962). The amino acid pool of *Escherchia coli. Bacteriological Reviews,* **26**, 292–335.

Busby, W. F. and Lewin, J. (1967). Silicate uptake and silica shell formation by synchronously dividing cells of the diatom *Navicular pelliculosa* (Breb) Hilse. *Journal of Phycology,* **3**, 127–131.

Caperon, J. (1968). Population growth response of *Isochrysis galbana* to variable nitrate environment. *Ecology,* **49**, 866–872.

Caperon, J. and Meyer, J. (1972a). Nitrogen-limited growth of marine phytoplankton. I. Changes in population characteristics with steady-state growth rate. *Deep-Sea Research,* **19**, 601–618.

Caperon, J. and Meyer, J. (1972b). Nitrogen-limited growth of marine phytoplankton. II. Uptake kinetics and their role in nutrient-limited growth of phytoplankton. *Deep-Sea Research,* **19**, 619–632.

Caperon, J. and Zieman, D. A. (1976). Synergistic effects of nitrate and ammonium ion on the growth and uptake kinetics of *Monochrysis lutheri* in continuous culture. *Marine Biology,* **36**, 73–84.

Chisholm, S. W. and Stross, R. G. (1976). Phosphate uptake kinetics in *Euglena gracilis* (Z) (Euglenophyceae) grown in light/dark cycles. II. Phased PO_4-limited cultures. *Journal of Phycology,* **12**, 217–222.

Chisholm, S. W., Stross, R. G. and Nobbs, P. A. (1975). Light/dark-phased cell division in *Euglena gracilis* (Z) (Euglenophyceae) in PO_4-limited continuous culture. *Journal of Phycology,* **11**, 367–373.

Cocito, C. and Bryson, V. (1958). Properties of colicine from *E. coli* Strain B. *Bacteriological Proceedings,* 1958, p. 38.

Coombs, J., Spanis, C. and Volcani, B. E. (1967). Studies on the biochemistry and fine structure of silica shell formation in diatoms. Photosynthesis and respiration in silicon-starvation synchrony of *Navicular pelliculosa* (Breb) Hilse. *Experimental Cell Research,* **47**, 315–328.

Cowie, D. B. (1962). Metabolic pools and the biosynthesis of protein. *In* "Amino Acid Pools" (Ed. J. T. Holden), pp. 633–645. Elsevier Publishing Company, Amsterdam, London, New York.

Darley, W. M. (1969). Silicon and the division cycle of the diatoms *Navicular pelliculosa* and *Cylindrotheca fusiformis. In* "Proceedings of the North American Paleontological Convention" (Ed. E. L. Yochelson), vol. 2, pp. 994–1009. Allen Press, Lawrence, Kansas, U.S.A.

Davis, C. O. (1976). Continuous culture of marine diatoms under silicate limitation. II. Effects of light intensity on growth and nutrient uptake of *Skeletonema costatum. Journal of Phycology,* **12**, 291–300.

Dawson, P. W. (1965). The intracellular amino acid pool of *Candida utilis* during growth in batch and continuous flow cultures. *Biochimica et Biophysica Acta,* **111**, 51–56.

deNoyelles, F. and O'Brien, W. J. (1974). The *in situ* chemostat—a self-contained continuous culturing and water sampling system. *Limnology and Oceanography,* **19**, 326–331.

Droop, M. R. (1961). Vitamin B_{12} and marine ecology: the response of *Monochrysis lutheri. Journal of the Marine Biological Association of the United Kingdom,* **41**, 69–76.

Droop, M. R. (1968). Vitamin B_{12} and marine ecology: IV. The kinetics of uptake, growth and inhibition in *Monochrysis lutheri. Journal of the Marine Biological Association of the United Kingdom,* **48**, 689–733.

Droop, M. R. (1970). Vitamin B_{12} and marine ecology: V. Continuous culture as an approach to nutritional kinetics. *Helgolander Wissenschaftliche Meeresuntersuchungen,* **20**, 629–636.

Droop, M. R. (1973a). Butrient limitation in osmotrophic protista. *American Zoologist,* **13**, 209–214.

Droop, M. R. (1973b). Some thoughts on nutrient limitation in algae. *Journal of Phycology,* **9**, 264–272.

Droop, M. R. (1974). The nutrient status of algal cells in continuous culture. *Journal of the Marine Biological Association of the United Kingdom,* **54**, 825–855.

Droop, M. R. (1975). The nutrient status of algal cells in batch culture. *Journal of Marine Biological Association of the United Kingdom,* **55**, 541–555.

Eppley, R. W. (1972). Temperature and phytoplankton growth in the sea. *Fishery Bulletin,* **70**, 1063–1085.

Eppley, R. W. and Coatsworth, J. L. (1966). Culture of the marine phytoplankter, *Dunaliella tertiolecta,* with light-dark cycles. *Archiv für Mikrobiologie,* **55**, 66–80.

Eppley, R. W. and Coatsworth, J. L. (1968). Uptake of nitrate and nitrite by *Ditylum brightwellii*—Kinetics and mechanisms. *Journal of Phycology,* **4**, 151–156.

Eppley, R. W. and Dyer, D. L. (1965). Predicting production in light-limited continuous culture of algae. *Applied Microbiology,* **13**, 833–837.

Eppley, R. W. and Renger, E. H. (1974). Nitrogen assimilation of an oceanic diatom in nitrogen-limited continuous culture. *Journal of Phycology,* **10**, 15–23.

Eppley, R. W. and Strickland, J. D. H. (1968). Kinetics of marine phytoplankton growth. *In* "Advances in Microbiology of the Sea" (Eds M. R. Droop and E. J. Ferguson Wood), pp. 23–62. Academic Press, London, New York.

Eppley, R. W. and Thomas, W. H. (1969). Comparison of half-saturation constants for growth and nitrate uptake of marine phytoplankton. *Journal of Phycology,* **5**, 375–379.

Eppley, R. W., Holmes, R. W. and Paasche, E. (1976). Periodicity in cell division and physiological behaviour of *Ditylum brightwellii*, a marine planktonic diatom, during growth in light–dark cycle. *Archiv für Mikrobiologie,* **56**, 305–323.

Eppley, R. W., Rogers, J. N. and McCarthy, J. J. (1969). Half-saturation constants for uptake of nitrate and ammonium by marine phytoplankton. *Limnology and Oceanography,* **14**, 912–920.

Eppley, R. W., Rogers, J. N., McCarthy, J. J. and Sournia, A. (1971). Light/dark periodicity in nitrogen and assimilation of the marine phytoplankters *Skeletonema costatum* and *Coccolitus huxleyi* in N-limited chemostat culture. *Journal of Phycology,* **7**, 150–154.

Fisher, N. S., Carpenter, E. J., Remsen, C. C. and Wurster, C. F. (1974). Effects on PCB on interspecific competition in natural and gnotobiotic phytoplankton communities in continuous and batch cultures. *Microbial Ecology,* **1**, 39–50.

Fitzgerald, G. P. (1968). Detection of limiting and surplus nitrogen in algae and aquatic weeds. *Journal of Phycology,* **4**, 121–126.

Fitzgerald, G. P. and Nelson, T. C. (1966). Extractive and enzymatic analysis for limiting or surplus phosphorus in algae. *Journal of Phycology,* **2**, 32–37.

Fogg, G. E. (1975). "Algal Cultures and Phytoplankton Ecology". University of Wisconsin Press, Madison.

Frisch, H. L. and Gotham, I. J. (1977). On periodic algal cyclostat populations. *Journal of Theoretical Biology,* **66**, 665–678.

Frisch, H. L. and Gotham, I. J. (1979). A simple model for periodic cyclostat growth. *Journal of Mathematical Biology,* **7**, 149–169.

Fuhs, G. W. (1969). Phosphorus content and rate of growth in the diatom *Cyclotella nana* and *Thalassiosira fluviatilis*. *Journal of Phycology,* **5**, 312–321.

Fuhs, G. W., Demerle, S. D., Canelli, E. and Chen, M. (1972). Characterization of phosphorus-limited plankton algae. (With reflections on the limiting nutrient concept.) *In* "Nutrient and Eutrophication" (Ed. G. E. Likens), Special Symposia, vol. 1, pp. 113–132. The American Society of Limnology and Oceanography, Allen Press, Lawrence, Kansas.

Giddings, J. M. (1975). Growth and chemical composition of *Scenedesmus abundans* in nitrogen-limited chemostat culture. Ph.D. thesis. Cornell University, Ithaca, New York.

Goldman, J. C. (1977a). Steady state growth of phytoplankton in continuous culture: Comparison of internal and external nutrient equations. *Journal of Phycology,* **13**, 251–258.

Goldman, J. C. (1977b). Temperature effects on phytoplankton growth in continuous culture. *Limnology and Oceanography,* **22**, 932–936.

Goldman, J. C. and Carpenter, E. J. (1974). A kinetic approach to the effect of temperature on algal growth. *Limnology and Oceanography,* **19**, 756–766.

Goldman, J. C. and Ryther, J. H. (1976). Temperature-influenced species competition in mass cultures of marine phytoplankton. *Biotechnology and Bioengineering,* **18**, 1125–1144.

Goldman, J. C. and McCarthy, J. J. (1978). Steady state growth and ammonium uptake of a fast growing marine diatom. *Limnology and Oceanography,* **23**, 605–703.

Goldman, J. C., Oswald, W. J. and Jenkins, D. (1974). The kinetics of inorganic carbon limited algal growth. *Journal of Water Pollution Control Federation,* **46**, 554–574.

Golterman, H. L., Bakels, C. C. and Jakobs-Mogelin, J. (1969). Availability of mud phosphate for the growth of algae. *Verhandlungen Internationale Vereinigung für Theoretische und Angewandte Limnologie*, **17**, 467–479.

Gons, H. J. and Mur, L. R. (1975). An energy balance for algal populations in light-limiting conditions. *Verhandlungen Internationale Vereinigung für Theoretische und Angewandte Limnologie*, **19**, 2729–2733.

Gotham, I. J. (1977). Polyphosphates and nutrient limited cell growth cycles in algal chemostats: A theoretical and physiological analysis. Ph.D. thesis. State University of New York at Albany, Albany, New York.

Guillard, R. R. L., Kilham, P. and Jackson, T. A. (1973). Kinetics of silicon-limited growth of the marine diatom *Thalassiosira pseudonana* Hasle and Heimdal (*Cyclotella nana* Hustedt). *Journal of Phycology*, **9**, 233–237.

Harold, F. M. (1966). Inorganic polyphosphate for growth of algae. *Bacteriological Reviews*, **30**, 772–795.

Harrison, W. G. (1976). Nitrate metabolism of the red tide dinoflagellate *Gonyalax polyedra* Stein. *Journal of Experimental Biology and Ecology*, **21**, 199–209.

Harrison, D. E. F. and Topiwala, H. H. (1974). Transient and oscillatory states of continuous culture. *In* "Advances in Biochemical Engineering", vol. 3 (Eds T. K. Ghose, A. Fiechter and N. Blackebrough), pp. 167–219. Springer-Verlag, Berlin and New York.

Harrison, P. J., Conway, H. L. and Dugdale, R. C. (1976). Marine diatoms grown in chemostats under silicate or ammonium limitation. I. Cellular chemical composition and steady-state growth kinetics of *Skeletonema costatum*. *Marine Biology*, **35**, 177–186.

Hastings, J. W. and Sweeney, B. M. (1964). Phased cell division in the marine dinoflagellates. *In* "Synchrony in Cell Division and Growth" (Ed. E. Zeuthen), pp. 307–321. Interscience Publishers, New York, London, Sydney.

Hattori, A. (1960). Studies on the metabolism of urea and other nitrogenous compounds in *Chlorella ellipsoidea*. III. Assimilation of urea. *Plant and Cell Physiology*, **1**, 107–116.

Healey, F. P. and Hendzel, L. L. (1975). Effects of phosphorus deficiency on two algae growing in chemostats. *Journal of Phycology*, **11**, 303–309.

Hellebust, J. A. (1975). Extracellular products. *In* "Algal Physiology and Biochemistry" (Ed. W. D. P. Stewart), pp. 838–863. Blackwell, Oxford.

Herbert, D. (1958). Some principles of continuous culture. 7th International Congress of Microbiology. Symposium "Recent Progress in Microbiology" (Ed. G. Tunevall), pp. 381–396. Almqvist and Wiksell, Stockholm.

Herbert, D., Elsworth, R. and Telling, R. C. (1956). The continuous culture of bacteria; a theoretical and experimental study. *Journal of General Microbiology*, **14**, 601–622.

Holden, J. T. (ed.) (1962). "Amino Acid Pools". Elsevier Publishing Company, Amsterdam, London, New York.

Holm-Hansen, O., Sutcliffe, Jr., W. H. and Sharp, J. (1968). Measurement of deoxyribonucleic acid in the ocean and its ecological significance. *Limnology and Oceanography*, **13**, 507–514.

Holt, M. G. and Smayda, T. (1974). The effect of daylength and light intensity on the growth rate of the marine diatom *Detonula confervacea* (Cleve) Gran. *Journal of Phycology*, **10**, 231–237.

Hutchinson, G. E. (1961). The paradox of the plankton. *American Naturalist*, **95**, 137–145.

Jannasch, H. W. (1965). Continuous culture in microbial ecology. *Laboratory Practice,* **14**, 1162–1167.
Jannasch, H. W. (1974). Steady state and the chemostat in ecology. *Limnology and Oceanography,* **19**, 716–720.
Jannasch, H. W. and Mateles, R. I. (1974). Experimental bacterial ecology studied in continuous culture. *In* "Advances in Microbial Physiology" (Eds A. H. Rose and D. W. Tempest), vol. 11, pp. 165–212. Academic Press, London, New York.
Jeanjean, R. (1969). Influence de carence en phosphore sur les vitesses d'adsorption du phosphate par les Chlorelles. *Bulletin de la Societé Francaise de Physiologie Végétale,* **15**, 159–171.
Jones, K. J., Tett, P., Wallis, A. C. and Wood, B. J. B. (1978a). The use of small, continuous multispecies cultures and cultures to investigate the ecology of phytoplankton in a Scottish Sea loch. *Mitteilungen der Internationalen Vereinigung für Theoretische und Angewandte Limnologie,* **18**, 71–77.
Jones, K. J., Tett, P., Wallis, A. C. and Wood, B. J. B. (1978b). Investigation of a nutrient-growth model using continuous culture of natural phytoplankton. *Journal of the Marine Biological Association of the United Kingdom,* **58**, 923–941.
Kayser, H. (1976). Waste-water assay with continuous algal cultures: the effect of mercuric acetate on the growth of some marine dinoflagellates. *Marine Biology,* **36**, 61–72.
Ketchum, B. H. (1939). The absorption of phosphate and nitrate by illuminated culture of *Nitzschia closterium. American Journal of Botany,* **26**, 399–407.
Ketchum, B. H. (1969). Eutrophication of estuaries. *In* "Eutrophication: Causes, Consequences and Correctives", Publication 1700, pp. 197–209. National Academy of Science/National Research Council.
Kilham, P. (1971). A hypothesis concerning silica and freshwater planktonic diatoms. *Limnology and Oceanography,* **16**, 10–18.
Kilham, S. S., Kott, C. L. and Tilman, D. (1977). Phosphate and silicate kinetics for the Lake Michigan diatom *Diatoma elongatum. Journal of Great Lakes Research,* **3**, 93–99.
Knauss, H. J. and Porter, J. W. (1954). The absorption of inorganic ions by *Chlorella pyrenoidosa. Plant Physiology, Lancaster,* **29**, 229–234.
Kornberg, H. L. (1973). Fine control of sugar uptake by *Escherichia coli. Symposia of the Society for Experimental Biology,* **27**, 175–193.
Kubitschek, H. E. (1970). "Introduction to Research with Continuous Cultures". Prentice Hall, Englewood Cliffs, New Jersey.
Kuhl, A. (1960). Die Biologie der kondensierten anorganischen Phosphate. *Ergebnisse der Biologie,* **23**, 144–185.
Kulaev, I. S. (1975). Biochemistry of inorganic polyphosphates. *Reviews of Physiology, Biochemistry and Experimental Pharmacology,* **73**, 131–158.
Laws, E. and Caperon, J. (1976). Carbon and Nitrogen metabolism by *Monychrysis lutheri*: measurement of growth-rate-dependent respiration rates. *Marine Biology,* **36**, 85–97.
Lorenzen, H. (1970). Synchronous cultures. *In* "Photobiology of Microorganisms" (Ed. P. Haldall), pp. 187–212. Wiley-Interscience, London, New York.
Maaløe, O. and Kjeldgaard, N. O. (1966). "Control of Macromolecular Synthesis". W. A. Benjamin, New York, Amsterdam.
Maddux, W. S. and Jones, R. F. (1964). Some interactions of temperature, light intensity and nutrient concentrations during the continuous culture of *Nitzschia closterium* and *Tetraselmis* sp. *Limnology and Oceanography,* **9**, 79–86.

Meers, J. L. (1974). Growth of bacteria in mixed cultures. *In* "Microbial Ecology" (Eds A. Laskin and H. Lechevalier), pp. 136–181. CRC Press.

Miyachi, S., Kanai, R., Mihara, S., Miyachi, S. and Aoki, S. (1964). Metabolic role of inorganic polyphosphate in *Chlorella* cells. *Biochimica et Biophysics Acta*, **93**, 625–634.

Monod, J. (1942). "Recherches sur la Croissance des Cultures Bactériennes". 2nd ed. Hermann, Paris.

Müller, H. (1970). Das Wachstum von *Nitzschia actinastroides* (Lemm.) v. Goor im chemostaten bei limitierender Phosphatkonzentration. *Berichte der Deutschen Botanischen Gesellschaft*, **83**, 537–544.

Müller, H. (1972). Wachstum und Phosphat bedarf von *Nitzschia actinastroides* (Lemm.) v. Goor in statischer und homokontinuierlicher Kulter unter Phosphat-limitierung. *Archiv für Hydrogiologie (Supplementband)*, **38**, 399–484.

Munson, R. J. (1970). Turbidostats. *In* "Methods of Microbiology" (Eds J. R. Norris and D. W. Ribbons), vol. 2, pp. 349–376. Academic Press, London and New York.

Mur, L. R., Gons, H. J. and van Liere, L. (1978). Competition of the green alga *Scenedesmus* and the blue-green alga *Oscillatoria*. *Mitteilungen der Internationalen Vereinigung für Theoretische und Angewandte Limnologie*, **21**, 473–479.

Myers, J. (1962). Laboratory cultures. *In* "Physiology and Biochemistry of Algae" (Ed. R. A. Lewin), pp. 595–615. Academic Press, New York and London.

Myers, J. and Graham, J. R. (1975). Photosynthetic unit size during the synchronous life cycle of *Scenedesmus*. *Plant Physiology Lancaster*, **55**, 686–688.

Niemeyer, R. (1976). Cyclic condensed metaphosphates and linear polyphosphates in brown and red algae. *Archives of Microbiology*, **108**, 243–247.

Novick, A. and Slizard, L. (1950). Experiments with the chemostat on spontaneous mutation of bacteria. *Proceedings of National Academy of Science, Washington*, **36**, 708–719.

Paasche, E. (1973a). Silicon and the ecology of marine plankton diatoms. I. *Thalassiosira pseudonana* (*Cyclotella nana*) grown in a chemostat with silicate as limiting nutrient. *Marine Biology*, **19**, 117–126.

Paasche, E. (1973b). Silicon and the ecology of marine plankton diatoms. II. Silicate-uptake kinetics in five diatom species. *Marine Biology*, **19**, 262–269.

Pearsall, W. H. (1932). Phytoplankton in English Lakes. II. The composition of the phytoplankton in relation to dissolved substances. *Journal of Ecology*, **20**, 241–262.

Perry, M. J. (1976). Phosphate utilization by an oceanic diatom in phosphorus-limited chemostat culture and in the oligotrophic waters of the central North Pacific. *Limnology and Oceanography*, **21**, 88–107.

Petersen, R. (1975). The paradox of the plankton: an equilibrium hypothesis. *American Naturalist*, **109**, 35–49.

Peterson, B. J., Barlow, J. P. and Savage, A. (1974). The physiological state with respect to phosphorus of Cayuga Lake phytoplankton. *Limnology and Oceanography*, **19**, 396–408.

Pickett, J. M. (1975). Growth of *Chlorella* in a nitrate-limited chemostat. *Plant Physiology, Lancaster*, **55**, 223–225.

Pipes, W. O. and Koutsoyannis, S. P. (1962). Light-limited growth of *Chlorella* in continuous culture. *Applied Microbiology*, **10**, 1–5.

Pirt, S. J. (1975). "Principles of Microbe and Cell Cultivation". John Wiley, New York.

Postgate, J. R. and Hunter, J. R. (1962). The survival of starved bacteria. *Journal of General Microbiology*, **29**, 233–263.

Price, C. A. and Quigley, J. W. (1966). A method for determining quantitative zinc requirements for growth. *Soil Science*, **101**, 11–16.

Redfield, A. C. (1958). The biological control of chemical factors in the environment. *American Scientist*, **46**, 205–221.
Rhee, G-Y. (1972). Competition between an alga and an aquatic bacterium for phosphate. *Limnology and Oceanography*, **17**, 505–514.
Rhee, G-Y. (1973). A continuous culture study of phosphate uptake, growth rate and polyphosphates in *Scenedesmus* sp. *Journal of Phycology*, **9**, 495–506.
Rhee, G-Y. (1974). Phosphate uptake under nitrate limitation by *Scenedesmus* sp. and its ecological implications. *Journal of Phycology*, **10**, 470–475.
Rhee, G-Y. (1978). Effects of N/P atomic ratios and nitrate limitation on algal growth, cell composition, and nitrate uptake: a study of dual nutrient limitation. *Limnology and Oceanography*, **23**, 10–25.
Rhee, G-Y. and Gotham, I. J. (1979). Continuous culture studies of environmental factors on phytoplankton growth: Effects of temperature and temperature-nutrient interactions. Submitted to *Limnology and Oceanography*.
Richardson, B., Orcutt, D. N., Schwertner, H. A., Martinez, C. L. and Wickline, H. E. (1969). Effects of nitrogen limitation on the growth and composition of unicellular algae in continuous culture. *Applied Microbiology*, **18**, 245–250.
Ryther, J. and Dunstan, W. M. (1971). Nitrogen, phosphorus and eutrophication in the coastal marine environment. *Science, New York*, **171**, 1008–1013.
Sager, R. (1975). Nuclear and cytoplasmic inheritance in green algae. *In* "Algal Physiology and Biochemistry" (Ed. W. D. P. Stewart), pp. 314–345. Blackwell, Oxford.
Serruya, C. (1974). Nitrogen and phosphorus balances in Lake Kinneret. *Verhandlungen Internationale Vereinigung für Theoretische und Angewandte Limnologie*, **19**, 489–508.
Shelef, G., Oswald, W. J. and Golueke, C. C. (1970). Assaying algal growth with respect to nitrate concentration by a continuous flow turbidostat. *In* "Proceedings of 5th International Conference on Water Pollution Research" (Ed. S. H. Jenkins), pp. III 25/1–III 25/9. Pergamon Press, Oxford.
Solomonson, L. P. and Spehar, A. M. (1977). Model for the regulation of nitrate assimilation. *Nature, London*, **265**, 373–375.
Sorokin, C. (1960). Kinetic studies of temperature effects on the cellular level. *Biochimica et Biophysica Acta*, **38**, 197–204.
Sournia, A. (1974). Circadian periodicities in natural populations of marine phytoplankton. *Advances in Marine Biology*, **12**, 325–389.
Stent, G. W. and Brenner, S. (1961). A genetic locus for the regulation of ribonucleic acid synthesis. *Proceedings of the National Academy of Science of the United States*, **47**, 2005–2014.
Stross, R. G. and Pemrick, S. M. (1974). Nutrient uptake kinetics in phytoplankton: a basis for niche separation. *Journal of Phycology*, **10**, 164–169.
Sundberg, I. and Nilshammer-Holmvall, M. (1975). The diurnal variation in relation to deposition of starch, lipid, and polyphosphate in synchronized cells of *Scenedesmus*. *Zeitschrift für Pflanzen physiologie*, **76**, 270–279.
Sutcliffe, Jr., W. H. (1965). Growth estimates from ribonucleic acid content in some small organisms. *Limnology and Oceanography*, **13**, R253–R258.
Swift, D. G. and Taylor, W. R. (1974). Growth of vitamin B_{12}- limited cultures: *Thalassiosira pseudonana, Monochrysis lutheri and Isochrysis galvana*. *Journal of Phycology*, **10**, 385–391.
Syrett, P. J. (1953). The assimilation of ammonia by nitrogen-starved cells of *Chlorella vulgaris*. I. The correlation of assimilation with respiration. *Annals of Botany*, **17**, 1–19.
Taylor, P. A. and Williams, P. J. LeB. (1975). Theoretical studies on the coexistence of

competing species under continuous flow conditions. *Canadian Journal of Microbiology*, **21**, 90–98.

Tempest, D. W. (1969). The continuous cultivation of microorganisms. I. Theory of chemostat. *In* "Methods in Microbiology" (Eds J. R. Norris and D. W. Ribbons), vol. 2, pp. 259–276. Academic Press, London and New York.

Tempest, D. W. (1970). The place of continuous culture in microbiological research. *In* "Advances in Microbial Physiology" (Eds A. H. Rose and J. F. Wilkinson), vol. 4, pp. 223–250. Academic Press, London and New York.

Tempest, D. W. and Hunter, J. R. (1965). The influence of temperature and pH value on the macromolecular composition of magnesium-limited and glycerol-limited *Aerobacter aerogenes* growing in a chemostat. *Journal of General Microbiology*, **41**, 267–273.

Tempest, D. W. and Neijssel, O. M. (1976). Microbial adaptation to low nutrient environments. *In* "Continuous Culture 6: Applications and New Fields" (Eds A. C. R. Dean, D. C. Elwood, C. G. T. Evans and J. Melling), pp. 281–296. Ellis Horwood, Chichester.

Tempest, D. W., Herbert, D. and Phipps, P. J. (1967). Studies on the growth of *Aerobacter aerogenes* at low dilution rates in a chemostat. *In* "Microbial Physiology and Continuous Culture" (Eds E. O. Powell, C. G. T. Evans, R. E. Strange and D. W. Tempest), pp. 240–253. Her Majesty's Stationery Office, London.

Terry, O. and Edmunds, L. N. Jr. (1969). Semicontinuous culture and monitoring systems for temperature-sychronized *Euglena*. *Biotechnology and Bioengineering*, **11**, 745–756.

Tett, P., Cottrell, J. C., Trew, D. O. and Wood, B. J. B. (1975). Phosphorus quota and the chlorophyll:carbon ratio in marine phytoplankton. *Limnology and Oceanography*, **20**, 587–603.

Thomas, W. H. and Dodson, A. N. (1968). Effects of phosphate concentrations on cell division rates and yield of a tropical oceanic diatom. *Biological Bulletin of The Woods Hole Oceanographic Institution*, **134**, 199–208.

Thomas, W. H. and Dodson, A. N. (1972). On nitrogen deficiency in tropical Pacific Oceanic phytoplankton. II. Photosynthetic and cellular characteristics of a chemostat-grown diatom. *Limnology and Oceanography*, **17**, 515–523.

Tilman, D. (1977). Resource competition between planktonic algae: an experimental and theoretical approach. *Ecology*, **58**, 338–348.

Tilman, D. and Kilham, S. S. (1976). Phosphates and silicate growth and uptake kinetics of the diatom. *Asterionella formosa* and *Cyclotella meneghiniana* in batch and semi-continuous cultures. *Journal of Phycology*, **12**, 375–383.

Titman, D. (1976). Ecological competition between algae: experimental confirmation of resource-based competition theory. *Science, New York*, **192**, 463–465.

Van Baalen, C. (1965). Quantitative surface plating of coccoid blue-green algae. *Journal of Phycology*, **1**, 19–22.

Veldkamp, H. (1976). Mixed culture study with the chemostats. *In* "Continuous Culture 6: Application and New Fields" (Eds A. C. R. Dean, D. C. Elwood, C. G. T. Evans and J. Melling), pp. 315–328. Ellis Horwood, Chichester.

Veldkamp, H. (1977). Ecological studies with chemostats. *In* "Advances in Microbial Ecology" (Ed. M. Alexander), vol. 1, pp. 59–94. Plenum Press, New York and London.

Veldkamp, H. and Jannasch, H. W. (1972). Mixed culture studies with the chemostat. *Journal of Applied Chemistry and Biotechnology*, **22**, 105–123.

Verduin, J. (1964). Principles of primary productivity: photosynthesis under completely natural conditions. *In* "Algae and Man" (Ed. D. F. Jackson), pp. 221–238. Plenum Press, New York.

Watson, T. G. (1972). The present status and future prospects of the turbidostat. *Journal of Applied Chemistry and Biotechnology*, **22**, 229–243.

Williams, F. M. (1971). Dynamics of microbial populations. *In* "Systems Analysis" (Ed. B. C. Patten), pp. 197–267. Academic Press, New York and London.

Growth rate of infusorian populations

MILOŠ LEGNER

Hydrobiological Laboratory, Botanical Institute, Czechoslovak Academy of Sciences, Prague

1 Introduction

Ciliata may form an essential part of the microbial component in aquatic communities. Many ciliate species live in benthic sediments, bacterial and algal mats or in a close proximity to them (reviews by Noland and Gojdics, 1967; Elliott and Bamforth, 1975). Water bodies with decaying plant material as, e.g. tree leaves or remnants of aquatic vegetation, are inhabited by a number of ciliate species (Bick, 1958; Wilbert, 1968); some of them reach considerable cell numbers especially in the interstitial environment of natant vegetation (Legner, 1964).

The biomass of ciliates usually increases with increasing amount of decomposable organic material in water (Legner, 1975a), as many species are capable of consuming heterotrophic bacteria in considerable quantities. As a result ciliates are frequent in various types of water bodies polluted with sewage, whether it be flowing or stagnant water (reviewed by Bick and Kunze, 1971). The effect of microphagous species on the removal of water turbidity caused by suspended bacteria has been reported (Curds *et al.*, 1968; Jones, 1973). Some data suggest that ciliates may stimulate growth of bacteria (Legner, 1973) and bacterial heterotrophic activity (Straškrabová-Prokešová and Legner, 1966). In addition to their effect on food organisms, the ciliates are a potential source of energy for predators in higher trophic levels. Hence, the opinion seems to be warranted that the importance of ciliates in aquatic ecosystems has not been appreciated sufficiently up to now.

Good estimates of ciliate biomass would be helpful in considering the role of ciliate populations in an ecosystem. Along with the value of biomass, the growth rate is utilizable as the basis for the calculation of production (Edmondson and Winberg, 1971). In the past, the growth rate of ciliates has been studied from several aspects. We shall attempt briefly to survey the present knowledge and to point out some possible approaches enabling deeper insight into the problem. The purpose of this paper is not to review all the literature relevant to this topic, but rather to outline possible ways to the understanding of ciliate population growth and to the estimation of its rate.

2 Growth rate concept

Generally, the multiplication of ciliate cells in pure culture under static (batch culture) conditions is considered to be comparable to that of other unicellular organisms (Hall, 1967). Similarly in continuous-flow culture conditions the kinetics of ciliate population growth is mostly supposed to correspond with that of bacterial culture (Curds and Bazin, 1977). Later on some special features of ciliate growth, distinct from the conventional scheme, will be discussed. To explain some basic terms, these peculiarities will be ignored for the sake of simplicity.

2.1 BASIC TERMS

The specific growth rate, μ, is understood as a measure of population biomass increase, which has the following relationship to the time of biomass doubling, τ_d:

$$\mu = \frac{\ln 2}{\tau_d}. \tag{1}$$

The specific growth rate of a population may but need not have the same value as the division rate, ν, of cells. The relation of ν to the generation time τ of cells analogous to equation (1) has been thoroughly dealt with by Powell (1956, cf. section 4.3). A third term we need for consideration of the population growth is the cell growth rate k_x, which is understood as a relative increase of individual cell mass. If not otherwise specified, we shall use k, the mean rate of cell biomass increase over entire generation time, instead of this value. All the listed parameters of unicellular growth have the dimension of h^{-1} and would assume the same value only if the cell mass exactly doubled during the interdivision interval and there were no variability in cell generation time.

Exact doubling of cell mass between divisions is considered as a characteristic feature of balanced growth (James, 1974). It seems probable that balanced growth is not normally reached in exponentially growing ciliate cultures. Earlier work, when the dry mass of *Paramecium* cells was measured (Kimball *et al.*, 1959), as well as the more recent report on doubling rate for cellular RNA of *Tetrahymena* (Plesner and Hartman, 1973) brought evidence that cell growth rate was slower than the division rate. Similarly, Curds and Cockburn (1971) reported that the mean cell volume of *Tetrahymena* decreased with time for at least five days of growth in a continuous-flow culture. These authors observed in continuous culture further decrease in cell volume accompanied by a gradual increase in division rate until a growth rate was reached never observed in batch culture.

Hall (1967) has, by anaology with bacterial cultures, differentiated lag phase, exponential phase, phase of negative growth acceleration, and maximal stationary phase, in protozoan batch culture. Szyszko *et al.* (1968) brought evidence that in addition to a generally known exponential (fast exponential) phase there is a further phase of slow exponential growth.

The cells in the former phase have generation times shorter than a day (ultradian growth) whereas the cells in the slow exponential phase have generation times longer than a day (infradian growth). Populations growing alternatively in one of the two phases differ in a susceptibility to synchronization stimuli (Meinert *et al.*, 1975).

2.2 NUTRIENT-LIMITED GROWTH

The question of nutrient limitation is central to the growth of cultures. Unlike bacteria, only a few species of ciliates have been kept on chemically defined media (Lilly and Stillwell, 1965; Soldo and van Wagtendonk, 1969; Meskill, 1970). The studies on limitation by a single nutrient are even scarcer (Lykkesfeldt, 1974). A set of papers reported growth of the ciliate *Tetrahymena pyriformis* in axenic media (growth in absence of other organisms; reviewed by Holz, 1973). They usually describe the growth in excess of nutrients (nutrient non-limited). However, the data on the functional form of nutrient limitation in ciliates may be obtained almost exclusively from the reports of continuous monoxenic (with one other organism) cultures of microphagous ciliates fed on bacteria (Hamilton and Preslan, 1970; Curds and Cockburn, 1971). Results of these authors agree in essence with Monod's (1942) function of saturation curve,

$$\mu = \frac{\mu_{max} S}{K_s + S}, \qquad (2)$$

when the concentration of bacterial mass is substituted for S, the concentration of limiting substrate. The constant K_s corresponds to the concentration of bacterial mass supporting growth of ciliates at a growth rate which is half the maximum attainable one (μ_{max}). A similar relationship was supposed when other prey organisms (algae) were limiting for *Tintinnopsis beroidea* (Gold, 1973).

2.3 NATURAL POPULATIONS

For application of these results to population living in aquatic ecosystems, the question of primary interest is what happens if several nutrients (or species of food organisms) approach the limiting amount in the environment. In algae, for instance, a threshold rather than multiplicative limitation is probably the case, i.e. only one limiting nutrient controls the population growth under given conditions, other ones displaying no effect (Droop, 1974). This problem should be a matter of further investigation in ciliates.

Cells both with nutrient-limited and non-limited growth may be encountered in natural populations. Generally, nutrient-limited growth is very likely the more frequent. One more physiological state of cells, however, is probably very important in the natural habitat, i.e. living under starving conditions when growth is arrested (cf. Jannasch and Mateles, 1974).

If we do not consider, for simplicity, the biotic interrelations in a community, the population in a natural habitat differs essentially from both batch and continuous culture (see Jannasch, 1974). In batch culture the population starts with non-limited exponential growth and gradually changes the closed system by depleting it of nutrients and excreting its metabolities. These changes act initially probably in favour of but later on obviously to the detriment of the population (cf. Grebecki, 1961). In continuous culture, the population initially adapts itself and then remains adapted to particular constant conditions. In an ideal case, its adaptive mechanisms are unused for a long time.

In the natural habitat, in contrast to the closed system of batch culture, the population is usually not able to change its environment so drastically, and sometimes even a situation may occur of relatively steady input of nutrients and output of external metabolites along with a part of the population. The environmental conditions, however, never keep constant, with the exception, perhaps, of the deep ocean and underground waters. Populations in nature, when growing, are probably in a permanent transient state, all the time adapting themselves to the continuously changing environment.

3 Evaluation of population biomass

3.1 SPACE DISTRIBUTION OF CILIATE POPULATIONS

Like other protozoan groups, most ciliate species are bound to the environment at the surface of various submerged bodies or sediments, preferring sites with slow water movement (Noland and Gojdics, 1967). They were found to attain substantially higher biomass at the walls of an experimental system simulating a river bed supplied with a moderate amount of artificial "sewage" than in the suspended component of the system (Legner, 1975b; Legner *et al.*, 1976).

One of the mechanisms of increasing ciliate biomass at submerged surfaces is thigmotaxis, i.e. a behaviour leading to a slowing down of active movement when in contact with a solid substratum (Noland and Gojdics, 1967).

The factors causing cells to aggregate at submerged surfaces may be different in different species. Sessile forms obviously find there a base for attaching their stalks. Various herbivores graze on bacterial and algal growths and carnivorous species prey on a gathered biomass of other protozoa. A preference for slowly moving water environment is "reasonable" also in free-swimming microphagous species that generate minute water currents to bring food to their cytostomes. A rapid water movement might obviously interfere with these feeding mechanisms. It was shown that *Glaucoma chattoni* in bacterized culture, and *Tetrahymena pyriformis* in axenic culture, attained lower population biomass when the culture fluid was stirred and aerated (Legner, 1973; Cameron, 1973).

The zonation of ciliates in aquatic environments may render proper quantification of biomass difficult. Collection of samples should be performed with respect to the most abundantly inhabited sites in such a way as to allow an estimation of the actual population biomass in the investigated part of a water-body. Special attention should be paid to determining the quantity of free-swimming cells concentrated at submerged surfaces (Legner, 1975b).

3.2 MEASUREMENT OF POPULATION GROWTH

For a calculation of the specific growth rate in a batch culture or any other closed system, it is sufficient to estimate the values of the population biomass at two different time instants. Then the specific growth rate is obtained from the formula,

$$\mu = \frac{\ln X_{t_2} - \ln X_{t_1}}{t_2 - t_1}, \tag{3}$$

where X_{t_1}, X_{t_2} are biomasses at the time t_1 and t_2, respectively and ln is the natural logarithm. The weak point of this type of growth rate measurement is the accuracy of the biomass estimation. When the population is not homogeneously distributed in a given space representative samples have to be taken of the whole system to compensate for possible migration (e.g. Legner, 1975a). Consequently, the biomass may be related to a unit volume of the inhabited space.

In continuous culture, with the ciliate homogeneously distributed over the culture vessel, simple relations should operate between steady-state values (Herbert *et al.*, 1956):

$$\frac{dX}{dt} = \mu X - DX = 0, \tag{4}$$

where D, the dilution rate, is the proportion between flow rate F and volume V of the culture vessel ($D = F/V$). The above equation (4) implies that $\mu = D$, i.e. the specific growth rate equals the dilution rate. Even in a multi-stage cascade of culture vessels a simple formula may be used to calculate the specific growth rate μ_n at the nth stage of the system (after Řičica *et al.*, 1967):

$$\mu_n = D\frac{X_n - X_{n-1}}{X_n}. \tag{5}$$

However, the specific growth rate may be calculated from changes in biomass in a homogeneous system prior to the attainment of a steady state. The formula is similar to that of the batch system (Jannasch, 1969):

$$\mu = D + \frac{1}{t}\ln\frac{X_{t_2}}{X_{t_1}}. \tag{6}$$

Although several homo-continuous systems of ciliates have been described since Curd's first reports (Curds, 1969; Curds and Cockburn, 1971), it seems that such a culture is not feasible for many ciliate species (Straškrabová and Legner, 1969; Legner, 1973). When a heterogeneous continuous-flow system is considered, the situation becomes more complicated. First, it is necessary to describe such a type of culture in terms of differential equations (Legner *et al.*, 1976). For a single stage we may write:

$$\frac{dX}{dt} = \mu' X - DX, \tag{7}$$

$$\frac{dX_T}{dt} = \mu X_T - DX. \tag{8}$$

The X and X_T are biomass of the suspended component and total biomass of the culture, respectively, both related to the unit volume of the culture vessel. Simultaneously X_{t_1}, X_{Tt_1} and then X_{t_2}, X_{Tt_2} are to be estimated at time instants t_1, t_2. The value of the specific growth rate μ', concerning only the suspended component, is readily calculable from the equation (6). When substituting $X_{t_1}, X_{\mathrm{T}t_1}$, for X, X_T and an arbitrary value for μ, the system of the differential equations may be solved by means of a computer program. Additional iterative sub-program is necessary for correction of the μ value, until a reasonable fit is obtained of the computed $X_{\mathrm{T}t_2}$ to the value actually found in the culture.

This procedure is applicable to a multi-stage cascade of heterogeneous cultures as well as to other types of heterogeneous systems, including natural ones. The conditions of the proper use of this method are that: (i) the total, as well as input and output biomass, is measurable; (ii) the volume of the system and respective flow rates are defined. A use of a mathematical treatment of a three-stage cascade was previously described (Legner *et al.*, 1976).

In an actual population an additional characteristic would usually be necessary for a complete description of population growth, the death rate being either intrinsic or caused by some external factor, e.g. by predation. In ciliates this value is not readily measurable in terms of total population biomass. For that reason it would be better to term the value of the specific growth rate computed from the changes of biomass as the "rate of biomass change" (see also Legner *et al.*, 1976).

The use of the above method is thus limited because there is a lack of information on the population state. One may suppose that a population growing exponentially but depressed by a heavy predation might have another type of cell than, for example, cells that die through starvation (see section 4).

4 Structural approach

4.1 CILIATE BIOLOGY

4.1.1 *Cell cycle*

With exception of one genus, ciliates have two type of nuclei which differ in size, morphology, chemical composition, mode of division and function.

Macronuclei function as somatic nuclei of the ciliate cell and synthesize RNA actively. In most ciliate species they are polypoid. It has been suggested that the degree of polyploidy depends on the relation between volume of cytoplasm and volume of micronecleus, the highest polyploidy occurring in large species with small micronuclei (Raikov,

1969). In three primitive families, Trachelocercidae, Loxodidae, and Geleidae (the order Karyorelictida after Corliss; the nomenclature corresponding with his scheme is currently used in the following text: cf. Corliss, 1977), the macronuclei do not divide, but arise during each cell division cycle from micronuclei (Raikov, 1969). In all the other ciliate groups the macronuclei divide, normally into two. Suctorida are one exception, where simultaneous multiple division may occur.

Micronuclei are mostly diploid; if polyploidy occurs, then, within the same species, its degree is usually directly proportional to the DNA titre (Raikov, 1969). Micronuclei are commonly reported not to synthesize the RNA (McDonald, 1973).

Ciliate nuclei, like other eucaryotic nuclei, show four main phases during the cell cycle (Prescott and Stone, 1967): the interval between the end of the nuclear division and the start of DNA replication (G_1), the period of DNA replication (S), the interval between the end of DNA replication and the start of nuclear division (G_2), and the period of nuclear division (D).

a. *Macronucleus*. The timing of the cell cycle has been studied most extensively in *Tetrahymena pyriformis* (Hymenostomatida, Tetrahymenina). The macronuclear synthetic phase occurs in the intermediate part of the generation cycle; the division of macronucleus occurs during cell division. The changes in nucleic acid content in the course of the cell cycle of *Tetrahymena* were recently reviewed by McDonald (1973). In axenic cultures the length of G_1 and S phases were observed to change with the composition of the nutrient medium. These changes were accompanied by variations in mean generation time (Cameron, 1973; Lykkesfeldt, 1974). One may wonder if this is also the case for populations living in a natural habitat; for example, mammalian cells in tissue culture were observed to have an extremely variable G_1 phase but the duration of $S + G_2 + D$ was not appreciably altered (Smith and Martin, 1974). As reported by Nilsson and Zeuthen (1974), the synthetic phase is readily detectable in the phase contast microscopy of living *Tetrahymena pyriformis*. During that period, in contrast to other cell cycle phases, the macronucleus appears "dark", due to the fine chromatin granulation. Prior to cell division a new feeding organelle is formed by the posterior daughter cell (opisthe), the old oral organelle being retained by the proter, the anterior cell (Frankel and Williams, 1973).

In *Paramecium aurelia* (Hymenostomatida, Peniculina), the S period in macronucleus starts later than in *Tetrahymena,* approximately in the middle of the generation cycle and lasts almost to the beginning of the division (Prescott and Stone, 1967).

In the large and highly organized cells of Heterotrichida, the macronuclear G_1 phase also varies in length under different conditions, the length of the S phase being different in different species (Minutoli and Hirshfield, 1968; De Terra, 1970). The macronucleus, often nodulated, condenses into a compact mass at the end of the synthetic phase prior to division. In *Stentor,* the timing of denodulation has been demonstrated to depend on cortical events. In a normal cell cycle the macronucleus condenses and ceases to synthesize DNA shortly before formation of a new adoral zone of membranelles (AZM). The same nuclear behaviour could be induced by removing the AZM from the cell (De Terra, 1973; Paulin and Brooks, 1975).

In Hypotrichida, as well as in Oligotrichida, the macronuclear S phase is connected with the appearance of replication bands, readily visible sites of DNA synthesis traversing the macronucleus. In hypotrichs like *Euplotes,* with single large macronucleus of elongated shape, two bands appear on the macronuclear tips, advance to the centre, fuse and finally disappear. For a species with many small macronuclei, single bands were reported to occur synchronously in all the macronuclei of a cell (Ruthmann, 1972). The presence of a single band has been reported in Oligotrichida (Grain, 1972; Salvano, 1975). In Hypotrichida, the macronuclear G_2 phase is usually absent; the elongated macronucleus condenses before division, as in other Spirotricha. The events of nuclear division are closely correlated with those of the morphogenesis of cell cortex. During the macronuclear S phase, which, for instance, in *Euplotes eurystomus* occupies about three quarters of the normal cell generation cycle, a new oral apparatus is usually formed for the opisthe, but old cirri are replaced by two new sets (cf. Hanson, 1967; Diller, 1975).

Owing to the absence of mitosis at macronuclear division, the amount of DNA distributed to daughter cells is very uneven (McDonald, 1958; DeBault and Ringertz, 1967). At the same time, controlling mechanisms operate keeping the macronuclear DNA content constant within certain limits. If the amount of DNA is decreased below a certain threshold the length of the cell cycle may be extended until a sufficient quantity of DNA is synthesized (Berger, 1973); in addition, a further DNA synthesis period may appear (Cleffmann,

1974). The extrusion of a part of macronuclear mass at division occurring in some ciliate species (reviewed by Morat, 1970) was reported to compensate for the excessive macronuclear DNA content (Cleffmann, 1974). Another mechanism for decreasing the DNA of macronuclei is simply the omission of DNA synthesis from one cycle (Zeuthen, 1963).

b. *Micronucleus.* The DNA in micronuclei is distributed to the daughters exactly, by a true mitosis. Two main types of micronuclear division cycle occur among ciliates. Cells of the first type lack the micronuclear G_1 phase and the S phase starts immediately after nuclear division or even in the course of it. *Tetrahymena pyriformis* and *Euplotes eurystomus* mentioned above belong to this group. The second type, by contrast, lack the G_2 phase and the synthesis of DNA takes place immediately before the micronuclear division. Species of the genera *Paramecium*, *Stylonychia* and *Urostyla* have been reported to possess this type of micronuclear cycle (reviewed by Golikova, 1974).

c. *Synchronized cells.* Insight into the cell cycle of ciliates has been advanced considerably by the study of synchronized cultures. It has been demonstrated that the cell division cycle can be divorced from the macronuclear replication cycle; thus, DNA synthesis, macronuclear and micronuclear division, oral morphogenesis, and cell division can be made mutually independent under various conditions (reviewed by Cleffmann, 1974).

Initially synchronous multiplication was usually induced by blocking the division for a certain period (Scherbaum and Zeuthen, 1954; Scherbaum, 1956), for instance by a temperature-induced block. The cells are arrested successively in a pre-divisional phase and grow abnormally in size at an exponential rate independent of the cell-cycle phase (Schmid, 1967a). At the same time they accumulate excess proteins destined for later incorporation into morphological structures, e.g. oral apparatus (Zeuthen, 1974). When the division block is released the cells start to divide synchronously at division rates usually higher than in a normal exponentially growing population (Zeuthern and Rasmussen, 1972).

Single or multiple changes of temperature or light and the application of metaphase-inhibiting agents (colchicine, colcemid, vinblastine) have often been used for blocking division (reviewed by Mitchison, 1971).

More recently, repetitive synchronization has been successfully attempted (Zeuthen, 1974). These experiments revealed that S-phase and G_2-phase activities overlap at the end of the DNA replication period and that a properly timed shock allows the S phase to be finished but delays the beginning of G_2. It was suggested that the actual cell cycle starts after the end of the S phase, and proceeds through morphogenesis, enzyme steps and cell division and is finished after DNA replication has been completed.

It has been shown that cell synchrony in ciliates, as in other eucaryotic cells, is related to their circadian rhythms. A high degree of synchrony in *Tetrahymena* has been attained by the combination of circadian thermal and feeding cycles, the relative position of the stimuli in the cycle being important. In oxygen-limited cultures a regular dying of part of the synchronized population (circadian chronotypic death) was found, which has not been observed in asynchronous cultures under the same conditions. These studies have led to the suggestion that cell requirements for environmental conditions change with physiological changes of the cell as it passes through the cycle (Meinert *et al.*, 1975).

4.1.2 *Life cycles*

When using the methods of growth rate calculation from the biomass data, the investigator tacitly assumes that population growth is exponential. As most vegetative ciliate cells divide into two almost equal daughters, this condition is usually fulfilled during vegetative growth. The budders, e.g. Suctorida, are exceptional. From their sessile mother cells buds commonly separate, either several at the same time or one after another (Raikov, 1969). Typically unequal separation of cell mass occurs during the preconjugation divisions characteristic of sessile Peritrichida and some Suctorida, giving rise to a free-swimming microconjugant (Raikov, 1972).

In some species, the growth rate of vegetative cells changes with number of generations completed since the last sexual event. After conjugation, the clone passes through periods of immaturity and maturity, when the division rate is rather high. Only during the latter period is the clone able to conjugate. If this does not occur, the division rate gradually decreases until the clone dies. In some strains of *Paramecium aurelia* no immaturity exists, in *Paramecium bursaria*, however, it can last for hundreds of cell divisions (reviewed by Preer, 1969). Immaturity phenomena were also found in *Tetrahymena* (Elliott, 1973).

The essence of the sexual process is a nuclear reorganization, when an old macronucleus is replaced by a new one. The complicated nuclear behaviour during mating has been reviewed by Raikov (1972). From the viewpoint of population structure, it is worth noting that most ciliate cells investigated so far have been reported to be in macronuclear G_1 phase when starting to mate (see Luporini and Dini, 1975). In addition to a normally occurring conjugation, when an exchange of genetic material between mating cells occurs, there exists autogamy, a nuclear reorganization without mating, and autogamy in pairs (also called "cytogamy").

The mating cells either fuse partially, and thereafter separate from each other as exconjugants (temporary conjugation), or unite into one individual (total conjugation). Peritrichida usually adopt the latter course (Grell, 1967; Raikov, 1972). This type of conjugation is an alternative to those involving mating types, i.e. clones that are able to mate with one another, being mostly self-sterile. An intraclonal mating, described in some clones, is termed selfing (Allen, 1967).

Conjugation may have a diurnal periodicity (Miyake and Nobili, 1974), while, on the other hand, in some natural populations it was found to be seasonal (Raikov, 1972). Feeding is frequently arrested during conjugation, and in some species the exconjugants do not eat for several days afterwards owing to reorganization of their oral apparatus (Kloetzel, 1975; Sikora and Kuznicki, 1973). A longer generation time was reported during the cell cycle following conjugation. This was suggested to be connected with a reduced rate of DNA synthesis in fragments of the old macronucleus (Berger, 1973). A mass appearance of sexual phenomena in a population may thus cause serious perturbations of the rate of population biomass growth.

Alternate formation of morphologically different cells may be another source of changes in biomass growth rate. In *Tetrahymena patula* transforming of microstome forms into macrostomes may be induced by adding prey organisms to its population. A transformation in a reverse direction is always accompanied by cell division (Trager, 1963).

A mass formation of cysts or a collective excystment may seriously affect population biomass changes. In some species, thick-walled resting cysts arise under adverse conditions. Members of the family Colpodidae commonly divide in temporary reproductive cysts (Tartar, 1967). Some ciliate species have both reproductive and resting cysts

(Corliss, 1973). While encysting, the ciliate cortex dedifferentiates to a varying extent according to species, and consequently may require a varying degree of morphogenesis during excystment (Hashimoto, 1963).

4.2 CELL VOLUME AND POPULATION GROWTH

4.2.1 *Effect of food concentration*

Nutrition is obviously of primary importance in growth of cell mass and therefore also in control of cell size. Kimball *et al.* (1959) working on *Paramecium aurelia*, and Hamilton and Preslan (1969) on *Uronema* sp., found that when starved cells are given excess food after a short lag they begin to grow at a normal rate. Since the cell division in this situation is delayed in comparison with normal cells, the previously starved cells grow progressively larger, but during subsequent generations they regain the volume of normal exponentially dividing cells. The reverse situation is observed when an exponentially growing population is deprived of food. While the division rate is maintained for some time, the cells stop their growth and small cells result (Kimball *et al.*, 1959; Cameron and Terebey, 1967; Hamilton and Preslan, 1969).

Curds and Cockburn (1971) studied the food/cell size/division rate relationship in a monoxenic ciliate culture. They demonstrated that when the cells of *Tetrahymena pyriformis* are allowed slowly to adapt their volume in continuous culture, it is possible to attain much higher specific growth rate than the maximum estimated from batch culture. They suggested two functions describing the dilution rate/mean cell volume relationship. For the maximum division rate ν_{max} (h^{-1}) of the cells of respective mean cell volumes m (10^3 μm^3) they found the empirical relationship

$$\nu_{max} = \frac{7 \cdot 97}{m}, \tag{9}$$

while the mean cell volume of steady-state population $\bar{m}$ was assumed to obey the equation

$$\bar{m} = \frac{m_{max} D}{D_k + D} \tag{10}$$

where m_{max} was maximum cell volume, D was dilution rate and D_k the dilution rate at which $\bar{m} = m_{max}/2$. The graphical representation of the functions (9) and (10) were curves surrounding the great majority of their experimental points. The mean cell volumes increased with an increase in the dilution rate at steady state whereas they were reciprocally related to their respective maximum specific growth (division) rates. The intersect of the two functions corresponded to the maximum growth (division) rate attainable by the strain investigated. It corresponded to one particular value of the mean cell volume.

Adopting the Caperon's (1967) treatment of food uptake by a unicellular population, Curds and Cockburn (1971) derived an equation relating mean cell volume, specific growth rate of the population and food concentration:

$$D = \frac{(K_3Y/m)S}{K_3/K_1m + S}, \tag{11}$$

where S is food concentration, K_1, K_3 rate coefficients for adsorption and release of the food at the adsorption site (mouth), Y is the yield coefficient, m the mean cell volume (mass) and D the dilution rate (which under steady-state conditions equals the specific growth rate of the population). The equation may be represented graphically as a curved surface.

The finding of the relationship between a definite mean cell volume and the optimum growth is impressive. Nevertheless, such near functional relationship is likely to be valid only for steady-state conditions, when the population growth is actually both nutrient limited and balanced (i.e. with the division rate proportional to the rate of biomass increase). Prior to the establishment of a steady state, cell division, and consequently mean cell volume, may also partly be controlled by other factors.

Taylor *et al.* (1976) were able to distinguish between axenic and monoxenic exponentially growing cells of *Tetrahymena pyriformis*. They showed that under axenic conditions the cells usually increased in volume when division was blocked before they had become food limited.

Similarly, some observations from continuous cultures cannot be explained by means of the Curds' relationship (equation 11). Hamilton and Preslan (1970) increased the mean cell size of *Uronema* by increasing the input concentration of bacteria while keeping the dilution rate constant. They explained this as the effect of increasing population density, though it might also be attributed to less than perfect steady-state conditions. I observed this phenomenon under conditions of incompletely mixed continuous culture. As the ciliates accumulated in the culture vessel, the rate of biomass output was lower than the rate of nutrient (bacteria) input and the excess nutrient mass per cell was obviously higher at a higher input concentration. An opposite situation was recently recorded by Donaghay (1974) who found the final minimal cell size in the chemostat to depend on the previous dilution rate. By increasing the dilution rate the final minimal cell size was reduced. Donaghay's data do not fit the Curds and Cockburn (1971) model, for they lie on the outer side of the curve for the steady-state populations.

4.2.2 *Effect of physical conditions*

A prominent physical factor capable of influencing both cell size and population growth rate in the natural environment is water temperature.

Up to now little attention has been devoted to this factor in field studies. Gold and Morales (1975) followed size distribution of the loricae of marine planktonic Tintinnina at different seasons. In both *Tintinnopsis levigata* and *T. tubulosoides* they found longer loricae in winter, during the time of low temperature (3–11°C) and shorter in the autumn when temperature was high (14–20°C). The variation in length was more pronounced in the former species; in the latter some overlap occurred. Gold and Morales (1975) demonstrated that the mean lorica size of *T. tubulosoides* cultivated at 10°C is smaller than that of the same species in the plankton collected at 5·5°C.

Most findings from pure cultures agree with the above data. For exponentially growing cultures of *Tetrahymena pyriformis* (not nutrient-limited), sufficient evidence has been gathered that both minimum mean cell volume and maximum division rate have temperature "optima" (James and Read, 1957; Prescott, 1957; Schmid, 1967a, 1967b). The temperature optimum for division rate was reported to be 29·0°C and 32·5°C, respectively, for two different strains of *Tetrahymena pyriformis* (Prescott, 1957). It was suggested (Schmid, 1967a) that the optimum temperature for individual cell growth is higher than that for division rate. Changes of cell volume caused by temperature took place in the entire population range simultaneously, the standard deviation of the distribution remaining constant (Schmid, 1967b).

Osmotic pressure is another factor that is important for the cell volume/growth rate relationship. *Tetrahymena* cells, when subjected to a sudden hypo-osmotic or hyper-osmotic change of the environment, immediately adjust their internal osmolarity by a volume change (swelling in hypo-osmotic and shrinkage in hyper-osmotic medium). Then, by changing the amount of intracellular solutes the cells return almost to their previous volumes within approximately an hour (Dunham and Kropp, 1973).

pH is a further factor that may affect the ciliate cell volume. Unlike the influence of the osmotic pressure, a constant functional dependence of the cell volume on pH has been reported for *Tetrahymena* cells. The changes of cell volume caused by the changes of pH lasted several hours and were more or less reversible. A significant difference was not found between living and dead cells (Lengerová-Kučerová, 1950).

Under natural conditions a variety of factors simultaneously affect cell volume. Hence, when interpreting data in terms of a cell volume/growth rate relationship, all the potential sources of interference should be evaluated carefully.

4.3 INDIVIDUAL CELL HISTORIES

4.3.1 *Theoretical considerations*

In the preceding section (4.2), no importance was attributed to the fact that most data on cell size were obtained by averaging a number of individuals. However, the size distribution of cells in a population is determined not only by their variability but also by the fact that they divide into approximate halves during the cell cycle. This also indicates that the assumption of normal distribution of cell size in a population is not fully warranted and estimation of the median instead of mean cell size would be advisable (Snedecor and Cochran, 1967).

a. *All cells cycling.* When the manifestation of individual cell histories in a population is to be treated theoretically, it is necessary first to simplify things as much as possible. Therefore, an ideal population of vegetative cells is assumed in which all cells are in one or other phase of the cell cycle, leading sooner or later to cell division. Cells of this population approach their cell divisions randomly with a perfect asynchrony. The total number of cells increase exponentially, the increase may be characterized by a constant division rate ν. The same rate of increase may characterize any existing group of cells in a population defined by the limits of their size, mass or quantity of some constituent, as well as by the phase of the cell cycle.

This ideal concept is complicated by at least two sources of variability, i.e. variable cell generation time and variable cell mass or some component thereof.

Powell (1956) described age distribution of a steady-state population with variable cell generation times by a function

$$n(s) = 2\nu e^{-s}\int_{s}^{\infty} f(\tau)d\tau \tag{12}$$

where τ is the variable generation time (length of the cell cycle), $0 \leq s \leq \tau$ is the cell age, and ν is the division rate. Under these conditions the division rate cannot be written explicitly, but must be determined by

$$2\int_{0}^{\infty} e^{-\nu\tau}f(\tau)d\tau = 1. \tag{13}$$

It follows from this equation that the cell generation time would equal $\ln 2/\nu$ only if it was invariable in length. Normally, when the *mean* generation time is used, there exists a difference, though usually small.

Collins and Richmond (1962) included the cell mass variability in the assessment of the growth rate k of an individual cell, but neglected the variability of τ, putting $\nu = \ln 2/\tau_d$ (τ_d is the doubling time). They used three frequency functions of the cell mass m, $\psi(m)$ of newborn cells just after division, $\phi(m)$ of cells at division, and $\lambda(m)$ of the whole steady-state population, and suggested a formula

$$k_x = \frac{\nu[2\int_0^{m_x}\psi(m)dm - \int_0^{m_x}\phi(m)dm - \int_0^{m_x}\lambda(m)dm]}{\lambda(m_x)} \tag{14}$$

for the individual cell mass growth rate k_x of a definite cell mass m_x. For the cell mass it is possible to substitute the cell volume or the mass of any cell constitutent.

To avoid some complications, Williams (1971) introduced relative age $a = s/\tau$; $0 \leqslant a \leqslant 1$ into the Powell's (1956) formula (equation 13) and obtained the relative age distribution of a steady-state population independent of the generation time:

$$n(a/da) = (2 \ln 2)e^{-a\ln 2}da. \tag{15}$$

The integral form of this function has been previously used in experimental work (Walker, 1954; Scherbaum and Rasch, 1957) and quoted as the "cumulative phase index" (CPI) (Schmid, 1967b). Williams

(1971) showed that in the absence of cell mass variability, this function is governing the cell mass distribution: $m = f(a)$. Then, the cell mass density function $h(m)$ is exactly the relative age density function divided by the increase of individual cell mass ($\mathrm{d}m/\mathrm{d}a$):

$$h(m) = \frac{1}{\mathrm{d}m/\mathrm{d}a}\,(2\ \ln\ 2)\mathrm{e}^{-f^{-1}(m)\ln 2}. \tag{16}$$

Again, in a steady-state population it is possible to apply this function to any constituent, either chemical compound or organelle. One may substitute an optimal cell growth function including "pulse" events as, for example, the occurrence of any definable developmental stage. In the case of exponential cell growth, for instance, this density function reduces to the expression (Williams, 1971):

$$h(m) = \frac{2m_0}{m^2}\,. \tag{17}$$

Williams (1971) further suggested a density function $h^*(m)$ of cell size with variability. The ideal cell size function $h(m)$ is then introduced as an *expected* size density function $h(m_{\mathrm{E}})$

$$h^*(m) = \int_1^2 h(m_{\mathrm{E}})\ p(m_{\mathrm{E}}, m)\mathrm{d}m_{\mathrm{E}} \tag{18}$$

where $p(m_{\mathrm{E}},m)$ is some probability density function.

It was shown in algae (Williams, 1971) that the distribution curves normalized for equal means and areas have constant shape for various steady-state growth rates of the same population. It has not been proved that this rule is valid also for nutrient-limited steady-state growth of ciliates. Evaluation of some existing data from batch culture (e.g. Dive, 1975) might help to answer this question.

For transient growth responses of populations, Williams (1971) expected that differences in individual cell growth mode (e.g. exponential versus linear) may be manifested in different population growth dynamics. With ciliate vegetative cells, one may assume exponential growth of an individual cell's biomass (Kimball *et al.*, 1959; Schmid, 1967b). The effect of different individual cell growth modes may manifest itself, however, during changes from vegetative fission to mating or other phases of the life-cycle where the individual cell growth is not necessarily exponential.

b. *Presence of non-cycling cells.* The manifestation of ciliate cell history in a size distribution of cell population might be greatly obscured when a part of cells formed at division are not able to continue in the cell cycle. Such cells probably rest in the presynthesis phase (G_1). Prescott and Carrier (1964) arrested an entire population of *Euplotes eurystomus* by letting a mass culture starve for about 5 days at 24–25°C. This phenomenon is obviously a general one, as results from slowly growing tissue cultures indicate. It was supposed that *some* cells do not continue through G_1 but proceed to a distinct state (G_0) and remain non-cycling (reviewed by Smith and Martin, 1974).

Powell (1956) considered the possibility of there being a fraction $(1 - \alpha)$ of "non-viable" cells. The mathematical treatment of these should be analogous to the treatment of non-cycling (but otherwise living) cells. There is, however, a difference in the outcome of the process.

Powell considers that a positive growth rate is possible only when $\alpha > 1/2$, whereas for $\alpha = 1/2$ the growth rate is zero and at $\alpha < 1/2$ the population dies out. The proportion of non-viable cells in a population is twice their percentage among newly formed cells, i.e. $2(1 - \alpha)$, in both batch and chemostat cultures. Part of the population at other than the vegetative stage of the life-cycle may behave as a "non-viable" fraction. An example of this is the increasing number of cysts in an otherwise stable population (Legner, unpublished results). On the other hand, if non-cycling cells are considered viable, a positive population growth is to be expected in all the three cases. For $\alpha = 1/2$, a special case, linear growth (arithmetical increase) of the overall population is predicted.

Powell (1956) assumed a higher proportion of non-growing cells in higher medium concentrations and over-acid environment. In ciliates, one may also expect such a population behaviour with starving cells incapable of cell growth or even diminishing in cell mass. The cell mass of some ciliate species decreases considerably on starvation. A 13-fold difference was found in mean cell volume between well fed and starved *Tetrahymena* (Cameron, 1973). *Dileptus* was reported to shrink to 1/100th its original volume, and *Bursaria* to less than a fifth of its length (reviewed by Tartar, 1961). In other eucaryotic cells, the median cell size of the non-cycling population fraction has actually been shown to be smaller than that of the whole population, even though the size distributions of non-cycling and cycling cells were overlapping (Yen *et al.*, 1975).

A linear population growth has been found in a mammalian tissue culture that could not be explained by partial formation of non-cycling cells as discussed above. This phenomenon has been interpreted (Smith and Martin, 1974) by the assumption that after division *all* cells enter a state of non-cycling. A cell may remain in this state (A state) for any length of time, but there is always a probability (P) that it re-enters the B phase (collective term for S, G_2, D phase and a part of G_1 phase). It was shown that the probability P is of an exponential nature, being analogous to that governing the decay of a radioactive element with a constant half-life. The proportion of cycling cells in a population is a function of P; it is expected to be unity if $P = 1$, and zero if $P = 0$. Just as mean generation time is affected by nutrient concentration, so the half-life of the A-state cells was shown to be shortened by increasing the concentration of stimulating substances supplied to them.

4.3.2 *Experimental data*

a. *Distribution of generation times.* Generation times have been investigated in the past (McDonald, 1958; Prescott, 1959; Nachtwey and Cameron, 1968) and a log normal distribution was reported for their distribution in a ciliate population (Schmid, 1967b; Nachtwey and Cameron, 1968). It was found (Nachtwey and Cameron, 1968) that in this case by a higher variability the median cell generation time is longer than the population doubling time, the mean generation time being slightly longer than the former.

More recent investigations (Jauker and Cleffmann, 1970) showed that the individual cells in an exponentially multiplying population of *Tetrahymena pyriformis* alternate long and short generation times rather regularly. In a semicontinuous culture of *Tetrahymena pyriformis* (1-hour intervals of a partial medium exchange), circadian rhythm of cell division persisted for 6 days after a light–dark synchronization (Edmunds, 1974). Therefore, the assumption that ciliate generation times are distributed randomly around a mean value is doubtful (Cleffmann, 1974).

b. *Distribution of cell mass.* The mass distribution of cells of defined relative age was studied carefully (Schmid, 1967a) and a remarkable constancy in the coefficient of variation was reported irrespective of cell size. When division was blocked by heat shocks, the mean cell volume increased in a well-fed cell population with the same cell growth rate as the volumes of cells at the extremes of the distribution. Schmid (1967a) concluded that the volume distribution of the cells of the same relative age, i.e. the "momentary size variation"—MV (size)—of Scherbaum and Rasch (1957), is a relatively stable parameter obviously controlled by efficient mechanisms. There have been no reports as yet on the variability of this parameter among different ciliate species. Generally, we can expect an increase of variability in populations living under incompletely homogeneous conditions, bearing in mind the dependence of cell volume on the amount of available food and the uneven distribution of food resources in a spatially inhomogeneous environment.

Schmid (1967b) has also shown that the size distribution of *Tetrahymena pyriformis* cells in an exponentially growing population tended to log normal as a result of a superposition of individual log normal "momentary size variations". The log normal distribution of cell size has been generally reported for the ciliate exponential phase (Cameron, 1973). A similar distribution was reported from mammalian tissue culture (Rosenberg and Gregg, 1969). The results of experiments on both types of cells also tends to confirm the exponential mode of interphase cell growth. Thus Rosenberg and Gregg noted that mammalian cells blocked in division grow in size by the same rate constant as do numbers in multiplying population. The kinetics of cell volume growth appeared to be independent of the cell cycle stages.

4.4 HYPOTHETICAL RELATION OF STRUCTURE AND GROWTH

It has been suggested that morphological characteristics of ciliates correlate with the growth rate of their populations and related phenomena. Taylor *et al.* (1976), using principal component analysis, showed that the slow-growing monoxenic cells of *Tetrahymena pyriformis* limited by food tended to increase the size of their buccal cavity. Suhr-Jessen *et al.* (1977) found that the cells in an axenic steady-state culture of *Tetrahymena pyriformis* have larger length/width ratio at a higher growth rate.

One may wonder if a similar relation would be found between population growth and the distribution of a particular cell property over the population. Obviously, data based on analysis of a population at a single instant hardly allow determination of the exact value of any dynamic parameter, including specific growth rate. Rather, an approximate value may be obtained in terms of a relative growth rate (Evans, 1976), i.e. the proportion of the maximum specific growth rate. Furthermore, even specific growth rate will not necessarily always be sufficient to characterize population growth. Data on the population structure could supplement it with information on the physiological state of population, when interpreted adequately.

Several hypothetical statements may be inferred from present knowledge of the relations of cell size to growth rate and cell age distribution to population structure and of the timing of events of the cell cycle (briefly reviewed in the preceding text, sections 4.1 to 4.3). The population structure, so far characterized mainly as cell size distribution, can be related to some types of population growth or population state. In addition to a variety of situations, which may occur in nature and are not predictable in the terms of present knowledge, five different types of populations may be recognized.

i. First, we may consider a population consisting only of cycling cells with a narrow momentary size variation. If no synchrony occurs, the distribution of relative cell ages (equation 15) should be reflected in the overall size distribution. Provided that a deterministric mathematical model were a good approximation (Schmid, 1967b), the frequency in the size class corresponding to the half-size of cells just dividing should approximately double the frequency of the class of dividing cells. If a stochastic model is adopted (equation 16) this ratio would be different depending on the mode of individual cell growth (Williams, 1971). In an actual population, the compound size distribution of both relative cell ages and momentary size variation may be log-normal (Schmid, 1967b; Cameron, 1973).

This type of distribution may be expected in a population which multiplies exponentially near the value of the maximum specific growth rate, being unlimited by nutrition and other external conditions, e.g. fast exponential growth in the sense of Szyszko *et al.* (1968) and Meinert *et al.* (1975). Since this state is usually preceded by a well-fed non-growing population (the transition from lag phase to the exponential one, cf. Kimball *et al.* 1959), it is reasonable to expect that the part of the MV (size) attributable to differential availability of food to individual cells would be relatively small. An accidental synchronization of cells in the transition from lag to exponential phase seems unlikely (Scherbaum, 1957) and therefore need not be considered. Growth of such a population is not expected to be balanced, since the mean cell volume would be decreasing. This might be the main way by which transitional cells differ from balanced steady-state cells limited by food, where the mean cell volume is constant and considerably smaller (Curds and Cockburn, 1971).

ii. As a second type, a population again consisting entirely of cycling cells may be considered. In this the MV (size) is assumed to be wide, in contrast to the preceding population type. It may be suggested that the cause of such a size distribution is the growth of a nutrient-limited population when the availability of nutrients to individual cells is not quite uniform. In consequence, the mean cell volume should be smaller than in the preceding type. However, it remains to be proved experimentally whether this is a characteristic of, for example, a later exponential phase of a food-limited batch culture or even of some ciliate steady-state populations.

iii. The third type of population might display a comparable or wider overall size variability than the preceding one if in addition to proliferating cells, it also includes a considerable fraction of cells not advancing in the cell cycle. If these cells are quiescent owing to starvation they are likely to be smaller than the others (cf. Prescott and Carrier, 1964).

This type may be represented by a food-limited population growing under heterogeneous conditions. It might occur in the late exponential phase of a batch culture, in a heterocontinuous culture, or in a natural habitat. Again, the mean cell volume should be smaller than in the first type.

iv. For the fourth type let us consider a population of cells whose growth was stopped owing to depletion of food resources. The cells are starving and in the same phase of the cell cycle (Prescott and Carrier, 1964). In consequence, the mean cell volume is generally lower than in the three preceding cases and the size distribution corresponds to a relatively narrow MV (size). Late stationary phase cells or early lag phase cells would represent this type of population.

v. The last population type cannot be defined in terms of cell size distribution alone because it is assumed that growth of this type is retarded by inhibition of cell division. These cells should be larger than the exponential cells. In the case of a complete division block they should maintain the size distribution of the state directly preceding (cf. the results of Schmid, 1967a, with division-blocked cells). Thus the cells in a transitional phase between the lag and the exponential probably maintain the pattern of the size distribution of lag phase cells, though their mean cell volume is much larger (cf. Kimball *et al.*, 1959). A special type of a division inhibition demonstrated by a morphogenetic marker was described by Taylor *et al.* (1976).

Our recent results (Legner, 1979) indicate that at least hypotheses (i), (ii), and (iv) might be valid, although direct experimental evidence is lacking thus far. Hypothesis (iii) conforms to the experimental results of Yen *et al.* (1975) from tissue cultures, which showed that the fraction of cells with small cell size are characterized by long generation times.

Since an estimation of structure characteristics is generally feasible, the above statements might serve as a starting point for a rough characterization of populations, including those occurring in natural habitats. This will help to formulate hypotheses more precisely and eliminate unwarranted assumptions.

4.5 METHODS

Several types of data may be acquired in order to test the hypotheses formulated in the preceding section (4.4) or otherwise characterize a ciliate population. Essentially, the methods used so far to obtain data may be classified, according to the subject of analysis, into three groups as follows: (i) the evaluation of the mass of a cell or its constituent, (ii) the evaluation of the proportion of cells with some morphological or physiological marker, (iii) the evaluation of temporal changes in the population structure.

4.5.1 *Cell mass or the quantity of any constituent*

a. *Distribution of individual biomass.* Obtaining the information on the actual mass distribution within a population represents the first step in population structure analysis. Several methods may be adopted.

The estimation of cell size (cell volume) is a convenient approximation of cell mass. The measurement of the dimensions of ciliate cells with the aid of a microscope or from micrographs with subsequent calculation of cell volumes has been used by a number of authors (e.g. Scherbaum and Rasch, 1957; Curds and Cockburn, 1971; Legner, 1973). Automation of this procedure is also feasible; the possibility of evaluating microscopic preparations by means of scanning equipment combined with pattern recognition techniques has been demonstrated

(Melamed and Kamentsky, 1969; Rogers and Kimzey, 1972; Kimzey, 1977). Size distribution analysis of ciliate cultures has been obtained by means of conductometric particle counter (Schmid, 1967a; Szyszko *et al.*, 1968; Hamilton and Preslan, 1969; Morrison and Tomkins, 1973; Dive, 1975). For constructing reliable distribution histograms the automated methods are preferable owing to the high statistical significance of their results (Sheldon and Parsons, 1967).

The dry weight of individual ciliate cells has been successfully estimated by interference microscopy (Kimball *et al.* 1959; Curds and Cockburn, 1971) and X-ray absorption microscopy (Kimball *et al.*, 1959; Zech, 1966). Fluorescence microscopy might also be employed for the measurement of cell mass distribution of a ciliate population. The protein of mammalian cells has been reliably stained with fluorescein isothiocyanate (Crissman and Steinkamp, 1973). In addition to the possibility of quantifying these cytological properties by automated scanning equipment (see above), flow-through cytophotometry seems to be the most promising method for obtaining distributions of cell constituents. It is based on imageless optical measurement in individual cells and the treatment of the data as electronic pulses just as in the Coulter-type analysers (Kamentsky and Melamed, 1969). In combination with a laser beam as a source of light this is a very rapid process (Stöhr and Goerttler, 1974).

b. *Distribution of DNA.* Except for its width, which may be indicative of the rate of growth, according to the hypotheses discussed previously, cell mass distribution supplied limited information on population physiological state generally and on population growth in particular. Provided that the steady-state distribution of the phases of the cell-cycle is known for a given species, data on the frequencies of cells occurring in particular phases would broaden the extent of information in two directions: (i) an increased proportion of cells in S, G_2, or D phase would indicate some kind of division block (cf. Taylor *et al.*, 1976); (ii) an increased proportion of G_1 phase cells may suggest an increase in the proportion of cells resting out of the cell cycle (e.g. owing to limited food source, cf. Prescott and Carrier, 1964).

In eucaryotes, the distribution of the DNA per cell is generally expected to correspond with the distribution of cell cycle phases in the population. In a rapidly growing population a bimodal distribution curve is usually obtained with a peak characterizing G_1 nuclei and a second lower peak corresponding to higher DNA content in G_2 and dividing nuclei. The depression in the frequency of cells between the peaks corresponds to the S phase (Crissman and Steinkamp, 1973). Computer programs have been worked out (Berkhan, 1975) to estimate automatically the fractions of cells in each phase. Microspectrophotometry has been used for DNA quantification in ciliate cells, for the most part employing the Feulgen reaction (McDonald, 1958; Kimball and Barka, 1959; Morat, 1970). Recently, fluorochromes have been employed in metazoan cells, e.g. fluorescent Schiff reagents (Yataganas *et al.*, 1975) and phenanthridium derivatives (Crissman and Steinkamp, 1973; Stöhr, 1975). Acridine orange has been reported to be able to separate DNA and RNA by fluorescence at two different wavelengths (Andreeff and Haag, 1975). This metachromasia would probably be helpful in the differential estimation of non-cycling cells (cf. Darzynkiewicz *et al.*, 1975). The evaluation of population samples has mostly been carried out with a flow-cytophotometer. A preliminary note reports on the use of this technique in parasitic protozoa (Jackson *et al.*, 1977).

Ultraviolet absorption of the nucleus may also be used for DNA estimation, as has been shown in the case of other protozoa (Rogers and Kimzey, 1972).

Thus far, little attention has been paid to the distribution of DNA content in ciliate populations. The amount of micronuclear DNA in the cells of *Colpidium campylum* varied very slightly, with no overlap of dividing and newly formed nuclei (Morat, 1970). Nevertheless, one could hardly expect that the distribution curve of micronuclear DNA in a ciliate population would be bimodal; according to present knowledge, either G_1 or G_2 phase is negligible (G_1 in the case of *C. campylum*, after Morat, 1970). The momentary variation of DNA content—MV (DNA)—in macronuclei is rather wide owing to the varying degree of polyploidy. In cultures of *C. campylum* (Morat, 1970) and *Tetrahymena pyriformis* (Jeffery *et al.*, 1973), the coefficient of variation of G_1 macronuclei was found to be about 0·20. The pattern of G_1 distribution in *C. campylum* (incidentally, bimodal) was reproduced in G_2 macronuclei (Morat, 1970). No double peaks have been observed thus far in the overall distribution of the macronuclear DNA content of asynchronously multiplying cells (Morat, 1970; Jeffery *et al.*, 1973), probably on account of the great MV (DNA) of the maronucleus. This need not, however, prevent the detection of an increased proportion of cells in a particular phase by means of DNA distribution.

4.5.2 *Use of markers*

As has been discussed earlier (section 4.3), the pattern of the overall cell mass distribution is obviously determined by several factors. In order to separate the effect of individual cell growth mode from that of MV, it is necessary to know at least one of the two corresponding distributions. The assessment of MV may be attempted along the lines of Scherbaum and Rasch (1957) and Schmid (1967a, 1967b), who used the variability of cells at the same stage of the cell cycle. Ideally, the stage of the cell cycle should be as short as possible. Various cell properties might be used as markers.

a. *Cell division.* In the past, the division furrow has been used most frequently as a marker of cytokinesis (cell division). The proportion of dividing cells in a population is called the division index I_D. This value should be a function of both the division rate and the length of the division phase t_D. In place of the function used by Scherbaum (1957), one modified after Nachtwey and Cameron (1968) is recommended for the calculation of this relation:

$$\nu = \frac{\ln(I_D + 1)}{t_D}, \tag{19}$$

since it may be readily derived from the integral form of the distribution function of relative cell ages (equation 15).

In practice, the length of the D phase and consequently the division index varied during the exponential growth of a ciliate culture while the division rate ν was constant (Scherbaum, 1957). Nevertheless, the duration of the division phase is short in comparison with the mean generation time; the distribution of the cell mass of dividing cells is thus a good approximation to the momentary variation. There is a methodological difficulty connected with short D-phase duration, because the proportion of dividing cells in a slowly growing population is very low and their microscopic scoring is time-consuming.

b. *Morphogenesis of cell cortex.* Morphogenetic stages of the cell cortex may be utilized for the marking of a particular cell cycle phase. Taylor *et al.* (1976) used the proliferation of kinetosomes along the 1st meridian in *Tetrahymena* as the indication of a predivision period (the exact coincidence with macronuclear S phase was not given—cf. Frankel and Williams, 1973). A very good marker of late macronuclear S phase is the proper formation of the oral apparatus by the opisthe (cf. Zeuthen,

1974). Likewise other phases of cortical morphogenesis might be used; for instance in *Hypotrichida*, the formation of new frontoventral cirri is a conspicuous event (cf. Deroux and Tuffrau, 1965; Diller, 1975).

c. *Macronuclear events and quiescent cells.* Some clearly specified phase of the macronuclear cycle may also be used as a cytological marker; for example, the condensation of the macronucleus at the end of the S phase in Spirotricha, or a distinct S phase macronucleus in *Tetrahymena* as seen by phase-contrast microscopy, or the replication bands of the Oligotrichida and Hypotrichida in macronuclear S phase (cf. section 4.1.1).

In addition, the MV might also be derived from the variability of cells in a non-growing population that is starving and consequently in G_1 phase (cf. Prescott and Carrier, 1964). Similarly, the distribution of an encysted population should approximate to the MV.

d. *Evaluation of data.* Since it is assumed that the MV is a well-regulated property of the ciliate population, a mathematical simulation of the expected distribution pattern is feasible. The possible methods were demonstrated by Schmid (1967b) and Williams (1971). Subsequently the analysis of population structure might be made in two ways. First, the theoretical and actual steady-state distributions may be compared and the parameters of the theoretical one readjusted accordingly. Second, this verified theoretical distribution might be compared with the distribution of a population from transient conditions. The percentage of non-cycling cells might be thus evaluated after the subtraction of the steady-state distribution from that investigated. Similar manipulations are possible with the distribution of a cell constituent, e.g. DNA. Again, a steady-state distribution may be derived and the difference between it and actual populations tested.

4.5.3 *Temporal changes in structure*

The methods discussed so far represent the analysis of a single population sample aimed at assessing a dynamic parameter, the growth rate. A still better approximation of population dynamics is attained when changes in the structure of population are evaluated. Autoradiography may be employed for following the changes in a set of several successive population samples. Convenient techniques are: (i) labelling of cells in some cell cycle phase by a radioactive percursor of a specific substance (in practice used for S phase); (ii) blocking of cells in a cell cycle phase by a blocking agent (used mainly for blocking in D phase).

a. *Labelled dividing cells.* The labelling of cell nuclei with radioactive isotopes has been used for the timing of cell cycle phases and may also be applied to the measuring of generation time (Sisken, 1964; Mitchison, 1971; Brock, 1971). Usually tritiated thymidine is added to the cell population in a proper form. This compound, being a precursor of thymine, is incorporated only into nuclei synthesizing DNA (S-phase nuclei), thus labelling them with tritium.

A short "pulse" of ^3H-thymidine supplied to an asynchronously multiplying population is incorporated into a fraction of the population which, after passing through the G_2 phase, appears as labelled dividing cells. In the autoradiograms of successively collected samples, these cells are recognized by the metallic silver grains developed around their dividing nuclei. The labelled dividing cells reach their maximum when the majority of former S-phase cells enter division. Then a decrease occurs, as the cells that were in G_1 phase at the time of labelling begin to divide. After repeating the cycle, a slightly lower peak appears. Smith and Martin (1974) showed that the time interval between the peaks corresponds to the minimum generation time. The presence of non-cycling (A-state) cells does not cause a delay of the second peak, but contributes to its decrease and a greater spread.

An estimate of the proportion of non-cycling cells allows one to correct the value of the mean generation time.

b. *Division block.* Blocking of cells in D phase was originally introduced as a means of improving the statistical significance of autoradiogram evaluation by increasing the proportion of dividing cells (Puck and Steffen, 1963). Division block is usually effected by colcemid, which is added along with the ^3H-thymidine. Then the changes are followed in the proportion of labelled cells, dividing cells, and labelled dividing cells. Assessment of the generation time is again possible, but the approximate length of the four phases of the nuclear cycle may also be estimated at the same time. Macdonald (1973) dealt thoroughly with the statistical analysis of this method, assuming all the cells were cycling. The method can be modified to take into account the proportion of non-cycling cells.

c. *Labelling of ciliates.* Ciliate cells may incorporate thymidine either in a dissolved form (e.g *Tetrahymena pyriformis*, cf. Stone and Cameron, 1964; Suhr-Jessen *et al.*, 1977) or bound to the organism's diet. For instance *Euplotes eurystomus* was fed labelled *Tetrahymena* cells (Prescott and Carrier, 1964) and *Paramecium aurelia* was supplied with labelled bacteria (Berger, 1971). Both *Tetrahymena* and bacteria were used for labelling *Blepharisma* (Minutoli and Hirschfield, 1968).

Generally, the doses of the radioactive thymidine in living cells should be as low as possible, to avoid artifacts in the cell cycle (Mitchison, 1971). In ciliates, difficulties are often encountered even at low concentrations owing to the complexity of their organization. McDonald (1973) reported that *Tetrahymena pyriformis* retained a water-soluble derivative of ^{3}H-thymidine after ingestion in G_2 phase through division and incorporated it into DNA in the next S phase. Also, the large size of ciliate cells may be a source of technical problems, as the low-energy beta particles emitted by tritium may be absorbed in the ciliate cytoplasm instead of reaching the autoradiographic emulsion. One may avoid this by flattening the cells in the preparation (Stone and Cameron, 1964). Another possibility is to disrupt cells chemically (e.g. with Triton X-100 supplemented with spermidine sulphate) while keeping their nuclei intact (Ron and Urieli, 1977).

5 Special approaches

Some problems concerned with the growth of ciliate populations cannot be solved solely on the basis of the methods discussed previously. For instance the distribution of individual generation times and its relation to other growth phenomena, the estimation of the maximum specific growth rate, and the relationships between a defined population growth and population structure, all require special approaches. Certain techniques, though devised originally for other purposes, may prove to be helpful for the study of these problems.

5.1 MEASUREMENT OF INDIVIDUAL GENERATION TIMES

The isolation of individual cells for measuring generation times is tedious, but information on the distribution of generation times in a population of free-swimming ciliates could hardly be obtained by other means. Nachtwey and Cameron (1968) have recommended the selection of dividing ciliate cells from a mass culture by means of a mouth-tube-controlled micropipette. Each dividing cell was placed in a drop of culture fluid and observed at short time intervals to record the time of daughter cell separation and of the repeated division of each daughter. Distribution histograms or cumulative distribution curves can be made from these data. This procedure is mainly useful in rapidly growing cultures, where a sufficient number of dividing cells are available.

For an exponentially growing population, the probit transformation of generation time distribution on a log-normal scale is usually close to a straight line. Nachtwey and Cameron (1968) suggested a simplified method for this: inspection of a larger number of drops at longer intervals and recording the number of cells that have divided before a given time. The mean generation time can be estimated, but the probit transformation also allows the calculation of the median generation time and the standard deviation.

Cell isolation data may also be evaluated by the method used for larger populations: by counting the progeny of a single cell in several replicates and subsequently by calculating the division rates (equation 3) of these sub-populations. This method reduces the biasing effect of cells with short generation times (Nachtwey and Cameron, 1968), but it suppresses information on the variability of generation times.

This modification of the isolation culture technique has been used for the estimation of ciliate division rates in suspensions of different bacterial strains (Curds and Vandyke, 1966). Petri dishes with glass rings placed into an agar layer served as special culture vessels enabling direct counting of ciliate cells under a low-power microscope. The growth rates were calculated from counts at the start (2 isolated cells) and after 24 hours using a formula corresponding essentially to equation (3). The means were calculated from several replicate estimates.

The importance of generation time variability may be illustrated by data reviewed by Nachtwey and Cameron (1968). *Tetrahymena* was reported to have a greater coefficient of variation in a simple medium than in a complex one. The authors suggested that this might be caused by forcing the cells in a simple medium to produce more precursor substances, and thus to increase the probability of errors in the approach to division.

5.2 USE OF CONTINUOUS CULTURE TO GROWTH STUDIES

In recent years, the continuous culture technique has been applied to the study of ciliate population dynamics. Partially successful attempts have been made to employ continuous methods in their original form, namely those of either turbidostat or chemostat.

5.2.1 *Turbidostat-type cultures*

The turbidostat is usually understood as a completely mixed continuous culture, where the flow rate is made dependent on the population biomass concentration. Essentially, light either transmitted or scattered by a culture is detected and used to control a pump or a solenoid valve, so that deviations from some preset light value result in the change of the flow of culture fluid (Munson, 1970). In the turbidostat, even under non-steady state conditions the specific growth rate is equal to the dilution rate. It allows cultures to multiply at a rate approaching the maximum specific growth rate, provided that all the nutrients in the medium are in excess. At the same time, the population density may be controlled over a considerable range of values (Watson, 1972).

Ciliate populations have not been studied frequently in the turbidostat. In an attempt to investigate phenomana related to the ultradian and the infradian growth, axenic cultures of *Tetrahymena pyriformis* were maintained on enriched proteose peptone medium in a continuous culture system of this type (Szyszko *et al.*, 1968; Wille and Ehret, 1968). A similar system was operated by Meinert *et al.* (1975).

5.2.2 *Chemostat-type cultures*

In the chemostat, under the conditions of complete culture mixing, the cell mass approaches a steady-state value dependent on the limiting nutrient concentration in the inflowing medium, while the specific growth rate of the culture stabilizes at a value equal the dilution rate. This holds unless the dilution rate is set higher than the critical value at which washout of the culture occurs. The kinetic aspects of the chemostat culture growth have been widely dealt with in both microbiologically and ecologically oriented papers (Hebert *et al.*, 1956; Droop, 1974; Jannasch, 1974; Curds and Bazin, 1977); hence further discussion of these topics would be redundant.

a. *Species-defined systems.* A chemostat-type culture of *Tetrahymena pyriformis*, was maintained axenically on a proteose-peptone medium by Suhr-Jessen *et al.* (1977) and dependence of cell morphology and the timing of cell cycle phases on the value of steady-state growth rate were analysed. Since these authors used a complex medium where the limiting substance was not specified, the system should not be considered as the chemostat in the strict microbiological sense. On the other hand, they have verified their assumption of culture homogeneity.

The investigator attempting to use a chemostat technique with a monoxenic ciliate culture has to consider the food organism as a limiting nutrient. The Monod-type limitation has usually been assumed by analogy with batch cultures.

Hamilton and Preslan (1970) kept *Uronema* sp., a pelagic marine ciliate on the bacterium *Serratia marinorubra*. Being aware of the possibility of bacterial growth in the presence of the ciliate they dispersed washed bacterial cells in sterile sea water with the addition of antibiotics. A steady-state culture was established unless the concentration of bacteria in the inflowing medium fell below a certain threshold value. Additional experiments with *Uronema marinum* in a second-stage vessel fed on a chemostat culture on *Vibrio* sp. resulted in periodic oscillations of the ciliate's cell size and feeding rate (Ashby, 1976).

Curds and Cockburn (1971) fed a steady-state population of *Tetrahymena pyriformis*, situated in the second stage of a two-stage chemostat, monoxenically on *Klebsiella aerogenes* limited by saccharose in the first stage. The ciliates were expected not to utilize saccharose; at the same time, it was tacitly assumed that no bacterial growth occurred in the presence of *Tetrahymena*, since substrate utilization was complete in the first stage. Later, Legner (1973) presented evidence that bacteria might be stimulated to further growth in the presence of ciliates even though all their substrate has been exhausted before entering the ciliate stage.

The use of a single-stage continuous culture proved to be a sensitive tool for analysing the kinetics of the ciliate monoxenic growth. Instead of steady state, periodic oscillations of both ciliates and their food organisms were expected from the analysis of the model based on the assumption of Monod-type relation of bacterial concentration and ciliate growth rate (Canale, 1970; Curds, 1971; Curds and Bazin, 1977). Jost *et al.* (1973a, 1973b) showed that the Monod model does not apply to the relation of *Tetrahymena*—bacteria *Escherichia coli* and *Azotobacter vinelandii* in a single-stage continuous culture, especially at the periods of low bacterial concentrations. Bonomi and Fredrickson (1976) suggested that the inclusion of bacterial wall growth, unavailable to ciliates, in the model would greatly improve the performance of a Monod-type mathematical simulation.

Unavailability of a part of bacterial population to a ciliate culture was also proposed by Sudo *et al.* (1975), who observed bacterial flocculation in the presence of high population densities of a continuous culture of *Colpidium campylum* on *Alcaligenes faecalis*. Unfortunately, these authors did not supply any quantitative data on the flocculation. Instead, they argued that the assumption of an "unavailable" bacterial fraction included in their mathematical model enabled better agreement of theoretical values with experimental findings. These authors also suggested that their substrate for the bacteria was concurrently utilized by *Colpidium*, though with an uptake rate much lower than that of the bacteria.

Autotrophic microbes have on occasions been utilized as a source of nutrition in ciliate continuous culture studies. Gold (1973) maintained continuous growth of *Tintinnopsis beroidea* on a mixture of four algal species. Taub (1969, 1977) used a two-stage continuous culture system with the ciliate *Tetrahymena vorax*, green alga *Chlamydomonas reinhardtii* and bacterial species *Pseudomonas fluorescens* and *Escherichia coli* in both stages. Glass beads were included to prevent the growth of organisms adhering to the bottom of the culture vessels. The rather complicated interrelations among the respective microbial species were mathematically modelled. Bader *et al.* (1976) studied a system of the blue-green alga *Anacystis nidulans* and the ciliate *Colpoda steinii*. The encystment of the ciliate and the attachment of both the cysts and a part of algal population to the walls of culture vessel were shown to be complicating factors giving rise to population dynamics that were difficult to explain in terms of the Monod model. It follows from these studies that the interactions of ciliates and their food with submerged surfaces should be quantified more precisely (cf. Legner, 1975b; Legner *et al.*, 1976).

b. *Incompletely defined system.* Some features of ciliate biology complicate the conditions in continuous clonal cultures on undefined mixtures of bacteria (Straškrabová and Legner, 1969; Legner, 1973). The common bacteria-feeding ciliate species *Colpidium campylum*, *Glaucoma chattoni* and *Cyclidium glaucoma* accumulate in regions of the culture vessels and are able to escape the drifting due to the flow of culture fluid. This behaviour characteristic provides the ciliates an opportunity for considerable self-control of their growth rate. *Colpidium campylum* kept a specific growth rate 0·01–0·02 h^{-1} at different dilution rates over the range 0·03–0·23 h^{-1} in either second or fourth stage of continuous culture units and with two different organic media as a source of nutrition for their food bacteria (Legner, 1973). It may be suggested that the infradian mode of growth was a response of this ciliate to a continuous supply of food when spatial zonation of culture was allowed.

Experiments in a continuous-flow cascade of three 225 litre stirred aquaria inoculated with a mixture of microbes from natural habitat showed that a considerable part of a free-swimming ciliate populations accumulated close to the walls and other submerged surfaces, even though they probably fed mainly on suspended bacteria (Legner, 1975b; Legner *et al.*, 1976). Indirect evidence was given that the free-swimming ciliates also attacked bacteria adhering to the walls (Legner *et al.*, 1976), which does not support the assumption of Bonomi and Frederickson (1976) mentioned earlier.

With some free-swimming ciliate species, it is not feasible to prevent the spatial heterogeneity of a population by a vigorous movement of the culture fluid. Legner (1973) observed a considerable lowering of the biomass of a semicontinuously reared *Glaucoma chattoni* when air bubbling was applied in order to keep the culture in a homogeneous suspension. The detrimental effect of an excessive movement of culture medium was also reported by Straškrabová and Legner (1969), Gold (1973) and Sudo *et al.* (1975).

5.2.3 *Estimation of population characteristics*

a. *Maximum specific growth rate.* In as far as Monod's (1942) function of the saturation curve (equation 2) can adequately describe the relationship between the limiting food and the ciliate growth rate, the maximum specific growth rate (μ_{max}), which is by definition the asymptote of this function, may be computed if several points of the curve are estimated experimentally (cf. Jannasch, 1974; Curds and Bazin, 1977). Moreover an experimental approximation of μ_{max} might still be feasible, even if the functional dependence only holds for part of the full range of possible values.

By slowly increasing the dilution rate in a chemostat-type monoxenic culture Curds and Cockburn (1971) were able to maintain *Tetrahymena pyriformis* at a specific growth rate considerably higher than the μ_{max} value previously calculated from batch culture data (Curds and Cockburn, 1968). The relation between the ciliate growth rate and mean cell volume (cf. section 4.2) was thought to account for this.

As mentioned earlier, the μ_{max} may nearly be attained in continuous cultures of the turbidostat type. It might for example be estimated for *Tetrahymena pyriformis* on an axenic medium in the apparatus used by Wille and Ehret (1968). The difficulties encountered with the maintenance of other ciliate species in homogeneous culture and the complications arising from light scatter caused by food organisms in monoxenic culture preclude the general application of this method.

A method drawn from bacterial ecology may be applicable. Jannasch (1969) estimated the growth rate of bacterial strains in a chemostat culture unit where sterilized sea water was dosed at a dilution rate higher than the critical one. A monotonic decrease of cell numbers occurred owing to the washout. A formula derived from his equation (6) cited earlier was used for the calculation. If all the nutrients were in excess, the calculated growth rate would correspond to the μ_{max}. In order to apply this approach adequately, a complete mixing of the culture must be accomplished. With ciliates, serious problems are to be expected among species habitually congregating at surfaces.

A possible way of estimating the μ_{max} of a submerged-surface bound species such as *Glaucoma scintillans* has been suggested by Legner (1979). In an enrichment experiment, the specific growth rate of this species, the first ciliate inhabitant of a continuous-flow system, was calculated from the changes of its biomass both in plankton and in periphyton using equation (8). The highest value recorded was taken to be μ_{max}. This was attained during the first day of population growth; subsequently the growth gradually decreased. The value obtained by this type of estimate would not be directly comparable with the μ_{max} obtained after a prolonged chemostat cultivation. Nevertheless, it characterizes the capabilities of a given wild strain measured in an open system immediately after the collection from the natural habitat.

b. *Steady-state population structure.* In addition to the possibilities of continuous culture discussed above, this technique might be helpful in the development of structure-based methods of growth rate assessment. The chemostat-type continuous culture can realize the limiting steady-state distribution of cell properties (cf. Williams, 1971). Suhr-Jessen *et al.* (1977) have evaluated the advantage of this in the study of various cell cycle parameters of *Tetrahymena pyriformis* at different growth rates. In addition, a chain of chemostats, a multistage continuous culture, can be expected to reproduce the sequence of physiological state occurring in transient states of batch growth curve (Fencl *et al.*, 1972). Multistage culture has also been recommended for the study of mixed microbial cultures by Jannasch and Mateles (1974). In the case of ciliates, in addition to events of the cell cycle, such life-cycle phenomena as encystment, conjugation, and cell transformation may be made independent of time (cf. section 4.1). One important fact should not be overlooked when studying population structure in a multistage system, even under homogeneous conditions: the population in the second state is not homogeneous by definition (Watson, 1972) and this obviously may be extended to the *n*th stage. Some cells have their ancestry in the *n*th stage, while others are in a state of physiological transition between the conditons of the preceding stage(s) and the *n*th one.

5.2.4 *Phased cultures*

Phased culture technique (Dawson, 1972) may be of a special importance in the study of the relation of population structure to the growth rate and to the physiological state of cells. Basically the method employs instantaneous additions of fresh medium at intervals of the culture doubling time. The biomass concentration, which doubles during these intervals, is halved owing to the addition of an amount of nutrient medium equalling the volume of the culture. Cell divisions occur shortly before the dilution. When the culture is halved, the cells are allowed to proceed through their cell cycles until the next dilution with a high degree of synchrony. Consequently cell cycle phenomena can be studied at the population level. Although this type of culture grows most of the time under the conditions of a batch culture, it is a true open system, which should not be confused with semicontinuous ones (Dawson, 1972).

5.2.5 *Specificity of ciliate continuous culture*

The main advantage of continuous cultures for ecological research is that they eliminate the artificial lag phase and stationary phase phenomena known from batch cultures (Jannasch and Mateles, 1974). This obviously holds also for heterocontinuous systems, as they also are open. It needs to be shown experimentally what other benefits of chemostat-type ecological experiments may be expected from ciliate heterocontinuous systems. For example the considerable reproducibility of enrichment experiments even by low substrate (and food) levels could be tested, as also could the possibility of competition between microbial species being controlled not only by μ_{max}, but also by the saturation constant characterizing the affinity of the given species to a given food (Veldkamp and Jannasch, 1972).

It has been stressed several times that chemostats must not be confused with realistic reproduction of the conditions of natural ecosystems. In chemostats, unlike the natural situation, the nutrient transport into the system is closely coupled with removing cells from the system (Jannasch, 1974). When the results from moderately stirred (Legner, 1973) and well stirred (Curds and Cockburn, 1971) ciliate cultures are compared, it seems that the growth rate of ciliate continuous culture may either be controlled by intrinsic factors independent of the dilution rate or, on the other hand, conform to the dilution rate as in true chemostats, depending on the culture vessel design. Hence, the ciliate continuous culture may represent a special type of open system where food supply and cell removal are not as closely linked as in the chemostat. In addition to microbiological methods often used in the maintenance of ciliate continuous cultures, the investigators should take advantage of protozoological and cytological approaches with emphasis on the distribution of cell parameters in a population and on the spatial zonation of biomass in culture vessels. A quantitative dynamic picture of the cell population behaviour might be a step towards comparing laboratory open systems to natural ones. Some findings from artificial continuous flow systems with mixed microbial populations seem to support this suggestion, as they display features of both batch and continuous systems (Jannasch and Mateles, 1974; Legner *et al.*, 1976) that have been observed in the microbial growth of natural populations.

6 Perspectives and possible applications

The ciliates, owing to their great morphological and functional complexity, allow us to trace a variety of cytological processes by both qualitative and quantitative methods. This enables us to study extensively the dynamics of their populations based on the changes of population structure. In consequence, the structural parameters of ciliate populations would deserve greater attention from ecologists in field studies.

Though the present hypotheses (section 4.4) might be tested by the direct comparison of measured data, a better insight into the relationship between individual cell histories and population growth may be expected by a more precise formulation of the functional relationships. To this end, the construction of mathematical models might be considered as a valuable approach. A brief discussion of theoretical concepts partially dealt with in several preceding sections (4.2, 4.3, 4.5) will survey the directions investigated so far.

The understanding of the relationship between cell volume, growth rate and food concentration may be considered as a key to the dynamic interrelations of individual cell properties and the population growth rate. Curds and Cockburn (1971) were able to derive a mathematical description of this relationship (equation 11), having considered the yield coefficient and the proportion of cytostome size to individual cell mass as constant. This model will probably require some improvement in future, since, for example, recent data of Taylor *et al.* (1976) show that the cytostome does not diminish proportionally with cell size. As a valuable background for further development of the model, the concept of an internal nutrient "pool" (cell quota) may be used, which is compatible with experimental data on algal (Droop, 1975) and bacterial (Law *et al.*, 1976) population growth. These kinetics diverge from the Monod (1942) saturation curve (equation 2) in the sense that the growth rate is controlled by intracellular nutrient concentration (Droop, 1977). The advantage of the internal nutrient pool concept is that the model may be applied to transient situations, fitting data from both continuous (Droop, 1974) and batch (Droop, 1975) systems. This model may be further arranged so that the pool material is removed not only by cell growth but also by endogeneous metabolism and excretion (Law *et al.*, 1976; Droop, 1975). The "feedback" effect of a ciliate species on bacteria indicates that the latter is not negligible in ciliates (Legner, 1973).

The model of relative cell ages which has been developed by Powell (1956) and further treated by Williams (1971) may help to define precisely the steady-state distribution of a cell property in a population. The concept of MV (size) (Scherbaum and Rasch, 1957; Schmid, 1967b) may be included into the model as variability. This enables one to fit the theoretical steady-state distribution by means of experimental data. Williams (1971) succeeded in distinguishing between different cell growth modes (exponential versus linear) in two species of algae. One may suppose that this approach may lead to differentiation between populations also in other than steady-state growth.

Smith and Martin (1974) suggested that resting mammal cells return to the cell cycle with a certain probability; return is therefore exponential and the "half-life" of the resting cells constant (cf. section 4.3). The value of this probability has been shown to depend on environmental conditions. Thus far, no experimental evidence has been furnished to indicate that resting ciliate cells also follow this pattern. Whether they do or not should be tested, since an approach similar to that of Smith and Martin (1974) might be helpful in theoretical treatment, especially of slowly growing ciliate populations.

To a certain extent, each of the above approaches might be successful in interpreting some type of experimental data. In addition, a deterministic model of bacterial biomass growth (Herbert *et al.*, 1956) may be easily adapted for ciliate populations especially for predator–prey interactions (Canale, 1970; Curds, 1971). Its simplicity, however, is counterbalanced by the lack of structuralization which excludes application to the relations between cell distribution and growth. The correspondence of this model with real systems is rather general. For transient situations, which prevail in natural populations, a direct application is usually not feasible.

An adequate mathematical description of the multiple interrelations among population growth phenomena might be helpful in the interpretation of data and in looking for the gaps in our knowledge. It should serve to unify the features of several population-structure oriented concepts.

The above approaches are derived from a theoretical base and they bring some prejudice to the evaluation of results. But there are other possible methods, based solely on statistical analysis of measured data. A general approach is to compare the population structure under different well-defined conditions of growth. Obviously, it is advantageous to evaluate the variability of several cell parameters simultaneously. This may be solved by principal component analysis, as by Taylor *et al.* (1976), who compared eight rectilinear and nine proportional morphogenetic characteristics of the ciliate cells in different phases of axenic and monoxenic culture growth. Another obvious approach to population structure analysis is to derive the limiting cell age and the limiting cell cycle phase durations from successively collected data on the changing distribution of cell parameters. This problem was treated by Macdonald (1973) using the theory of multiphase branching processes. Although he demonstrated the use of the method on the data of Puck and Steffen (1963) from metazoan tissue cultures, it might be applied to ciliate populations.

The main advantage of the growth rate assessment based on the population structure would be that it might be related to the growth rate *in situ*. In a natural habitat, where ciliates not only prey upon their food organisms but also are consumed by other predators, structure analysis may be the only method of estimating the actual growth rate, since all the other methods underestimate this value.

The coupling of mathematical procedures with an automated population structure analysis is also likely to achieve a considerable acceleration of data interpretation. In contrast to the "retrospective" calculation the growth rate from the changes in biomass or cell counts, a method based on the relationship between the population structure and the population growth rate might enable assessment of the growth rate value almost immediately after an automated analysis of a population sample.

Progress in the methods of ciliate growth rate measurement may reflect in other fields of the limnological and oceanographical research. Generally, reasonable estimates of the production of the microbial component in aquatic communities are hardly possible without those of ciliate populations. Both the effect of ciliates on their food organisms and the function of ciliates as a potential source of nutrition for predators at higher trophic levels cannot be neglected. As has been mentioned at the beginning of this paper, along with biomass, specific growth rate is a basis for a good production estimate.

Several aspects of the application of ciliate population dynamics to solving practical problems have been discussed by Curds and Bazin (1977) and need not be repeated here. However, in addition to the conclusions of these authors, one possible application should be mentioned. The actual values of dynamic parameters assessed for ciliate populations in the natural habitat can be utilized in predicting the response of ciliates and probably of the entire microbial component to environmental changes. It may be of practical importance to forecast the changes brought about by contamination of the aquatic environment with the products of human activity. The understanding of ciliate population growth may thus favour progress in the evaluation of water quality, enabling rational prognoses of the microbial decomposition of organic "pollution" in the aquatic environment.

References

Allen, S. L. (1967). Chemical genetics of protozoa. *In* "Chemical Zoology", Volume 1, Protozoa (Ed. G. W. Kidder), pp. 617–694. Academic Press, New York and London.

Andreeff, M. and Haag, D. (1975). Fluorochromierung mit Acridinorange in der Impulscytophotometrie. *In* "Impulscytophotometrie" (Ed. M. Andreeff), pp. 31–37. Springer-Verlag, Berlin.

Ashby, R. E. (1976). Long term variations in a protozoan chemostat culture. *Journal of Experimental Marine Biology and Ecology*, **24**, 227–235.

Bader, F. G., Tsuchiya, H. M. and Fredrickson, A. G. (1976). Grazing of ciliates on blue-green algae: effects of ciliate encystment and related phenomena. *Biotechnology and Bioengineering*, **18**, 311–331.

Berger, J. D. (1971). Kinetics of incorporation of DNA precursors from ingested bacteria into macronuclear DNA of *Paramecium aurelia*. *Journal of Protozoology*, **18**, 419–429.

Berger, J. D. (1973). Regulation of DNA content and length of the cell cycle in *Paramecium aurelia*. *In* "Progress in Protozoology" (Eds P. de Puytorac and J. Grain), p. 39. Université de Clermont, Clermont-Ferrand. (Abstract.)

Berkhan, E. (1975). Digitale Ausgabe und Auswertung der Messergebnisse des Impulscytophotometers. *In* "Impulscytophotometrie" (Ed. M. Andreeff), pp. 50–54. Springer-Verlag, Berlin.

Bick, H. (1958). Ökologische Untersuchungen an Ciliaten fallaubreicher Kleingewässer. *Archiv für Hydrobiologie*, **54**, 506–542.

Bick, H. and Kunze, S. (1971). Eine Zusammenstellung von autökologischen Befunden an Süsswasserciliaten. *Internationale Revue der gesamten Hydrobiologie*, **56**, 337–384.

Bonomi, A. and Fredrickson, A. G. (1976). Protozoan feeding and bacterial wall growth. *Biotechnology and Bioengineering*, **18**, 239–252.

Brock, T. D. (1971). Microbial growth rates in nature. *Bacteriological Reviews*, **35**, 39–58.

Cameron, I. L. (1973). Growth characteristics of Tetrahymena. *In* "Biology of Tetrahymena" (Ed. A. M. Elliott), pp. 199–226. Dowden, Hutchinson and Ross, Stroudsburg, Pennsylvania.

Cameron, I. L. and Terebey, N. (1967). Cell division in the absence of cell growth in *Tetrahymena pyriformis*. *Journal of Protozoology*, **14**, Supplement, 7. (Abstract.)

Canale, R. P. (1970). An analysis of models describing predator–prey interactions. *Biotechnology and Bioengineering*, **12**, 353–378.

Caperon, J. (1967). Population growth in micro-organisms limited by food supply. *Ecology*, **48**, 715–722.

Cleffmann, G. (1974). The cell cycle. *In* "Actualités Protozoologiques" (Eds P. de Puytorac and J. Grain), vol. 1, pp. 315–325. Université de Clermont, Clermont-Ferrand.

Collins, J. F. and Richmond, M. H. (1962). Rate of growth of *Bacillus cereus* between divisions. *Journal of General Microbiology*, **28**, 15–33.

Corliss, J. O. (1973). History, taxonomy, ecology, and evolution of species of Tetrahymena. *In* "Biology of Tetrahymena" (Ed. A. M. Elliott), pp. 1–55. Dowden, Hutchinson and Ross, Stroudsburg, Pennsylvania.

Corliss, J. O. (1977). Annotated assignment of families and genera to the orders and classes currently comprising the Corlissian scheme of higher classification for the phylum Ciliophora. *Transactions of the American Microscopical Society*, **96**, 104–140.

Crissman, H. A. and Steinkamp, J. A. (1973). Rapid, simultaneous measurement of DNA, protein, and cell volume in single cells from large mammalian cell populations. *Journal of Cell Biology*, **59**, 766–771.

Curds, C. R. (1969). Quantitative feeding studies on *Tetrahymena pyriformis* using continuous-culture techniques. *In* "Progress in Protozoology" (Eds A. A. Strelkov, K. M. Sukhanova and I. B. Raikov), p. 188. Nauka, Leningrad. (Abstract.)

Curds, C. R. (1971). A computer-simulation study of predator–prey relationships in a single-stage continuous-culture system. *Water Research*, **5**, 793–812.

Curds, C. R. and Bazin, M. J. (1977). Protozoan predation in batch and continuous culture. *In* "Advances in Aquatic Microbiology" (Eds M. R. Droop and H. W. Jannasch), vol. 1, pp. 115–176. Academic Press, London.

Curds, C. R. and Cockburn, A. (1968). Studies on the growth and feeding of *Tetrahymena pyriformis* in axenic and monoxenic culture. *Journal of General Microbiology*, **54**, 343–358.

Curds, C. R. and Cockburn, A. (1971). Continuous monoxenic culture of *Tetrahymena pyriformis*. *Journal of General Microbiology*, **66**, 95–108.

Curds, C. R. and Vandyke, J. M. (1966). The feeding habits and growth rates of some fresh-water ciliates found in activated-sludge plants. *Journal of Applied Ecology*, **3**, 127–137.

Curds, C. R., Cockburn, A. and Vandyke, J. M. (1968). An experimental study of the role of the cilated protozoa in the activated-sludge process. *Water Pollution Control*, **67**, 312–329.

Darzynkiewicz, Z., Traganos, F., Sharpless, T. and Melamed, M. R. (1975). Conformation of RNA *in situ* as studied by acridine orange staining and automated cytofluorometry. *Experimental Cell Research*, **95**, 143–153.

Dawson, P. S. S. (1972). Continuously synchronized growth. *Journal of Applied Chemistry and Biotechnology*, **22**, 79–103.

DeBault, L. E. and Ringertz, N. R. (1967). A comparison of normal and cold synchronized cell division in Tetrahymena. *Experimental Cell Research*, **45**, 509–518.

Deroux, G. and Tuffrau, M. (1965). *Aspidisca orthopogon*, n. sp. révision de certains mécanismes de la morphogenèse à l'aide d'une modification de la technique au protargol. *Cahiers de Biologie Marine*, **6**, 293–310.

De Terra, N. (1970). Cytoplasmic control of macronuclear events in the cell cycle of Stentor. *In* "Control of Organelle Development" (Ed. P. L. Miller), pp. 345–368. Cambridge University Press, London.

De Terra, N. (1973). Evidence for cortical control of macronuclear DNA synthesis in Stentor. *In* "Progress in Protozoology" (Eds P. de Puytorac and J. Grain), p. 412. Université de Clermont, Clermont-Ferrand. (Abstract.)

Diller, W. F. (1975). Nuclear behavior and morphogenetic changes in fission and conjugation in *Aspidisca costata* (Dujardin). *Journal of Protozoology*, **22**, 221–229.

Dive, D. (1975). Influence de la concentration bacteriénne sur la croissance de *Colpidium campylum*. *Journal of Protozoology*, **22**, 545–550.

Donaghay, P. L. (1974). Cell size selectivity in chemostatic systems *In* "Abstracts of Papers Submitted for the 37th Annual Meeting ASLO". American Society of Limnology and Oceanography, Seattle. (Abstract.)

Droop, M. R. (1974). The nutrient status of algal cells in continuous culture. *Journal of the Marine Biological Association of the United Kingdom*, **54**, 825–855.

Droop, M. R. (1975). The nutrient status of algal cells in batch culture. *Journal of the Marine Biological Association of the United Kingdom*, **55**, 541–555.
Droop, M. R. (1977). An approach to quantitative nutrition of phytoplankton. *Journal of Protozoology*, **24**, 528–532.
Dunham, P. B. and Kropp, D. L. (1973). Regulation of solutes and water in Tetrahymena. *In* "Biology of Tetrahymena" (Ed. A. M. Elliott), pp. 165–198. Dowden, Hutchinson and Ross, Stroudsburg, Pennsylvania.
Edmonson, W. T. and Winberg, G. G. (eds) (1971). "A Manual on Methods for the Assessment of Secondary Productivity in Fresh Waters". International Biological Programme, Blackwell Scientific Publications, Oxford and Edinburgh.
Edmunds, L. J., Jr. (1974). Effect of light on cell division in exponentially increasing cultures of Tetrahymena grown at low temperatures. *In* "Actualités Protozoologiques" (Eds P. de Puytorac and J. Grain), vol. 1, p. 360. Université de Clermont, Clermont-Ferrand. (Abstract.)
Elliott, A. M. (1973). Life cycle and distribution of *Tetrahymena*. *In* "Biology of *Tetrahymena*" (Ed. A. M. Elliott), pp. 259–286. Dowden, Hutchinson and Ross, Stroudsburg, Pennsylvania.
Elliott, P. B. and Bamforth, S. S. (1975). Interstitial protozoa and algae of Louisiana salt marshes. *Journal of Protozoology*, **22**, 514–519.
Evans, C. G. T. (1976). The concept of relative growth rate. *In* "Continuous Culture 6: Applications and New Fields" (Eds A. C. R. Dean, D. C. Ellwood, C. G. T. Evans, J. Melling), pp. 346–348. Ellis Horwood, Chichester.
Fencl, Z., Řičica, J. and Kodešová, J. (1972). The use of the multi-stage chemostat for microbial product formation. *Journal of Applied Chemistry and Biotechnology*, **22**, 405–416.
Frankel, J. and Williams, N. E. (1973). Cortical development in *Tetrahymena*. *In* "Biology of *Tetrahymena*" (Ed. A. M. Elliott), pp. 375–409. Dowden, Hutchinson and Ross, Stroudsburg, Pennsylvania.
Gold, K. (1973). Methods for growing Tintinnida in continuous culture. *American Zoologist*, **13**, 203–208.
Gold, K. and Morales, E. A. (1975). Seasonal changes in lorica sizes and the species of Tintinnida in the New York bight. *Journal of Protozoology*, **22**, 520–528.
Golikova, M. N. (1974). Cytophotometrical study of DNA amount in micronucleus of *Paramecium bursaria*. *Acta Protozoologica*, **13**, 109–120. (Russian.)
Grain, J. (1972). Etude ultrastructurale *d'Halteria grandinella* O.F.M. (Cilié Oligotriche) et considérations phylogénétiques. *Protistologica*, **8**, 179–197.
Grebecki, A. (1961). Experimental studies on the selection and adaptability in *Paramecium caudatum*. *Acta Biologiae Experimentalis*, **21**, 35–52.
Grell, K. G. (1967). Sexual reproduction in Protozoa. *In* "Research in Protozoology" (Ed T.-T. Chen), vol. 2, pp. 147–213. Pergamon Press, Oxford.
Hall, R. P. (1967). Nutrition and growth of Protozoa. *In* "Research in Protozoology" (Ed. T.-T. Chen), vol. 1, pp. 337–404. Pergamon Press, Oxford.
Hamilton, R. D. and Preslan, J. E. (1969). Cultural characteristics of a pelagic marine hymenostome ciliate (*Uronema* sp.). *Journal of Experimental Marine Biology and Ecology*, **4**, 90–99.
Hamilton, R. D. and Preslan, J. E. (1970). Observations on the continuous culture of a planktonic phagotrophic protozoon. *Journal of Experimental Marine Biology and Ecology*, **5**, 94–104.
Hanson, E. D. (1967). Protozoan development. *In* "Chemical Zoology", Volume 1, Protozoa (Ed. G. W. Kidder), pp. 395–539. Academic Press, New York and London.

Hashimoto, K. (1963). Formation of ciliature in excystment and induced re-encystment of *Oxytricha fallax* Stein. *Journal of Protozoology*, **10**, 156–166.

Herbert, D., Elsworth, R. and Telling, R. C. (1956). The continuous culture of bacteria; a theoretical and experimental study. *Journal of General Microbiology*, **14**, 601–622.

Holz, G. G., Jr. (1973). The nutrition of *Tetrahymena*: essential nutrients, feeding, and digestion. *In* "Biology of *Tetrahymena*" (Ed. A. M. Elliott), pp. 89–98. Dowden, Hutchinson and Ross, Stroudsburg, Pennsylvania.

Jackson, P. R., Fisher, F. M., Jr., Winkler, D. G. and Kimzey, S. L. (1977). Laser studies with protozoans: Cytofluorograf analysis of *Plasmodium yoelii*, *Trypanosoma gambiense* and *Trypanosoma equiperdum*. *In* "Abstracts of Papers Read at the Fifth International Congress on Protozoology" (Ed. S. H. Hutner), p. 142. Society of Protozoologists, New York City. (Abstract.)

James, T. W. (1974). A cell cycle nomogram: graphs of mathematical relationship allow estimates of the timing of cycle events in eucaryotic cells. *In* "Cell Cycle Controls" (Eds G. M. Padilla, I. L. Cameron, A. Zimmerman), pp. 31–42. Academic Press, London and New York.

James, T. W. and Read, C. P. (1957). The effect of incubation temperature on the cell size of *Tetrahymena*. *Experimental Cell Research*, **13**, 510–513.

Jannasch, H. W. (1969). Estimations of bacterial growth rates in natural waters. *Journal of Bacteriology*, **99**, 156–160.

Jannasch, H. W. (1974). Steady state and the chemostat in ecology. *Limnology and Oceanography*, **19**, 716–720.

Jannasch, H. W. and Mateles, R. I. (1974). Experimental bacterial ecology studied in continuous culture. *In* "Advances in Microbial Physiology" (Eds A. H. Rose and D. W. Tempest), vol. 11, pp. 165–212. Academic Press, London and New York.

Jauker, F. and Cleffmann, G. (1970). Oscillation of individual generation times in cell lines of *Tetrahymena pyriformis*. *Experimental Cell Research*, **62**, 477–480.

Jeffrey, W. R., Frankel, J., DeBault, L. E. and Jenkins, L. M. (1973). Analysis of the schedule of DNA replication in heat-synchronized *Tetrahymena*. *Journal of Cell Biology*, **59**, 1–11.

Jones, G. L. (1973). Role of Protozoa in waste purification systems. *Nature, London*, **243**, 546–547.

Jost, J. L., Drake, J. F., Fredrickson, A. G. and Tsuchyiya, H. M. (1973a). Interactions of *Tetrahymena pyriformis*, *Escherichia coli*, *Azotobacter vinelandii*, and glucose in a minimal medium. *Journal of Bacteriology*, **113**, 834–840.

Jost, J. L., Drake, J. F., Tsuchiya, H. M. and Fredrickson, A. G. (1973b). Microbial food chains and food webs. *Journal of Theoretical Biology*, **41**, 461–484.

Kamentsky, L. A. and Melamed, M. R. (1969). Rapid multiple mass constituent analysis of biological cells. *Annals of the New York Academy of Sciences*, **157**, 310–323.

Kimball, R. F. and Barka, T. (1959). Quantitative cytochemical studies on *Paramecium aurelia*. II. Feulgen microspectrophotometry of the macronucleus during exponential growth. *Experimental Cell Research*, **17**, 173–182.

Kimball, R. F., Caspersson, T. O., Svensson, G. and Carlson, L. (1959). Quantitative cytochemical studies on *Paramecium aurelia*. I. Growth in total dry weight measured by the scanning interference microscope and X-ray absorption methods. *Experimental Cell Research*, **17**, 160–172.

Kloetzel, J. A. (1975). Scanning electron microscopy of stomatogenesis accompanying conjugation in *Euplotes aediculatus*. *Journal of Protozoology*, **22**, 385–391.

Law, A. T., Robertson, B. R., Dunker, S. S. and Button, D. K. (1976). On describing

microbial growth kinetics from continuous culture data: some general considerations, observations, and concepts. *Microbial Ecology*, **2**, 261–283.
Legner, M. (1964). Annual observations on ciliates inhabiting the natant vegetation of two naturally polluted pools. *Acta Societatis Zoologicae Bohemoslovenicae*, **33**, 192–213.
Legner, M. (1973). Experimental approach to the role of Protozoa in aquatic ecosystems. *American Zoologist*, **13**, 177–192.
Legner, M. (1975a). Concentration of organic substances in water as a factor controlling the occurrence of some ciliate species. *Internationale Revue der gesamten Hydrobiologie*, **60**, 639–654.
Legner, M. (1975b). Growth of a mixed ciliate culture in a continuous-flow system. *Journal of Protozoology*, **22**, 66A. (Abstract.)
Legner, M. (1979). A characterization of water quality by the protozoan growth rate. *In* "Bioindicatores Deteriorisationis Regionis". Academia, Prague. (In press.)
Legner, M., Punčochář, P. and Straškrabová, V. (1976). Development of the microbial component of a river community. *In* "Continuous Culture 6: Applications and New Fields" (Eds A. C. R. Dean, D. C. Ellwood, C. G. T. Evans, J. Melling), pp. 329–344. Ellis Horwood, Chichester.
Lengerová-Kučerová, A. (1950). Cell volume of *Glaucoma pyriformis* as a function of hydrogen ion concentration in the medium. *Acta Societatis Zoologicae Bohemoslovenicae*, **14**, 207–228.
Lilly, D. M. and Stillwell, R. H. (1965). Probiotics: growth-promoting factors produced by microorganisms. *Science*, **147**, 747–748.
Luporini, P. and Dini, F. (1975). Relationships between cell cycle and conjugation in 3 hypotrichs. *Journal of Protozoology*, **22**, 541–544.
Lykkesfeldt, A. E. (1974). Effects of 5-bromodeoxyuridine on *Tetrahymena pyriformis* grown in low concentrations of tetrahydrofolic acid. *Comptes Rendus des Travaux du Laboratoire Carlsberg*, **40**, 91–100.
Macdonald, P. D. M. (1973). On the statistics of cell proliferation. *In* "The Mathematical Theory of the Dynamics of Biological Populations" (Eds M. S. Bartlett, R. W. Hiorns), pp. 303–314. Academic Press, London and New York.
McDonald, B. B. (1958). Quantitative aspects of deoxyribose nucleic acid (DNA) metabolism in an amicronucleate strain of *Tetrahymena*. *Biological Bulletin*, **114**, 71–94.
McDonald, B. B. (1973). Nucleic acids in *Tetrahymena* during vegetative growth and conjugation. *In* "Biology of *Tetrahymena*" (Ed. A. M. Elliott), pp. 287–306. Dowden, Hutchinson and Ross, Stroudsburg, Pennsylvania.
Meinert, J. C., Ehret, C. F. and Antipa, G. A. (1975). Circadian chronotypic death in heat-synchronized infradian mode cultures of *Tetrahymena pyriformis* W. *Microbial Ecology*, **2**, 201–214.
Melamed, M. R. and Kamentsky, L. A. (1969). An assessment of the potential role of automatic devices in cytology screening. *Obstetrical and Gynecological Survey*, **24**.
Meskill, V. P. (1970). Factors influencing the growth of *Glaucoma chattoni* in a chemically defined medium. *Journal of Protozoology*, **17**, 104–107.
Minutoli, F. and Hirshfield, H. (1968). DNA synthesis cycle in Blepharisma. *Journal of Protozoology*, **15**, 532–535.
Mitchison, J. M. (1971). "The Biology of the Cell Cycle". Cambridge University Press, Cambridge.
Miyake, A. and Nobili, R. (1974). Mating reaction and its daily rhythm in *Euplotes crassus*. *Journal of Protozoology*, **21**, 584–587.

Monod, J. (1942). "Recherches sur la Croissance des Cultures Bactériennes". Hermann, Paris.

Morat, G. (1970) Études cytophotométrique et autoradiographique de l'appareil nucléaire de *Colpidium campylum* au cours du cycle cellulaire. *Protistologica,* **6**, 83–95.

Morrison, G. A. and Tomkins, A. L. (1973). Determination of mean cell size of *Tetrahymena* in growing cultures. *Journal of General Microbiology,* **77**, 383–392.

Munson, R. J. (1970). Turbidostats. *In* "Methods in Microbiology" (Eds J. R. Norris and D. W. Ribbons), vol. 2, pp. 349–376. Academic Press, London and New York.

Nachtwey, D. S. and Cameron, I. L. (1968). Cell cycle analysis. *In* "Methods in Cell Physiology" (Ed D. M. Prescott), vol. 3, pp. 213–259. Academic Press, New York and London.

Nilsson, J. R. and Zeuthen, E. (1974). Microscopical studies on the macronucleus of heat synchronized *Tetrahymena pyriformis* GL. *Comptes Rendus des Travaux du Laboratoire Carlsberg,* **40**, 1–18.

Noland, L. E. and Gojdics, M. (1967). Ecology of free-living Protozoa. *In* "Research in Protozoology" (Ed. T.-T. Chen), vol. 2, pp. 215–266. Pergamon Press, Oxford.

Paulin, J. J. and Brooks, A. S. (1975). Macronuclear differentiation during oral regeneration in *Stentor coeruleus*. *Journal of Cell Science,* **19**, 531–541.

Plesner, P. and Hartman, B. (1973). Synthesis of ribosomal protein in *Tetrahymena* during exponential growth and cell division. *In* "Progress in Protozoology" (Eds P. de Puytorac and J. Grain), p. 325. Université de Clermont, Clermont-Ferrand. (Abstract.)

Powell, E. O. (1956). Growth rate and generation time of bacteria, with special reference to continuous culture. *Journal of General Microbiology,* **15**, 492–511.

Preer, J. R., Jr. (1969). Genetics of the Protozoa. *In* "Research in Protozoology" (Ed. T.-T. Chen), vol. 3, pp. 129–278. Pergamon Press, Oxford.

Prescott, D. M. (1957). Relation between multiplication rate and temperature in *Tetrahymena pyriformis,* strains HS and GL. *Journal of Protozoology,* **4**, 252–256.

Prescott, D. M. (1959). Variations in the individual generation times of *Tetrahymena gelei* HS. *Experimental Cell Research,* **16**, 279–284.

Prescott, D. M. and Carrier, R. F. (1964). Experimental procedures and cultural methods for *Euplotes eurystomus* and *Amoeba proteus*. *In* "Methods in Cell Physiology" (Ed. D. M. Prescott), vol. 1, pp. 85–95. Academic Press, New York and London.

Prescott, D. M. and Stone, G. E. (1967). Replication and function of the protozoan nucleus. *In* "Research in Protozoology" (Ed. T.-T. Chen), vol. 2, pp. 117–146. Pergamon Press, Oxford.

Puck, T. T. and Steffen, J. (1963). Life cycle analysis of mammalian cells. I. A method for localizing metabolic events within the life cycle, and its application to the action of colcemide and sublethal doses of X-irradiation. *Biophysical Journal,* **3**, 379–397.

Raikov, I. B. (1969). The macronucleus of ciliates. *In* "Research in Protozoology" (Ed. T.-T. Chen), vol. 3, pp. 1–128. Pergamon Press, Oxford.

Raikov, I. B. (1972). Nuclear phenomena during conjugation and autogamy in ciliates. *In* "Research in Protozoology" (Ed. T.-T. Chen), vol. 4, pp. 147–289. Pergamon Press, Oxford.

Řičica, J., Nečinová, S., Stejskalová, E. and Fencl, Z. (1967). Properties of microorganisms grown in excess of the substrate at different dilution rates in continuous multistream culture systems. *In* "Microbial Physiology and Continuous Culture" (Eds E. O. Powell, C. G. T. Evans, R. E. Strange, D. W. Tempest), pp. 196–208. Her Majesty's Stationery Office, London.

Rogers, T. D. and Kimzey, S. L. (1972). Rapid scanning microspectrophotometry of

colorless *Euglena gracilis* and *Astasia longa*. A basis for differentiation. *Journal of Protozoology*, **19**, 150–155.
Ron, A. and Urieli, S. (1977). Qualitative and quantitative studies on DNA and RNA synthesis in *Loxodes striatus* nuclei. *Journal of Protozoology*, **24**, 150–154.
Rosenberg, H. M. and Gregg, E. C. (1969). Kinetics of cell volume changes of murine lymphoma cells subjected to different agents in vitro. *Biophysical Journal*, **9**, 592–606.
Ruthmann, A. (1972). Division and formation of the macronuclei of *Keronopsis rubra*. *Journal of Protozoology*, **19**, 661–666.
Salvano, P. (1975). Comparaison du fonctionnement des bandes de réorganisation d'*Euplotes crassus* (Dujardin) et de *Strombidium sulcatum* Claparède et Lachmann après analyse microspectrographique en UV. *Journal of Protozoology*, **22**, 230–232.
Scherbaum, O. (1956). Cell growth in normal and synchronously dividing mass cultures of *Tetrahymena pyriformis*. *Experimental Cell Research*, **11**, 464–476.
Scherbaum, O. (1957). The division index and multiplication in a mass culture of Tetrahymena following inoculation. *Journal of Protozoology*, **4**, 257–259.
Scherbaum, O. and Rasch, G. (1957). Cell size distribution and single cell growth in *Tetrahymena pyriformis* GL. *Acta Pathologica et Microbiologica Scandinavica*, **41**, 161–182.
Scherbaum, O. and Zeuthen, E. (1954). Induction of synchronous cell division in mass cultures of *Tetrahymena pyriformis*. *Experimental Cell Research*, **6**, 221–227.
Schmid, P. (1967a). Temperature adaptation of the growth and division process of *Tetrahymena pyriformis*. I. Adaptation phase. *Experimental Cell Research*, **45**, 460–470.
Schmid, P. (1967b). Temperature adaptation of the growth and division process of *Tetrahymena pyriformis*. II. Relationship between cell growth and cell replication. *Experimental Cell Research*, **45**, 471–486.
Sheldon, R. W. and Parsons, T. R. (1967). "A Practical Manual on the Use of the Coulter Counter in Marine Research". Coulter Electronics Sales Company, Canada.
Sikora, J. and Kuznicki, L. (1973). Cytoplasmic movements during conjugation of *Paramecium aurelia*. *In* "Progress in Protozoology" (Eds P. de Puytorac and J. Grain), p. 380. Université de Clermont, Clermont-Ferrand. (Abstract.)
Sisken, J. E. (1964). Methods for measuring the length of the mitotic cycle and the timing of DNA synthesis for mammalian cells in culture. *In* "Methods in Cell Physiology" (Ed. D. M. Prescott), vol. 1, pp. 387–401. Academic Press, New York and London.
Smith, J. A. and Martin L. (1974). Regulation of cell proliferation. *In* "Cell Cycle Controls" (Eds G. M. Padilla, I. L. Cameron and A. Zimmerman), pp. 43–60. Academic Press, London and New York.
Snedecor, G. W. and Cochran, W. G. (1967). "Statistical Methods" (Sixth Edition). Iowa State University Press, Ames, Iowa.
Soldo, A. T. and van Wagtendonk, W. J. (1969). The nutrition of *Paramecium aurelia*, stock 299. *Journal of Protozoology*, **16**, 500–506.
Stöhr, M. (1975). On the correlation of chromosome number and DNA content in HeLa cell clones. *Acta Cytologica*, **19**, 299–305.
Stöhr, M. and Goerttler, K. (1974). Neue instrumentelle Möglichkeiten zur Optimierung der ultraschnellen Mikrofluorometrie. *Histochemistry*, **39**, 35–40.
Stone, G. E. and Cameron, I. L. (1964). Methods for using *Tetrahymena* in studies of the normal cell cycle. *In* "Methods in Cell Physiology" (Ed. D. M. Prescott), vol. 1, pp. 127–140. Academic Press, New York and London.
Straškrabová-Prokešová, V. and Legner, M. (1966). Interrelations between bacteria and Protozoa during glucose oxidation in water. *Internationale Revue der gesamten Hydrobiologie*, **51**, 279–293.

Straškrabová, V. and Legner, M. (1969). The quantitative relation of bacteria and ciliates to water pollution. *In* "Advances in Water Pollution Research" (Ed. S. H. Jenkins), pp. 57–67. Pergamon Press, Oxford and New York.

Sudo, R., Kobayashi, K. and Aiba, S. (1975). Some experiments and analysis of a predator–prey model: Interaction between *Colpidium campylum* and *Alcaligenes faecalis* in continuous and mixed culture. *Biotechnology and Bioengineering,* **17**, 167–184.

Suhr-Jessen, P. B., Stewart, J. M. and Rasmussen, L. (1977). Timing and regulation of nuclear and cortical events in the cell cycle of *Tetrahymena pyriformis. Journal of Protozoology,* **24**, 299–303.

Szyszko, A. H., Prazak, B. L., Ehret, C. F., Eisler, W. J., Jr. and Wille, J. J., Jr. (1968). A multi-unit sampling system and its use in the characterization of ultradian and infradian growth in *Tetrahymena. Journal of Protozoology,* **15**, 781–785.

Tartar, V. (1961). "The Biology of Stentor". Pergamon Press, Oxford.

Tartar, V. (1967). Morphogenesis in Protozoa. *In* "Research in Protozoology" (Ed. T.-T. Chen), vol. 2, pp. 1–116. Pergamon Press, Oxford.

Taub, F. B. (1969). Gnotobiotic models of freshwater communities. *Verhandlungen der Internationalen Vereinigung für theoretische und angewandte Limnologie,* **17**, 485–496.

Taub, F. B. (1977). A continuous gnotobiotic (species-defined) ecosystem. *In* "Aquatic Microbial Communities" (Ed. J. Cairns, Jr.), pp. 105–138. Garland Publishing, New York and London.

Taylor, W. D., Gates, M. A. and Berger, J. (1976). Morphological changes during the growth cycle of axenic and monoxenic *Tetrahymena pyriformis. Canadian Journal of Zoology,* **54**, 2011–2018.

Trager, W. (1963). Differentiation in Protozoa. *Journal of Protozoology,* **10**, 1–6.

Veldkamp, H. and Jannasch, H. W. (1972). Mixed culture studies with the chemostat. *Journal of Applied Chemistry and Biotechnology,* **22**, 105–123.

Walker, P. M. B. (1954). The mitotic index and interphase processes. *Journal of Experi nental Biology,* **31**, 8–15.

Watson, T. G. (1972). The present status and future prospects of the turbidostat. *Journal of Applied Chemistry and Biotechnology,* **22**, 229–243.

Wilbert, N. (1968). Ökologische Untersuchung der Aufwuchs- und Planktonciliaten eines eutrophen Weihers. *Archiv für Hydrobiologie,* **35**, Supplement, 411–518.

Wille, J. J., Jr. and Ehret, C. F. (1968). Light synchronization of an endogenous circadian rhythm of cell division in Tetrahymena. *Journal of Protozoology,* **15**, 785–789.

Williams, F. M. (1971). Dynamics of microbial populations. *In* "Systems Analysis and Simulation in Ecology" (Ed. B. C. Patten), vol. 1, pp. 197–267. Academic Press, New York and London.

Yataganas, X., Yasuharu, M., Traganos, F., Strife, A. and Clarkson, B. (1975). Evaluation of a Feulgen-type reaction in suspension using flow microfluorimetry and a cell separation technique. *Acta Cytologica,* **19**, 71–78.

Yen, A., Fried, J., Kitahara, T., Strife, A. and Clarkson, B. D. (1975). The kinetic significance of cell size. II. Size distributions of resting and proliferating cells during interphase. *Experimental Cell Research,* **95**, 303–310.

Zech, L. (1966). Dry weight and DNA content in sisters of *Bursaria truncatella* during the interdivision interval. *Experimental Cell Research,* **44**, 599–605.

Zeuthen, E. (1963). Independent cycles of cell division and of DNA synthesis in *Tetrahymena.* In "Cell Growth and Cell Division" (Ed. R. J. C. Harris), pp. 1–8. Academic Press, New York and London.

Zeuthen, E. (1974). A cellular model for repetitive and free-running synchrony in *Tetrahymena* and *Schizosaccharomyces. In* "Cell Cycle Controls" (Eds G.

M.Padilla, I. L. Cameron and A. Zimmerman), pp. 1–30. Academic Press, London and New York.

Zeuthen, E. and Rasmussen, L. (1972). Synchronized cell division in Protozoa. *In* "Research in Protozoology" (Ed. T.-T. Chen), vol. 4, pp. 9–145. Pergamon Press, Oxford.

A conceptual model of marine detrital decomposition and the organisms associated with the process

J. J. LEE

Department of Biology, City College of CUNY, New York, NY, USA

1 The importance of litter in carbon and energy flow

Most of the energy trapped photosynthetically by plants, particularly the terrestrial ones, is used to build structural polymers which, although representing (as they have through evolutionary time) tremendous potential energy, only slowly become available to other biota.

An appreciation of the proportion of fixed carbon that passes through decomposition processes can be gained from a few examples. A montane grassland had an annual above-ground net primary production of ~3·5 ton ha^{-1} of which 1·3 ton was assimilated by sheep and slugs, 1·4 ton entered the decomposer web as dead herbage and 0·8 entered as dung (Satchel, 1974). In a mixed deciduous forest with net above-ground primary production of ~0·8 ton ha^{-1}, only 0·1 ton was consumed by the herbivores. While ~⅓ of the production was channelled into wood standing crop, the other ⅔ of the total annual production (6·5 ton) entered the decomposition cycle as leaf litter and dead wood (Satchel, 1974).

Data from marine systems are similar. Coastal primary productivity lies between 0·4 and 1·1 kg cm^{-2}, a figure which includes beds of giant kelps, sea grasses, cord grasses and mangroves (Mann, 1972).

Many problems impede precise estimations of the carbon channelled through various marine detrital food webs but there seems to be a consensus that only a very small fraction of the macrophyte primary production enters food webs through grazing (e.g. Burkholder and Bornside, 1957; Darnell, 1967a, b; Odum and LeCruz, 1967; Mann, 1972; Odum and Heald, 1975; Thayer *et al.*, 1975). Among the more generally accepted estimates are those of Teal (1972) on cord grass grazing (~5 per cent) and Heald (1969) on mangrove grazing (6 per cent). Based on thorough sifting of the evidence, Mann (1972) estimated that in the near-shore <10 per cent of macrophyte primary production enters grazing food chains.

While the area of the coastal zones is small compared with the rest of the ocean, their productivity accounts for a very large proportion of the world's commercial fishes (Ryther, 1969; Russel-Hunter, 1970; Mann, 1972; McHugh, 1976). In the United States ~73·8 per cent of the total commercial landings are estuarine-dependent species, a figure obtained from averaging the catches of Gulf and Southern Coastal states like Louisiana, Mississippi and Virginia, where 99 per cent of the fish are estuarine dependent and states like California, where tropical tuna skew the average, since estuarine-dependent species made up only 4 per cent of the catch (McHugh, 1976). Detrital food webs are important even in the open ocean and the deep sea where undigested, insoluble fragments of phytoplankton and other organisms are recycled from faeces, inefficient grazing on blooms, dead organisms and, to a lesser extent, from other sources.

2 Systems for decomposition of refractory (fibrous) plant structure and channelling of products into secondary production

Very few simple systems have evolved for decomposition of the more refractory plant structural compounds. From a theoretical point of view, this is not surprising. Margalef's (1968) perspective on the strategy of ecosystem evolution, which suggests that maturing ecosystems accumulate biomass in persistent vegetative structures, thereby enlarging potential habitats for organisms with more specialized requirements, seems applicable to the complex decomposing compartments of ecosystems. With rare exceptions, which nicely prove the rule, all decomposer systems depend primarily on microbial (including fungal and protozoal) cellulases.

High plant tissue and litter consists of six main chemical constituents: cellulose; hemicellulose; lignin; water-soluble compounds such as sugar, amino acids and aliphatic acids; ether- and alcohol-soluble constituents such as fats, oils, waxes, resins and carotenoids, and proteins. Only cellulose and lignin, and to a smaller degree, hemicelluloses, defy the digestive systems of herbivores. Two general systems have evolved to degrade the more refractory plant structural compounds and channel their energy and resultant metabolities into secondary production: open-ecosystem detrital food webs and intestinal fermentation systems. Judging only from the extensive evidence obtained on evolution of the mutualistic relationships between xylophagus insects and the flagellates (or fungi in some insects), which inhabit their hindguts, it would appear that the intestinal systems have evolved from the external environmental ones (Honigberg, 1967).

3 Intestinal fermentation systems

Aside from the obvious economic incentives for the study of the intestinal fermentations of large herbivorous mammals, these animals also offer experimental advantages over the study of other cellulose-decomposing systems. A horse, cow, rabbit or a sheep is an entity easily isolated from the ecosystem. Materials entering and leaving the animal are easily recognized, and agents responsible for fermentations are readily accessible. The efficiency of processing can be readily measured. These factors account for the relatively advanced knowledge of such cellulose-digesting systems. Although symbionts and hosts appear very specialized and highly evolved (Hungate *et al.*, 1964;

Hungate, 1966; Honigberg, 1967), fundamental aspects of the intestinal process are helpful in comparing systems having the same kind of function in ecosystems.

Intestinal fermentations, as the term fermentation implies, are anaerobic. Between 10^{10} and 10^{11} bacteria and a lesser number but probably equivalent biomass of protozoa (10^5 to 10^6) characteristically inhabit each cubic centimetre of rumen contents. A typical sheep rumen of 4·7 litres has microorganisms occupying more than 20 per cent (0·93 litres) of the total space! A 55 kg sheep fed almost a kilogram of gound alfalfa pellets digested $\sim 2/3$ of its ration. Calculations suggested that 86 per cent could be digested in 42 h by the microbes if the food could be retained in the rumen (Hungate, 1975).

From the point of view of energetics, Owens and Isaacson (1977) suggest that increasing turnover of rumen contents appears to: (a) enhance bacterial protein production, (b) increase ruminal acetate and methane production, and (c) flush out that 30–60 per cent of the feed material which is more resistant to digestion. Providing that the animal can provide new forage at regular intervals they calculate that an increase in dilution rate from 2 per cent to 12 per cent per hour would double bacterial growth rates.

Present evidence suggests that the amount of cellulose digested by the fascinatingly complex entodiniomorph protozoa is probably very small compared to that digested by the bacteria. The primary role of the protozoa in the system seems to be one of rapidly removing sugars from the rumen liquor to prevent sugar build-up which would thereby slow the bacterial fermentation. Eliminating the protozoa in the rumen increased the bacterial populations, suggesting that the protozoa also serve as bacterial grazers (Hungate, 1975).

One key barrier litter (cellulose and lignin) digestion is the three-dimensional cross-linking of long chains of the polymer. The surface area of fibre exposed to enzymatic attack limits the rate of digestion. Grinding (e.g. rumination, wave action, mastication) increases surface area, breaks bonds between lignin, silica and other fibre components, and improves digestibility. This helps explain why grinding and pelleting of feed increases consumption and speeds passage and assimilation by many ruminants.

Enzymes of cellulolytic microorganisms, whether intestinal or free living (i.e. spp. of *Myrothecium*, *Cellvibrio*, *Pseudomonas*, *Cellulomonas*, *Verticillium*, *Rubinococcus*, *Trichoderma*) fall into three main categories:

1. C_1—endoglucanase(s): forms linear chains of anhydroglucose units;
2. C_x—p-1,4 glucanohydrolase(s) (cellobiohydrolase): cleaves cellulobiose units from longer chains;
3. β-glucosidase(s): cleaves cellobiose into glucose units.

A microbe may have more than one form of the above enzyme; it may be extracellular, mural (bound to the cell wall or cell membrane) or intracellular. Some of the extracellular enzymes bind directly to cellulose. This prevents the enzymes from being diluted or washed away. Some of the microorganisms with huge mural enzymes (MW $\sim 2 \times 10^6$) cement themselves directly to the cellulose (as depicted by Cheng *et al.*, 1977), which suggests that the intimacy of cellulolytic bacteria with the insoluble substrates may be important for the action of cell-bound cellulase. Cavities, presumably formed by enzymatic digestion, develop in the cellulose substratum under the bound bacterial–capsule complex (Cheng *et al.*, 1977).

As noted earlier, deployment of the various cellulases is subject to complete repression by readily assimilated carbon sources such as glucose (the chief end product of cellulose degradation), fructose, sucrose, trehalose, maltose, cellobiose, and starch (Yamene *et al.*, 1970; Berg *et al.*, 1972; Mandels, 1975; Benguin *et al.*, 1977). Cellobiose, an intermediate cleavage metabolite, inhibits both cellobiohydrolase and endoglucanase (Sternberg, 1976).

4 Terrestrial detrital-degradation systems

Detrital pathways, whether in soils or aquatic systems, are difficult subjects for research. It boggles the mind to conceive of the microflora of a single gram of clay, which, according to one estimate (Alexander, 1964), can often have a total surface area in excess of 100 m^2.

New ways to examine microbes and their activities *in situ* promise that progress will be made on many aspects of detrital degradation (Holm-Hansen, 1972; Waid *et al.*, 1973; Zimmerman and Meyer-Riel, 1974; Bancroft *et al.*, 1976; Flett *et al.*, 1976; Gough and Woelkerling, 1976; Hoppe, 1976; Moriarty, 1975, 1977). But it is not apparent to many (e.g. Harley, 1971; Clark, 1973; Parkinson, 1973; Stotsky, 1973), that progress will be slow.

In constrast with the gut flora, filamentous fungi and yeasts play major roles in the decomposition of cellulose and other plant structural compounds in open environmental systems, e.g. soils, leaf litter, sewage treatment systems, marine benthos (Goodfellow and Cross, 1974; Hones, 1974; Pugh, 1974). Another difference between intestinal microbial fermentative systems and some of their environmental analogues is the discontinuity, heterogeneity, variability and environmental sensitivity of the latter processes. Colonization of new materials by microbial propagules, be they cells, spores or fragments of mycelia, is poorly predictable; also, it is often impossible to define steps in the sequence (Alexander, 1971). At each phase in the sequence, we rationalize that the organisms present are particularly well suited for the conditions of the habitat as it exists at that time. Microbial populations appear in gradual and continuous progressions, peak, and then decline as they are succeeded by other species. The limiting factors may be chemical or physical changes that render the environment no longer suitable for growth or they may be biologically imposed stresses.

Decomposition, however, begins not in the soil but in the phylloplane. An extensive literature deals with organisms inhabiting the above-ground surfaces of plants (reviews: Pomeroy, 1970; Bell, 1974; Hensen, 1974; Miller, 1974; Stout, 1974). Although it is hard to generalize from the myriad habitats studied, yeasts and bacteria are the initial colonizers, using sugars exuded from leaf surfaces or released by aphid and other insect action from sub-cuticular tissues. The sugar content on the surfaces of some leaves (e.g. tobacco and coffee) is amazing: 115–224 mg l^{-1} of dew collected on the surface of the leaves. As leaves senesce the phylloplane microflora attack with cutinases, pectinases, and cellulases which eventually penetrate cuticles, attack middle lamellae and disintegrate cell walls. Above-ground humidity is an important factor at this stage of decompositon.

Once leaves fall on the ground, additional organisms attack them. In soils, species of *Pseudomonas*, other Gram-negative rods, and *Bacillus* spp. are among the primary colonizing bacteria. They metabolize amino acids, organic acids, and readily mobilized constitutents of plant remains; meanwhile filamentous fungi are attacking the larger structural molecules (Ecklund and Gyllenberg, 1973; Pugh, 1974). Except for chitin and other nitrogen-rich organic structural materials, where they are primary colonizers (Goodfellow and Cross, 1974), the actinomycetes, notably *Streptomyces* spp., appear to be members of the secondary populations of fungi. The most common initial fungal colonizers are species of *Cladosporium*, *Aureobasidium* and *Alternaria*. These are followed by a fairly large number of other, more specialized fungi. Brock (1966), in an excellent overview, suggests a sequence of colonizers for leaf litter: sugar-phycomycetes, then hemicellulose-utilizing fungi, cellulose utilizers, and lastly lignin digesters. To some extent the little nitrogen left seems to exclude some genera of fungi (e.g. *Botryotrichum, Chaetomium, Masoniella,* etc.) and favours others (*Actinomucor, Coniothyrium, Trichoderma*) (Pugh, 1974). Decomposition of lignin (polymers of phenolic acids) releases material (phenols and tannins), which are toxic or repellent to many soil organisms, thus favouring some species over others.

Soil animals, from protozoan size to large worms, may be very important for decomposition. They contribute by fragmenting and burying plant litter by coating egested materials with microflora and enzymes, by decomposing cellulose and lignin with the aid of internal microbial fermentation systems, and by harvesting large numbers of soil microflora (Atlavinyte and Pociene, 1973) and thereby keeping microbial populations fluctuating. Evidence for the importance of various groups of animals, notably microarthropods, to the decomposition process comes largely from exclusion experiments. Although experiments with litter in bags of various mesh sizes to exclude different groups of animals have led to equivocal results in some cases, litter-bag experiments can provide acceptable evidence of the importance of animal activities in decomposition (Harding and Stuttard, 1974). Naphthalene and DDT have also been used to exclude particular groups of animals from soil. The agents for litter fragmentation in temperate regions are mainly earthworms, diplopods, isopods, dipterous larvae, Collembola and oribatid mites (Lofty, 1974). For example, Witkamp and Crossly (1966) found that after one year only 45 per cent of litter treated with naphthalene to exclude most invertebrates was decomposed compared to 60 per cent in the controls; Von Perel *et al.* (Lofty, 1974) found that litter in bags that excluded earthworms decomposed 203 times slower than did controls.

Although micro-animals such as nematodes and protozoa abound in all kinds of litter-decomposing soils, e.g. nematodes in an oak forest litter, 15 g m^{-2} of $\sim 6 \times 10^6$ animals m^{-2} 20 cm deep (Twinn, 1974), their roles have not been as intensively studied as those of larger animals. Presumably their effects on decomposition comprise: (1) selective and rapid removal of various types of microorganisms from such communities; (2) removal of metabolites and by-products, which might inhibit decomposition if accumulated; (3) nutrient cycling, particularly releases of microbially bound N, S and P compounds; and (4) as reservoirs or packages of nutrients for larger consumers (carnivores) (Stout, 1974; Twinn, 1974).

Though Twinn (1974) is cautious in assigning an important role in the terrestrial decomposition process to nematodes, his suggestion that better definition of the nematode role might require parallel study of nematodes and microflora on the same material made us strive to complete the analysis by analogy with other ecosystems (e.g. Connell, 1970). Since soil nematodes are so abundant, might not they, by selective grazing, control numbers and types of microorganisms in litter-decomposing communities?

These small animals discussed are, in turn, often lower and intermediate steps in the food webs supporting the familiar terrestrial vertebrates.

5 Litter decomposition in aquatic habitats (particularly in the marine sublittoral)

The tremendous differences in aquatic habitats renders it surprising that the essential features of plant-litter decomposition in fresh waters and marine habitats are similar (reviews: Perkins, 1958; Ingold, 1971; Mann, 1972; Wiebe and Pomeroy, 1972; Barlocher and Kendrick, 1973; Willoughby, 1974; Jones, 1974; Berrie, 1976; Cook, 1976; Dick, 1976; Jones and Harrison, 1976; Newell, 1976; Saunders, 1977; Rich and Wetzel, 1978). Since the terrestrial sources of plant matter are the same whether decomposition takes place in soils or freshwater environments, senescent leaves and other plant structures enter the water with a flora of phylloplane-decomposing mycelial fungi, yeasts, actinomycetes and other bacteria, etc. The general events in soils occur also in benthic environments as soon as lotic litter becomes waterlogged and sinks. First there is rapid loss of soluble and easily mobilized plant materials either by leaching or by initial microbial action (e.g. Mann, 1972; Perkins, 1974; Willoughby, 1974). Though not often measured in this initial phase, the carbon to nitrogen ratio of the leaves and wood rises as nitrogeneous materials (e.g. nucleic acids, amino acids, and proteins) leach or are digested. Then, as colonization and microbial growth increases on freshly deposited litter, the nitrogen associated with their organic constituents lowers the carbon to nitrogen ratio. However, the recent data of Knauer and Ayers (1977) on decomposing *Thalassia testudinum* leaves does not seem to show this. Experimental evidence is sparse for most freshwater habitats but there is some. For example, Kaushik and Haynes (1968, 1971) used antifungal antibiotics to show that fungi are far more successful in initial colonization and decomposition of leaves than bacteria. Mason (Mason, 1976), working with the same techniques with *Phragmites communis* litter in a freshwater marsh, found that bacteria and fungi were about equally responsible for decomposing the *Phragmites* leaves. Fragmentation by whatever means (wind, wave, running stream, animal action) increases surfaces of the holocellulose, making it more accessible to attack, and speeds up the decomposition. Finally, as in soils, small animals feed upon the microorganisms. Examples that might be cited are stonefiles, chironomid larvae (Willoughby, 1974), amphipods (Marzolf, 1966; Hargrave, 1970; Barlocher and Kendrick, 1973, 1975a, b, 1976; Kostalos and Seymour, 1976), nematodes (Meyers *et al.*, 1964; Hopper and Meyers, 1966; Meyers and Hopper, 1967; Garlach, 1978) and ciliates (Fenchel and Jørgensen, 1977). There is evidence that grazing releases nutrients that can stimulate primary production of benthic algae (Cooper, 1973). The interesting experiments by Barlocher and Kendrick (1976) depict the influence of fungi on palatability of ash, maple and oak leaves for *Gammarus pseudolimnaeus*. They inoculated leaves with 10 different species of fungi and then placed notched identifiable leaves with mycelia of different species in dishes with the animal. Maple and oak leaves with *Anguillospora* mycelia were the most palatable for the *Gammarus*. *Humicola* mycelia rendered ash leaves most palatable. The three kinds of leaves were least palatable when *Tricladium* or *Tetracladium* grew on them.

The impression obtained from the published studies of litter decomposition in aquatic habitats is that many aspects of the process are obscure. Aquatic mycologists are beginning to find out which fungi grow on different substrates in different habitats (review: Jones, 1975). What underlies habitat selection? What promotes growth? How do they compete with each other and other microorganisms for substrates? How do they attack litter? What are the kinetics? Are they controlled by associated microflora? What factors limit development of the whole detrital community? What are the roles of animals, especially the numerous small ones, in aquatic decomposition and in detrital food webs? This last question is the focus of this discussion on detrital decomposition in shallow benthic marine environments.

The shoal benthic system, of which the detrital food web is an integral part, has few obvious or commonplace analogues. The organisms intimately associated with the process span more than 4 orders of magnitude in size ($\leqslant 0{\cdot}5\ \mu m - \geqslant 5$ cm). They are diverse, with the likely diversity of thousands of species in any one region and the possibility that perhaps hundreds of different species may be packed as densely as $\geqslant 1 \times 10^6$ organisms cm^{-3}. By almost every index employed, the organisms involved ($\sim 1\mu m$ to > 1 cm) are heterogeneously distributed (Lackey, 1961; Buzas, 1969; Lee *et al.*, 1969; Matera and Lee, 1972). As with soils and leaf-litter communities, structure is an important microhabitat determinant in the benthos. Physical factors such as circulation, particle grain size and sorting, and biological factors such as the growth of epistrate animals and plants, and grazing by meiofauna, are important in establishment, maintenance, and changes in an intricate mosaic of microhabitats.

The community is also functionally a mosaic; many organisms are interacting at different trophic levels at the same time. There are many good illustrations of this point. The amphipod, *Ampelisca abdita*, for instance, is both a detritus feeder and a deposit feeder as it uses its antennae and other appendages either to swirl up or grasp various types of food (Watling, 1975). Plant structural debris, associated microflora, micro- and meiofauna are ingested by this animal. An interesting quantitative approach to omnivory was recently proposed by Odum and Heald (1975). They rated animals numerically on the basis of identifiable gut contents. The herbivores, for instance, fell into three groups based upon the extent to which small animals were eaten. An animal was considered an herbivore if recognizable macro-animal material contributed < 30 per cent of the volume of the digestive tract contents. Omnivores and lower carnivores were also rated by similar logic. Our own tracer experiments suggest that Odum and Heald may have underestimated the quantity of meiofauna ingested by macro- and megafauna.

There seems little question that the trophic dynamics of the lower and intermediate steps in the detrital food web are especially complex. A vast diversity of microorganisms, microflora and fauna, and meiofauna intimately interwoven in time and space, collectively are the transformers of the plant structural materials into compounds utilizable by the animals at higher levels. Are all these organisms essential for the process ("species redundancy")? Why so many species? What adaptation do various species have for living in this extremely dynamic community? Are the dynamic changes in community structure essential for high productivity?

6 Models of the decomposition process

Links between the primary detritus-decomposing fungi and bacteria and other organisms in the marine detrital food web are not easily organized into conceptual schematic model diagrams. Odum and Heald (1975) made important advances, particularly in their recognition of a mixed trophic level. The detailed schematic conceptualizations in Coull's review (Coull, 1973) of meiofaunal–microbial interactions demonstrate how difficult it is to create a broadly encompassing scheme depicting key relationships at the lower and intermediate steps in the detrital food web.

Our model of relationships in the marine detrital food web reflects our interpretations of current evidence with the gaps filled in by extrapolation from other systems or by conjectures (Fig. 1). The close physical relationships of inhabitants of the microflora to each other and to the decomposing plant litter (Fig. 2) enter into our modelling even though they are not expressed in the diagram. We believe that materials released or mobilized by one organisms do not necessarily come into equilibrium with environmental pools or micro-pools. Thus, metabolites released by one organism may be taken up and turned over by organisms in juxtaposition without ever reaching easily detectable or equilibrium concentrations.

In shallow, relatively unpolluted benthic systems, there are usually three major external sources of energy: light, plant, litter and wind/wave/current/tidal action. When present, the physical actions on the habitat are important energy subsidies to the system, since they serve to grind and macerate plant litter and circulate materials. The contributions of the other two energy sources are very delicately interwoven.

Autolysis, leaching and microbial action quickly release the readily decomposed fractions of plant litter, leaving the more resistant holocellulose and lignins behind. The readily mobilized metabolities are taken up by bacteria, fungi, algae, protozoa and perhaps, to some extent, by those animals that can extract them from the environment. The standing

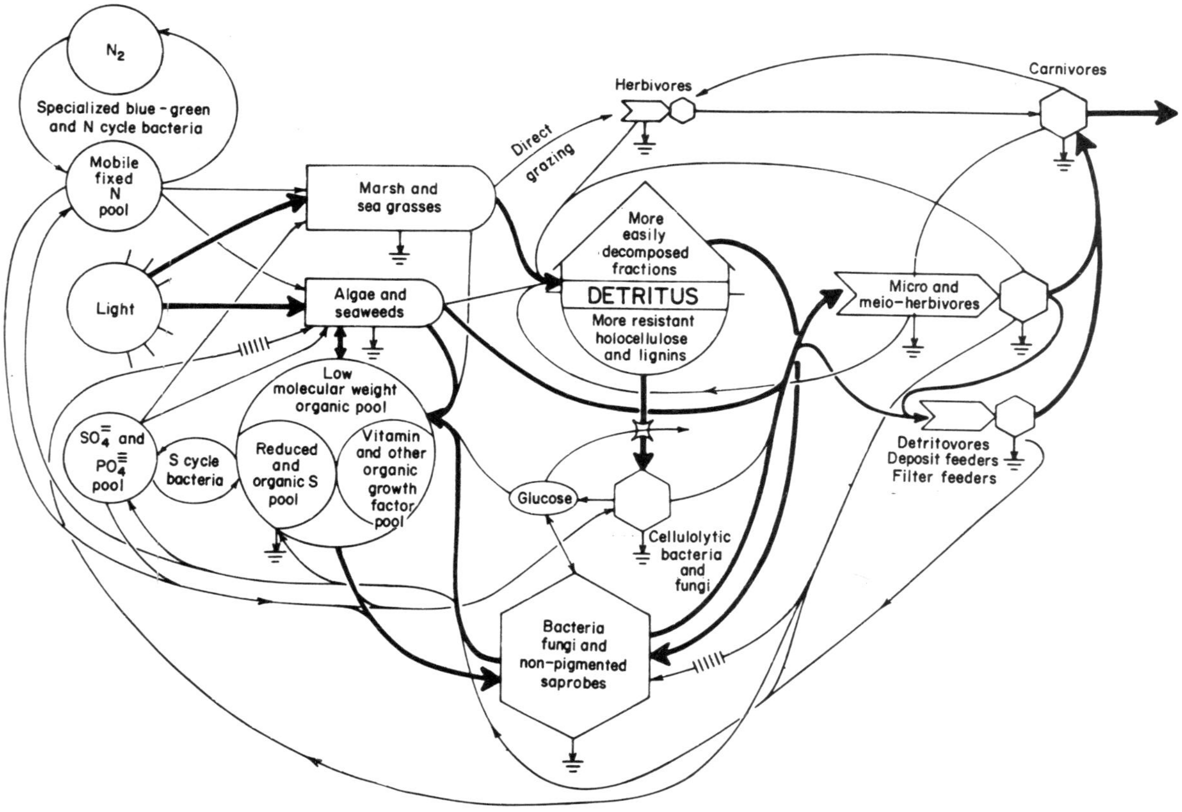

Fig. 1. A conceptual model of marine detrital decomposition and the organisms associated with the process.

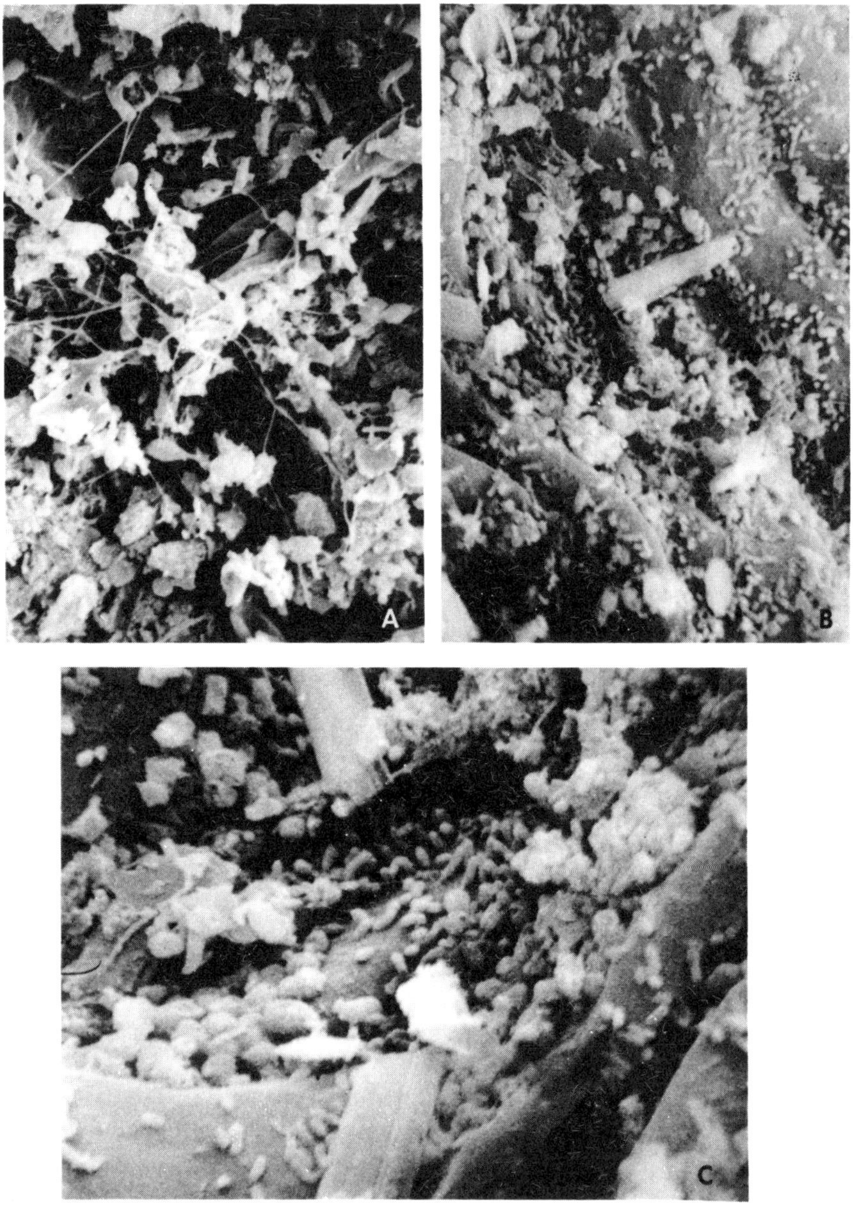

Fig. 2. SEM micrographs of typical surfaces of leaves of *Spartina alterniflora* decomposing on the sublittoral benthos in a US north eastern salt marsh. Note the intimate associations of the organisms to the substrate and to each other. (Photographs by Dr Marie E. McEnery, Department of Biology, City College of City University of New York.)

crop of beneficiaries are, in turn, consumed by various animals (detailed later). The microbial attack on holocellulose and lignins is a relatively slow process in the ecosystem. As noted, grinding and maceration are important in increasing the surface available for attack by C_1 endoglucanases. Formation and release of the microbial cellulases is sensitive to repression by even low levels of glucose, cellobiose, and other readily assimilated carbon sources. Representation of glucose in the key switching role in our model (Fig. 1) does not preclude important roles for other repressor or retro-inhibiting metabolities known for the cellulase system. It follows that the heterotrophic bacteria and algae can regulate the rate of decomposition of cellulose by keeping low the levels of low-molecular, readily assimilated, organic compounds. This aspect of the system is complex since many of the organisms release as well as take up metabolities. This is particularly true when the benthic microalgae are vigorously fixing carbon photosynthetically. Microbes can also control the growth rates of cellulolytic microorganisms by controlling the availability of key nutrients. The soluble nitrogen levels in the shallow benthos, for instance, are balanced by N_2 fixers and denitrifiers, as well as by organisms that remove, metabolize and/or excrete various forms of fixed nitrogen.

Many self-regulating mechanisms characteristically come into play in nutrient pools and associated microorganisms. In the N cycle—a good example—many of the microbial enzymes involved are induced or are repressed by ambient nutrient levels.

Several prominent factors shape the population structure of the microbial and microfloral assemblages:

1. initial colonization;
2. varying abilities of different species to be able to grow and reproduce on (or be inhibited by) the available energy sources—phosphorus, nitrogen and sulphur pools, micronutrients (vitamins, etc.);
3. physical conditions (i.e. temperature, pH, redox, salinity, light, etc.) and physiological responses of the organisms to these conditions;
4. indiscriminate or selective cropping of the microbial and microalgal assemblages.

The micro- and meiofauna in our scheme play the following roles:

1. they quantitatively and selectively crop the microbial and mic-

roalgal assemblages, thereby shaping the population structures of both assemblages, and also help keep both assemblages in relatively youthful and rapidly growing seres;
2. they regenerate macro- and micronutrients and act in mineralization processes;
3. they change the physical characteristics of their environment; and
4. they serve as large and qualitatively different packages of food for larger consumers.

The population structure of the micro- and meiofaunal assemblages is in turn shaped by:

1. their responses to the physical characteristics and nutrients available;
2. the population structure of the microbial and microalgal assemblage; and
3. their consumption by omnivores or carnivores.

In summary, key points of regulation of energy flow from the potential energy in holocellulose are:

1. Regulation of cellulolytic enzyme production and activities by end-product retroinhibition and repression— a phenomenon modulated by associated bacteria, fungi and algae.
2. Control of decomposer growth rates by:
    a. Regulation of nutrient availability and certain environmental gradients, phenomena modulated by the activities of associated bacteria, fungi, algae, micro- and meiofauna, and larger animals.
    b. Selective and non-selective cropping of decomposers by micro-, meio- and macrofauna which harvest or strip the decomposers from decomposing substrates, digest them, and release the remaining primary decomposers from auto-inhibiting growth restraints or free substrates for fresh colonization and rapid growth of new decomposers.

7 Some evidence for the model

The sketchiness of available evidence argues that our comprehensive scheme is premature. Yet some pattern is needed to incorporate seemingly disparate bits of information and to provide testable organizing hypotheses.

Some decomposer (in the holistic sense) communities are better understood than others. It is fairly well known, for instance, that heterotrophic capabilities are widespread among shallow-water benthic marine pennate diatoms. (Lee *et al.*, 1975; Hellebust and Lewin, 1977) while only a fraction of the carbon budget of many species seems satisfied by heterotrophy (Droop, 1974; Saks *et al.*, 1976). Some algal species seem highly adapted for the mixed trophic role that may be characteristic of the assemblage. Quite a variety of transport systems, both constitutive and adaptive, occur in different algal species. At least three different types of amino acids and several sugar and organic acid transport systems have been found in various diatoms. The systems vary greatly in the concentration ranges of their activities. (Hellebust and Lewin (1977) have reviewed the evidence). Many of the same species of algae seem to release a fraction of their photosynthetates as metabolities usable by other organisms in the community (Droop, 1974; Provasoli and Carlucci, 1974; Hellebust, 1977). The microphytes are often growing in juxtaposition to the degradative fungi and bacteria (Sieburth and Thomas, 1973). Undoubtedly many algal cells simultaneously withdraw metabolities from their surroundings, release other metabolites, and photosynthetically fix carbon. SEM studies of eelgrass indicated that the leaves are virtually carpeted by epiphytic diatoms as the leaves emerge in the spring (Sieburth and Thomas, 1973). One diatom, *Cocconeis scutellum,* actually impresses attachment pockets into the surface of the apparently soft and mucoid emerging leaves. They are also found on the surfaces of mature leaves in the fall. Surprisingly few bacteria were observed on the *Zostera* epithelium in the same study. Since the same pennate diatom species are also found in littoral and benthic piles of decaying *Zostera* and *Spartina,* there is reason to think that by virtue of their abundance on the leaves at time of abscission and their heterotrophic abilities, they indeed play the role suggested for them in our model. The relative importance of algae against bacteria as nutrient absorbers has been argued by some researchers (Wright and Hobbie, 1966; Wright and Shaw, 1975; Hellebust and Lewin, 1977). By one view, kinetics seem to favour bacterial uptake (Wright and Hobbie, 1966); evidence on the other side of the ledger is cited by Hellebust and Lewin (1977). Presumably both groups are active in the process, depending upon the stochastic, environmental and biotic interactions which have taken place. Interaction between algae and bacteria is well documented. Wright and Shaw (1975) found

⅔ of the 144 strains of bacteria they tested to take up a glycollic acid metabolite excreted by many algae. Less well known are the dynamic aspects of the activities of primary decomposers in the marine environment. Most of the work has been descriptive.

It appears, just as for rumen microflora, the polysaccharide slimes and pili are important for attachment of marine bacteria in detritus (e.g. Floodgate, 1972). Aside from Fenchel's experimental systems (Fenchel, 1973; Fenchel, 1977a, b), there is little other evidence that bacteria are early colonizers of those plant materials studied recently by new methods. If the actions of bacteria attached to plant fibres could be seen by scanning electron microscopy (SEM) and transmission electron microscopy (TEM) of rumen contents, there is little reason to believe that they would be missed in fine structure studies of marine materials. May's study (May, 1974) of *Spartina* by SEM is a good example. He found heavy infestation of fungal hyphae but relatively few bacteria. The Sieburth and Thomas (1973) paper was mentioned earlier in a different context. A note of caution: standing-drop measurements are often deceptive. Earlier work (Burkholder and Bornside, 1957; Odum and de la Cruz, 1967) suggested that bacteria are important in marsh-grass degradation. Bacteria, as is true with many small organisms, can have a very high reproductive and metabolic rate. By using Brock's autoradiographic technique (Brock and Brock, 1968), we found heavy tritiated thymidine labelling of yeast and bacteria on *Spartina* detritus incubated for 2 hours *in situ*. Only an occasional fungal mycelium was labelled by comparison. We can only infer from this spot check that bacterial and yeast reproductive rates were high in the detritus. Rather than being ecosystem process-oriented, mycological studies of *Spartina* detritus have dwelled largely on floristics (Johnson, 1956; Goodman, 1959; Webster and Lucas, 1961; earlier literature reviewed in Johnson and Sparrow, 1961; Jones, 1962a, b, 1963; Lloyd and Wilson, 1962; Dennis, 1968; Ellis, 1971; Kohlmeyer and Kohlmeyer, 1971; Johnson, 1977). Gessner and Goos (1973) related fungal colonies to loss of *Spartina* biomass from litter bags. Besides *Spartina* litter, fungi are also numerous in *Spartina* salt-marsh muds (Cowley, 1973; Pugh, 1974). Mangrove litter mycology has been given considerable attention by various researchers from the Institute of Marine Science of the University of Miami (excellent reviews are Fell and Master, 1973; Newell, 1976). These identify more than 60 genera of fungi associated with mangrove decay system.

Epifluorescence microscopy and SEM make it easier to count microorganisms colonizing detritus. According to one estimate the bacterial density on the surface of detritus can attain $5 \times 10^8 - 10^{10}$ bacteria g^{-1} (dry weight) detritus (Fenchel and Jørgensen, 1977). In various laboratory experiments with leached, dried and fragmented plant materials, Fenchel (1973, 1977a, b) has found a characteristic pattern of initial bacterial colonization of sterilized substrates 6–8 hours after incubation in sea or fresh water. The bacteria reproduced slowly (generation time 10–20 h) for 407 days before reaching stationary phase. During log growth $^{14}CO_2$ from the tagged plant substrate was released as the decomposition proceeded in almost direct proportion to increasing microbial population. Small non-pigmented flagellates appeared in the cultures soon after the bacteria and peaked when the bacterial populations stabilized. Ciliates apppeared later. The flagellate and ciliate populations were $\sim 5 \times 10^7 - 10^8$ g^{-1} (dry weight) detritus. In contrast with earlier evidence Sieburth and Thomas (1973), Fenchel and Jørgensen (1977) believe that diatoms, rhizopods, and small metazoans are late colonizers and that the role of the zooflagellates and ciliates in the decomposition process is to absorb nutrients and consume bacteria. The rate of consumption of bacteria by one ciliate. *Tetrahymena pyriformis*, varied from ~ 500–600 bacteria/individual/hour in log phase to 150 bacteria/individual/hour in stationary phase. Another role usually assigned to this group of small bacteriovores—recyling nutrients like phosphorus released from consumed bacteria (Johannes, 1965, 1968; Pomeroy, 1970)—has been questioned in the light of recent experimental evidence obtained by Fenchel and co-workers (Barsdate *et al.*, 1974; Fenchel and Jørgensen, 1977). The original assumption was that decomposition slows down because nutrients become locked up in the bacteria. Grazing animals release nutrients and stimulate bacterial growth. The challenge to this idea is based on Johannes' approach, where bacteria were grown in batch culture on easily utilized substrates and transferred to vessels in which the bacteria were deprived of carbon sources—a condition Fenchel believes does not stimulate a steady-state environmental situation. Their experimental microcosms (Barsdate *et al.*, 1974; Fenchel and Harrison, 1976; Fenchel, 1977a, b) are more natural. Pulse-labelling with ^{32}P showed rapid turnover of the bacterial P pool even in microcosms with bacteria alone. In one experiment ^{32}P was turned over hourly by the bacteria while the turnover of ^{32}P due to ciliate grazing was estimated to be only 5 per cent of the P cycling of the entire experimental system.

Some investigators suggest that protozoan grazing controls bacterial densities (Bick, 1964; Fenchel, 1969; Curds, 1973; Barsdate *et al.*, 1974; Fenchel and Harrison, 1976). Bacterial populations in experimental detrital decomposing systems are typically 2–5 times higher in ungrazed systems than in systems with flagellates and ciliates (Fenchel and Jørgensen, 1977). At room temperature these protozoa turn over the bacterial populations every 10–30 hours. From their experience and familiarity with the literature Fenchel and Jørgensen (1977) felt that protozoans are the most important bacterial grazers in the sediments, and also suggest that meiofauna may be important as bacterial consumers but that the evidence is still weak.

Another perspective on ciliate grazing was suggested by measurements of grazing and growth rates of the benthic *aufwuchs* hypotrich *Euplotes vannus* in the presence of two other ubiquitous and equally hardy small herbivores, *Chromadorina germanica* (nematode) and *Allogromia laticollaris* (foraminifera) (Muller and Lee, 1977).

8 The particular role of meiofauna and slightly larger animals in the decomposition process and the detrital food web

A logical first question one might ask is: Is there evidence that meiofauna have any effect on the decomposition process? Good evidence seems to come from our own studies and those of our colleagues (Lee *et al.*, 1975; Tenore, 1977; Tenore *et al.*, 1977).

In experiments with mixtures of meiofauna and the polychaetes *Capitella capitata* or *Nephthys incisa*, decomposition (oxidation) of detritus derived from eelgrass, *Zostera marina*, was up to three times faster in the presence of meiofauna plus microflora than with microflora alone or with microflora plus either polychaete. Furthermore, rate of incorporation of ^{14}C from the detritus by *Nephthys incisa* doubled when meiofauna was present relative to those in identical vessels but lacking meiofauna (Tenore *et al.*, 1977). There were no differences in cultures with *Capitella;* it did not seem to consume the meiofauna.

How do (or could) meiofauna affect the decomposition process? In the absence of any demonstration that any of the meiofauna in the detrital community has cellulases or internal microbial fermentation systems, we have two principle hypotheses, neither excluding the other.

1. Meiofauna, in concert with microfauna and slightly larger amphipods, crop significant numbers of the actual cellulose digesting microorganisms. By so doing they remove inhibitors which slow the activities of the microbial community.

   a. The above hypothesis could be made stronger: the large populations of meiofauna, by their selective and heavy grazing pressures, determine the population structure of the microbial community and thereby become an important factor in regulation of rates of detrital decomposition.

2. Meiofauna in concert with larger metazoa significantly alter water sediment and detritus characteristics by their feeding, metabolic and locomotor activities, which fragment, mix and aerate their habitats, and release and cycle nutrients. These provide microfloral environmental factors more conducive to detrital decomposition. What evidence supports either hypothesis?

Barsdate *et al.* (1974) found, in an experimental model detrital decomposition ecosystem, that mineralization of detritus was stimulated by protozoan grazing. Regardless of the controversy raised on cause or effect (Fenchel and Jørgensen, 1977), phosphorus cycling was faster in the presence of the grazing protozoa. Although criticized (Fenchel and Jørgensen, 1977), Johannes' idea (Johannes, 1968) that "nematodes (and perhaps other meiobenthic organisms) may share with the protozoa a major role in benthic nutrient regeneration" is still attractive. The evidence obtained by Gerlach (1978) also supports this view.

Experiments with the detritivorous amphipod *Orchestia grillus* on *Spartina alterniflora* litter (Lopez *et al.*, 1977) seem to support the mineral-recycling concepts articulated by Johannes (1965) and Pomeroy *et al.* (1972). *O. grillus* efficiently fed upon the microorganisms attached to the ingested *Spartina* litter, but did not digest the litter itself. Microbial activity was accelerated by grazing. Lopez *et al.* (1977) concluded that the observed increases in microbial activity may depend on ammonia excretion and higher diffusion rates due to the animals' movement, but is mainly a direct response to grazing. To my knowledge, there are no published data on nutrient regeneration by nematodes or other marine meiofauna.

As for the mechanical activities of the meiobenthos, there aren't many experimental data. Cullen (1973) and Lee *et al.* (1975) suggest that meiofauna might be important in effecting small-scale sedimentological processes. In sheltered areas where detritus accumulates, meiobenthic bioturbation may be important. Conceivably, the meiofauna may, by their locomotory activities, help keep detrital surfaces open to microbial colonization. This interesting possibility needs investigation.

Inferential evidence supports our hypothesis that meiofauna, by selective heavy grazing, might alter the population structure of the microbial community (Alexander, 1964). The mechanisms by which meiofauna differentially feed and digest are not easily categorized. However, the following categories seem reasonable:

1. Morphological adaptations that allow selection of food organisms according to size or toughness.
2. Differential digestive abilities, i.e. everything eaten is not digested.
3. Differential or seasonal utilization of resources by species that, for physiological reasons, are not always present in the population.
4. Morphological adaptations to exploit prey spatially separated in different microhabitats.
5. Special attraction for certain species of meiofauna to particular species of microflora (Lee *et al.*, 1977). In the case of relatively sessile meiofauna (i.e. formaminifera) special attraction of particular food species to the animal, e.g. "Circean effect" (Lee and Pierce, 1963; Lee, 1974).
6. Presence of growth-promoters or inhibitors (molecular information) in different species of microflora that differentially stimulate, encourage, permit, slow or block growth, reproduction or fecundity of meiofaunal species.
7. Ability of some small animals to adjust their behaviour or their metabolic processes to handle particular food.

Small animals can be very efficient in harvesting microorganisms from detritus-size particles. In an experiment with ^{14}C-labelled bacteria, Fenchel *et al.* (1975) showed that mouth parts of the shallow-water burrowing amphipod *Corphidium voluator* were adapted for removing bacteria from clay or silt particles. The animal was unable to feed on the labelled bacteria in the absence of particles.

Some evidence for resource partitioning or differential ingestion of food because of differences in buccal morphology has been described (Wieser, 1953; Richman and Rogers, 1969; Yates, 1971; Boucher, 1973; Twinn, 1974; Fenchel, 1975; Ivester and Coull, 1975, 1977; Nassogne, 1977; Tietjen and Lee, 1977). Differences in food particle size and shape have been suggested as a means whereby food also may be differentially ingested by different micrometazoan species (Ivester and Coull, 1975; Tietjen and Lee, 1977). Protozoa also have been observed to ingest differentially food particles of different sizes (Droop, 1966).

The role of particle size in food ingestion has been examined for suspension feeders (Mullin, 1963; Jørgensen, 1966; Richman and Rogers, 1969; Frost, 1972; Lemm and Frost, 1976; Nival and Nival, 1976). Far less information is available on particle-size selection by benthic deposit feeders (Fenchel *et. al.*, 1975; Hyllenberg, 1975). Such differential ingestion could have pronounced effects on remineralization rates of detritus.

The relationships among nematode feeding habits, diets, digestive enzymes and structure of digestive tracts have been examined. Jennings and Deutsch (1975) found the glycosidase, β-glucuronidase, acted both extra- and intracellularly during digestion of bacteria by *Monhystera denticulata*, a nematode that ingests bacteria and small algae (Tietjen and Lee, 1977). The enzyme was not present in *Chromadorina germanica*, which does not ingest bacteria. β-glucuronidase hydrolyses various components of mucopolysaccharides, and may be important in digesting the capsules and slime layers of bacteria (Jennings and Deutsch, 1975). In an overall fine-structure study of ingestion and digestion of the nematodes *Diplolaimella* sp. and *C. germanica*, Deutsch-Levy (1977) found that the two ate differently. *C. germanica* used teeth to pierce food and powerful muscles to suck in food which was digested intracellularly. *Diplolaimella* sp., in contrast, ate small algal and bacterial cells intact and digested the food enzymatically in the intestinal lumen. *Diplolaimella* produced luminal carbohydrases, presumably for digesting bacterial and algal cell envelopes; *C. germanica* lacked them. Both worms produced esterases but lacked exopeptidases, characteristic enzymes of carnovires.

Small marine herbivores and bacteriovores often grow at different rates of diets of similar sized food organisms (Provasoli *et al.*, 1959; Mullin, 1963; Droop, 1966; Muller and Lee, 1969; Lee and Muller, 1973; Muller, 1975; Lee *et al.*, 1976a; Pappenhofer, 1976; Rubin and Lee, 1976; Tietjen and Lee, 1977). This should not be surprising since bacterial and algal cells differ in chemical composition (Schliefer and Kandle, 1972), and since algal cells vary from species to species (Lewin, 1962; Chuecas and Tiley, 1969). What is surprising are the great ranges of ecological efficiencies associated with different diets. The data on two species of marine ciliates, *Uronema marinum* and *Euplotes vannus*, suggested differences on similar diets as large as 2–21 per cent (Rubin and Lee, 1976).

Molecular signals built into food organisms are "recognized" by the ciliate's genetic and metabolic machinery, which alters metabolic processing if it has the genetic capabilities to do so. Since energy is needed for the above processes, an organism can save energy if it can reduce the cost of converting ingested molecules into its own structure, i.e. given an animal's nutritional requirements, the closer its food meets these requirements the less energy is needed for biosynthesis. Thus food quality may be an important aspect of the energy-transformations in all ecosystems.

9 Meiofauna as food for larger organisms

There is an expanding literature on the consumption of meiofauna by members of the meiofaunal group itself and by larger organisms. Predatory species are known to occur in the following meiofaunal groups: Hydrozoa, Turbellaria, Nemertini, Nematoda, Tardigrada, Polychaeta and Halacarida (McIntyre, 1969, 1971; McIntyre and Murison, 1973). The influence of meiofaunal predators on the meiofauna itself is hard to estimate. Suggestions have been made that meiobenthic hydroids (Muus, 1966; Christensen, 1967; Nassogne, 1970; Heip and Smol, 1975) and turbellarians (Bilio, 1967; Coull and Dudley, 1976) may be especially important in regulating the numbers of their chief prey, which appear to be nematodes, oligochaetes and harpacticoid copepods. Predatory nematodes may also be important, although less is known of their role (McIntyre, 1969).

As for predation on meiofauna by larger organisms, there has been much speculation but few hard data exist. Gerlach (1978) has recently estimated that meiofauna and foraminifera contribute ~30 and 12 per cent of the living biomass in the sediment and that they represent approximately the same percentage of the diets of deposit feeding macrofauna. Juvenile flatfish are active predators on meiofauna and consume very large numbers of harpacticoid copepods, oligochaetes, ostracods and nematodes (Smidt, 1951; Bregnballe, 1961; Macer, 1967; Edwards and Steele, 1968; McIntyre, 1969; Nassogne, 1970). *Gobius* spp. have also been observed with harpacticoid copepods, nematodes, and foraminifera in their guts (Smidt, 1951; Tenore, Skidaway Institute, Georgia, personal communication). Harpacticoids are important food of juvenile salmonids (Kaczinski *et al.*, 1973; Feller and Kaczinski, 1975).

Lassere *et al.* (1976) observed that the mullet, *Chelon labiosus*, significantly depends on ingestion of harpacticoid copepods, nematodes, ostracods, oligochaetes, polychaetes, and other meiobenthic organisms. Of equal or greater interest, they observed an apparent inhibition of mullet growth by the meiofauna. In enclosed fish reservoires in the Bassin d'Arcachon, France, mullet growth is sustained by detritus. In summer, when meiofauna is abundant, mullet growth is low. They suggested that in summer the high densities of meiofauna outcompete mullet for detritus. Since the meiofauna in the Bassin have high metabolic rates, appreciable detrital potential energy may be diverted from mullet by the meiofaunal community respiration. McIntyre (1964) and Marshall (1970) reached similar conclusions. Among the invertebrate predators, a voracious eater of copepods, nematodes, ostracods and turbellarians is *Nereis diversicolor* (Rees, 1940; Perkins, 1958). Tietjen (1969) observed a decline in nematodes in later summer in two New England estuaries coincident with appearance of nematodes in the guts of juvenile nereid polychaetes. Other invertebrate predators of meiofauna include *Crangon crangon* (Smidt, 1951; Gerlach and Schrage, 1969) and crabs of the genus *Uca* (Teal, 1962; McIntyre, 1968). Warner and Woodley (1975) have observed benthic foraminifera in the guts of the ophiuroid *Ophiothrix fragilis*, and suggest that the organism may be feeding on benthic detritus. Tietjen (1971) mentioned that deep-sea ophiuroids might feed on meiofauna; the feeding mechanism of ophiuroids indeed makes such cropping of meiofauna plausible (Hyman, 1955). McIntyre (1964) and Marshall (1970) arrive at similar conclusions for macrobenthic filter feeders and meiofauna.

10 Concluding remarks

It is clear from the examination of the evidence and the construction of a conceptual model that, although we are presently able to identify the key assemblages of organisms and critical process that seem to be associated with decomposition in the shallow marine benthos, many aspects of the entire system need clarification. The untidy and complex interrelationships between the organisms involved constitute real challenges to our ingenuity and imagination. Judging from the rate of recent progress and development of many fresh approaches, the answers to the main questions raised in our analysis can be eagerly anticipated.

Acknowledgements

The author is appreciative of support from both the Department of Energy (E11-1 3254-37) and the National Science Foundation (GA 33388) who have contributed to the research on which this review is based. Much of the work was done in collaboration with Dr John H. Tietjen.

References

Alexander, M. (1964). Biochemical ecology of soil microorganisms. *Annual Reviews of Microbiology*, **18**, 217–252.
Alexander, M. (1971). "Microbial Ecology". John Wiley, New York.
Atlavinyte, O. and Pociene, D. (1973). The effect of earthworms and their activity on the amount of algae in the soil. *Pedociologica*, **13**, 445–455.
Bancroft, K. Paul, E. A. and Wiebe, W. J. (1976). The extraction and measurement of adenosine triphosphate from marine sediments. *Limnology and Oceanography*, **21**, 473–480.
Barlocher, F. and Kendrick, B. (1973). Fungi and food preferences of *Gammarus pseudolimnaeus*. *Archiv für Hydrobiologie*, **72**, 501–516.
Barlocher, F. and Kendrick, B. (1975a). Leaf conditioning by microorganisms. *Oecologia (Berlin)*, **20**, 359–362.
Barlocher, F. and Kendrick, B. (1975b). Assimilation efficiency of *Gammarus pseudolimnaeus* (Amphipoda) feeding on fungal mycelium or autumn-shed leaves. *Oikos*, **26**, 55–59.
Barlocher, F. and Kendrich, B. (1976). Hyphomycetes as intermediaries of energy flow in steams. *In* "Recent Advances in Aquatic Mycology" (Ed. E. B. Gareth Jones). John Wiley, New York.
Barsdate, R. J., Prentki, R. T. and Fenchel, T. (1974). Phosphorus cycle of model ecosystems: significance for decomposer food chains and effect of bacterial grazers. *Oikos*, **25**, 239–251.
Beguin, P., Eisen, H. and Roupas, A. (1977). Free and cellulose-bound cellulases in a *Cellulomonas* species. *Journal of General Microbiology*, **101**, 191–196.
Bell, M. K. (1974). Decomposition of herbaceous litter. *In* "Biology of Plant Litter Decomposition" (Eds C. H. Dickinson and G. J. F. Pugh), vol. I, pp. 37–67. Academic Press, London and New York.

Berg, B., von Hofsten, B. and Pettersson, G. (1972). Growth and cellulase formation by *Cellvibrio fulvus*. *Journal of Applied Microbiology*, **35**, 201–204.
Berrie, A. D. (1976). Detritus, micro-organisms and animals in fresh water. *In* "The Role of Terrestrial and Aquatic Organisms in Decomposition Processes" (Eds J. M. Anderson and A. Macfadyen), pp. 301–322. Blackwell Scientific Publications, Oxford.
Bick, H. (1964). "Die Sukzession der Organismen bei der Selbstreinigung von organisch unrunreinigtem Wasser unter verschiedenen Milieubedingungen". Düsseldorf.
Bilio, M. (1967). Hahrungsbeziehungen der Turbellarien in Kustensalzwiesen. *Helgoländer wissenschaftliche Meeresuntersuchungen*, **15**, 602–621.
Boucher, G. (1973). Premieres donnes ecologiques sur les nematodes libres marins d'une station de vose cotiere de Banyuls. *Vie et Milieu*, **23**, 69–100.
Bregnballe, F. (1961). Plaice and flounder as consumers of the microscopic bottom fauna. *Meddelelser fra Danmarks Fiskeri- og Havundersøgelser, København*, **3**, 133–182.
Brock, T. D. (1966). "Principles of Microbial Ecology". Prentice Hall, Englewood Cliff, New Jersey.
Brock, T. D. and Brock, M. L. (1968). Measurement of steady-state growth rates of a thermophilic alga directly in nature. *Journal of Bacteriology*, **95**, 811–915.
Burkholder, P. R. and Bornside, G. H. (1957). Decomposition of marsh grass by aerobic marine bacteria. *Bulletin of the Torrey Botanical Club*, **74**, 366–383.
Buzas, M. A. (1969). Foraminiferal species densities and environmental variables in an estuary. *Limnology and Oceanography*, **14**, 411–422.
Cheng, K. J., Akin, D. E. and Costerton, J. W. (1977). Rumen bacteria: interaction with particulate dietary components and response to dietary variation. *Federation Proceedings*, **36**, 193–197.
Christensen, H. E. (1967). Ecology of *Hydractinia echinata* (Fleming) (Hydroidea, Athecata). I. Feeding biology. *Ophelia*, **4**, 245–275.
Chuecas, L. and Riley, J. P. (1969). Component fatty acids of the total lipids of some marine phytoplankton. *Journal of the Marine Biological Association of the United Kingdom*, **49**, 97–116.
Clark, F. E. (1973). Problems and perspective in microbial ecology. *In* "Modern Methods in the Study of Microbial Ecology" (Ed. T. Rosswall), pp. 13–16. Uppsala, Sweden.
Connell, J. H. (1970). A predator–prey system in the marine intertidal region. I. *Balanus gladual* and several predatory species of *Thais*. *Ecological Monographs*, **40**, 49–78.
Cooke, W. B. (1976). Fungi in sewage. *In* "Recent Advances in Aquatic Mycology" (Ed. E. B. G. Jones), pp. 513–542. Halsted Press, New York.
Cooper, D. C. (1973). Enhancement of net primary productivity by herbivore grazing in aquatic laboratory microcosms. *Limnology and Oceanography*, **18**, 31–37.
Coull, B. C. (1973). Estuarine meiofauna: a review, trophic relationships and microbial interactions. *In* "Belle W. Baruch Library in Marine Science", Vol. 1: "Estuarine Microbial Ecology" (Eds L. H. Stevenson and R. R. Colwell), pp. 499–511. South Carolina Press, Columbia.
Coull, B. C. and Dudley, B. W. (1976). Delayed naupliar development of meiobenthic copepods. *Biological Bulletin of the Woods Hole Oceanographic Institution*, **150**, 38–46.
Cowley, G. T. (1973). Variations in soil fungus populations in a South Carolina salt marsh. *In* "Estuarine Microbial Ecology" (Eds L. H. Stevenson and R. R. Colwell), pp. 441–454. South Carolina Press, Columbia.
Cullen, D. J. (1973). Bioturbation of superficial marine sediments by interstitial meiobenthos. *Nature, London*, **242**, 323–324.

Curds, C. R. (1973). The role of protozoa in the activated-sludge process. *American Zoologist,* **13**, 161–169.
Darnell, R. M. (1976a). Organic detritus in relation to the estuarine ecosystem. *In* "Estuaries" (Ed. G. H. Lauff). American Association for the Advancement of Science, Publication 83, Washington, D.C.
Darnell, R. M. (1967b). The organic detritus problem. *In* "Estuaries" (Ed. G. H. Lauff). American Association for the Advancement of Science, Publication 83, Washington, D.C.
Dennis, R. W. (1968). "British Ascomycetes". J. Cramer, Lehre, Germany.
Deutsch-Levy, A. (1977). Gut structure and digestive physiology of the free-living marine nematodes, *Chromadorina germanica* (Butschli, 1864) and *Diplolaimella* sp. Ph.D. dissertation. City University of New York.
Dick, M. W. (1976). The ecology of aquatic phycomycetes. *In* "Recent Advances in Aquatic Mycology", pp. 513–542. Halsted Press, New York.
Droop, M. (1966). The role of algae in the nutrition of *Heteramoeba clara* Droop, with notes on *Oxyrrhis marina* Dujardin and *Philodina rolseola* Ehrenberg. *In* "Some Contemporary Studies in Marine Science" (Ed. H. Barnes), pp. 225–231. George Allen and Unwin, London.
Droop, M. R. (1974). Heterotrophy of carbon. *In* "Algal Physiology and Biochemistry" (Ed. W. D. P. Stewart), pp. 530–559. Blackwell, Oxford.
Edwards, R. R. C. and Steele, J. H. (1968). The ecology of O-group plaice and common dabs at Loch Ewe. I. Population and food. *Journal of Experimental Marine Biology and Ecology*, **2**, 215–238.
Eklund, E. and Gyllenberg, H. (1973). Bacteria. *In* "Biology of Plant Litter Decomposition" (Eds. C. H. Dickinson and G. L. Pugh), vol. 2, pp. 245–268. Academic Press, New York and London.
Ellis, M. B. (1971). "Dematiaceous Hyphomycetes". Commonwealth Mycological Institute, Kew, Surrey, England.
Fell, J. W. and Master, I. M. (1973). Fungi association with degradation of mangrove (*Rhizophora mangle* L.) leaves in South Florida. *In* "Estuarine Microbial Ecology" (Eds H. L. Stevenson and R. R. Colwell) pp. 445–466. University of South Carolina Press, Columbia, S. C.
Feller, R. J. and Kaczynski, V. W. (1975). Size selective predation by juvenile chum salmon (*Oncorhynchus keta*) on epibenthic prey in Puget Sound. *Journal of the Fisheris Research Board of Canada*, **32**, 1419–1429.
Fenchel, T. (1969). The ecology of marine microbenthos. IV. Structure and function of the benthic ecosystem, its chemical and physical factors and the microfauna communities with special reference to the ciliated protozoa. *Ophelia*, **6**, 1–182.
Fenchel, T. (1973). Aspects of the decomposition of seagrasses. Review for the Decomposition Working Group at the International Seagrass Workshop, Leiden, Netherlands. Oct. 22–26, 1973.
Fenchel, T. (1975). The quantitative importance of the benthic microfauna of an arctic tundra pond. *Hydrobiologia*, **46**, 445–464.
Fenchel, T. (1977a). Competition, coexistence and character displacement in mud snails "Hydrobiidae"). *In* "Ecology of Marine Benthos" (Ed. B. C. Coull), pp. 229–243. University of South Carolina Press, Columbia.
Fenchel, T. (1977b). The significance of bacterivorous protozoa in the microbial community of detrital particles. *In* "Freshwater Microbial Communities". 2nd Edition (Ed. J. Cairns). Garland Publishing Company, New York.
Fenchel, T. and Harrison, P. (1976). The significance of bacterial grazing and mineral cycling for the decompositon of particulate detritus. *In* "The Role of Terrestrial and

Aquatic Organisms in Decomposition Processes" (Ed. J. M. Anderson), pp. 285–299. Blackwell Scientific, Oxford.

Fenchel, T. and Jørgensen, B. (1977). Detritus food chains of aquatic ecosystem: the role of bacteria. *In* "Advances in Microbial Ecology" (Ed. M. Alexander), pp. 1–58. Plenum Press, New York.

Fenchel, R., Kofoed, L. H. and Lappalainen, A. (1975). Particle size selection of two deposit feeders: the amphipod, *Corophium volutator* and the prosobranch, *Hydrobia ulvae*. *Marine Biology,* **30**, 119–128.

Flett, R. J., Hamilton, R. D. and Campbell, N. E. R. (1976). Aquatic acetylene-reduction techniques: solutions to several problems. *Canadian Journal of Microbiology,* **22**, 43–51.

Floodgate, G. D. (1972). The mechanism of bacterial attachment to detritus in aquatic systems. *Memorie dell'Institut italiano di Idrobiologia,* **29** (Suppl.), 309–323.

Frost, B. W. (1972). Effects of size and concentration of food particles on the feeding behaviour of the marine planktonic copepod, *Calanus pacificus*. *Limnology and Oceanography,* **17**, 85–815.

Gerlach, S. A. (1978). Food chain relationships in subtidal silty and marine sediments and the role of meiofauna in stimulating bacterial productivity. *Oecologia (Berlin),* **33**, 55–69.

Gerlach, S. A. and Schrage, M. (1969). Freilibende Nematodes als Nahrung der Sandgarnele *Crangon crangon*. *Oecologia,* **2**, 362–375.

Gessner, R. V. and Goos, R. D. (1973). Fungi from decomposing *Spartina alterniflora*. *Canadian Journal of Botany,* **51**, 51–55.

Goodfellow, M. and Cross, T. (1974). Actinomycetes. *In* "Biology of Plant Litter Decomposition" (Eds C. H. Dickinson and G. J. F. Pugh), pp. 269–302. Academic Press, New York and London.

Goodman, P. J. (1959). The possible role of pathogenic fungi in die-back of *Spartina townsendii* agg. *Transactions of the British Mycological Society,* **42**, 409–415.

Gough, S. B. and Woelkerling, W. J. (1976). On the removal and quantification of algal aufwuchs from macrophyte hosts. *Hydrobiologia,* **48**, 203–207.

Harding, D. J. L. and Stuttard, R. A. (1974). Microarthropods. *In* "Biology of Plant Litter Decomposition" (Eds C. H. Dickinson and G. J. F. Pugh), vol. 2, pp. 489–532. Academic Press, New York and London.

Hargrave, B. T. (1970). The utilization of benthic microflora by *Hyalella azteca* (Amphipoda). *Journal of Animal Ecology,* **39**, 427–532.

Harley, J. L. (1971). Fungi in ecosystems. *Journal of Animal Ecology,* **40**, 1–16.

Heald, E. J. (1969). The production of organic detritus in a South Florida estuary. Dissertation, University of Miami, Florida.

Hellebust, J. A. (1974). Extracellular products. *In* "Algal Physiology and Biochemistry" (Ed. W. D. P. Stewart), pp. 838–863. Blackwell, Oxford.

Hellebust, J. A. and Lewin, J. (1977). Hetertrophic nutrition. *In* "The Biology of Diatoms" (Ed. D. Werner), pp. 169–197. Blackwell, Oxford.

Heip, C. and Smol, N. (1975). On the importance of *Protohydra leuckarti* as a predator of meiobenthic populations. *In* "Proceedings of the 10th European Symposium on Marine Biology" (Eds G. Persoone and E. Jaspers), vol. 2, pp. 285–296. Universal Press, Wesseren, Belgium.

Holm-Hansen, O. (1973). Determination of total microbial biomass by measurement of adenosine triphosphate. *In* "Estuarine Microbial Ecology" (Eds L. H. Stevenson and R. R. Colwell), pp. 73–89. University of South Carolina Press, Columbia.

Honigberg, B. M. (1967). Chemistry of parasitism among some protozoa. *In* "Chemical

Zoology. I. Protozoa" (Ed. G. Kidder), pp. 695–802, Academic Press, New York and London.

Hoppe, H. G. (1976). Determination and properties of actively metabolizing heter-trophic bacteria in the sea, investigated by means of microautoradiography. *Marine Biology*, **36**, 291–302.

Hopper, B. and Meyers, S. (1966). Aspects of the life cycles of marine nematodes. *Helgoländer wissenschaftliche Meeresunterscuhungen*, **13**, 444–449.

Hungate, R. E. (1966). "The Rumen and Its Microbes". Academic Press, New York and London.

Hungate, R. E. (1975). The rumen microbial exosystem. *Annual Reviews of Ecological Systems*, **6**, 39–66.

Hungate, R. E., Bryant, M. P. and Mah, R. A. (1964). The rumen bacteria and protoza. *Annual Reviews of Microbiology*, **18**, 131–166.

Hyllenberg, J. (1975). Selective feeding by *Abarenicola pacifica* with notes on *Abarenicola vagabunda* and a concept of gardening in lugworms. *Ophelia*, **14**, 113–137.

Hymen, L. H. (1955). "The Invertebrates". vol. 4. Echinodermata. McGraw Hill, New York.

Ingold, C. T. (1971). "Fungal Spores, their Liberations and Dispersal". Clarendon Press, Oxford.

Ivester, M. S. and Coull, B. C. (1975). Comparative study of ultrastructural morphology and some mouthparts of four hastoriid amphipods. *Canadian Journal of Zoology*, **53**, 408–417.

Ivester, M. S. and Coull, B. C. (1977). Niche fraction studies of two sympatric species of *Enhydrosoma* (Copepoda, Harpacticoida). *Microfauna, Meeresboden*, **61**, 137–152.

Jennings, J. B. and Deutsch, A. (1975). Occurrence and possible adaptive significance of β-glucuronidase and arylamidase (leucine aminopeptidase) activity in two species of marine nematodes. *Comparative Biochemistry and Physiology*, **52A**, 611–614.

Jensen, V. (1974). Decomposition and angiosperm tree leaf litter. *In* "Biology of Plant Litter Decomposition" (Eds C. H. Dickinson and G. J. F. Pugh), vol. 1, pp. 37–104. Academic Press, New York and London.

Johannes, R. E. (1965). The influence of marine protozoa on nutrient regeneration. *Limnology and Oceanography*, **12**, 189–195.

Johannes, R. E. (1968). Nutrient regeneration in lakes and oceans. *In* "Advances in Microbiology of the Sea" (Eds M. R. Droop and E. J. Ferguson Wood), vol. 1, pp. 203–213. Academic Press, London and New York.

Johnson, R. G. (1977). Vertical variation in particulate matter in the upper twenty centimetres of marine sediments. *Journal of Marine Research*, **35**, 273–282.

Johnson, T. W. (1956). Marine fungi. I. *Leptosphaeria* and *Pleospora*. *Mycologia*, **48**, 495–505.

Johnson, T. W. and Sparrow, F. K. (1961). "Fungi in Oceans and Estuaries". J. Cramer, Weinhein, Germany.

Jones, E. B. (1962a). Marine fungi. *Transactions of the British Mycological Society*, **45**, 93–114.

Jones, E. B. (1962b). *Haligena spartinae* sp. nov., a Pyrenomycete on Spartina townsendii. *Transactions of the British Phycological Society*, **45**, 245–248.

Jones, E. B. (1963). Marine fungi. II. Ascomycetes and Deuteromycetes from submerged wood and drift *Spartina*. *Transactions of the British Mycological Society*, **46**, 135–144.

Jones, E. B. G. (1974). Aquatic fungi: freshwater and marine. *In* "Biology of Plant Litter Decomposition" (Eds C. H. Dickinson and G. J. F. Pugh), pp. 337–383. Academic Press, New York and London.

Jones, E. B. G. and Harrison, J. L. (1976). Physiology of marine phycomycetes. *In* "Recent Advances in Aquatic Mycology" (Ed. E. B. G. Jones), pp. 261–278. Halsted Press, New York.

Jørgensen, B. (1966). "Biology of Suspension Feeding". Pergamon Press, New York.

Kaczynski, V. W., Feller, R. J., Calyton, J. and Cerke, R. J. (1973). Trophic analysis of juvenile pink and chum salmon (*Oncorhynchus gorbuscha* and *O. keta*) in Puget Sound. *Journal of the Fisheries Research Board of Canada*, **30**, 1003–1008.

Kaushik, N. K. and Haynes, H. B. N. (1968). Experimental study on the role of autumn-shed leaves in aquatic environments. *Journal of Ecology*, **56**, 229–245.

Kaushik, N. K. and Haynes, H. B. (1971). The fate of dead leaves that fall into streams. *Archiv für Hydrobiologie*, **68**, 465–575.

Knauer, G. A. and Ayers, A. V. (1977). Changes in carbon, nitrogen, adenosine triphosphate and chlorophyll-a in decomposing *Thalassia testudinum* leaves. *Limnology and Oceanography*, **22**, 408–414.

Kohlmeyer, J. and Kohlmeyer, E. (1971). "Synoptic Plates of Higher Marine Fungi". Verlag von J. Cramer, Germany.

Kostalos, M. and Seymour, R. L. (1976). Role of microbial enriched detritus in the nutrition of *Gammarus minus* (Amphipoda). *Oikos*, **27**, 512–516.

Lackey, J. B. (1961). Bottom sampling and environmental niches. *Limnology and Oceanography*, **6**, 271–279.

Lam, R. K. and Frost, B. W. (1976). Model of copepod filtering response to changes in size and concentration of food. *Limnology and Oceanography*, **21**, 490–500.

Lasserre, P., Renaud-Mornant, J. and Castel, J. (1976). Metabolic activities of meiofaunal communities in a semi-enclosed lagoon. Possibilities of trophic competition between meiofauna and mugilid fish. *In* "Proceedings of the 10th European Symposium on Marine Biology" (Eds G. Persoone and E. Jaspers), vol. 2, pp. 393–414. Universal Press, Wetteren, Belgium.

Lee, J. J. (1974). Towards understanding the niche of foraminifera. *In* "Foraminifera" (Eds R. H. Hedley and G. Adams), vol. 1, pp. 208–257. Academic Press, New York and London.

Lee, J. J. and Muller, W. A. (1973). Trophic dynamics and niches of salt marsh foraminifera. *American Zoologist*, **13**, 215–223.

Lee, J. J. and Pierce, S. (1963). Growth and physiology of foraminifera in the laboratory. Part 4—monoxenic culture of an allogromiid with notes on its morphology. *Journal of Protozoology*, **19**, 404–411.

Lee, J. J., Muller, W. A., Stone, R. J., McEenery, M. and Zucker, W. (1969). Standing crop of foraminifera in sublittoral epiphytic communities of a Long Island salt marsh. *Journal of Marine Biology*, **4**, 44–61.

Lee, J. J., Tietjen, J. H. and Garrison, J. R. (1976a). Seasonal switching in the nutritional requirements of *Nitocra typica* Boeck, an harpacticoid copepod from salt marsh aufuwuchs communities. *Transactions of the American Microscopical Society*, **96**, 628–637.

Lee, J. J., McEnery, M., Kennedy, E. and Rubin, H. (1975). A nutritional analysis of a sublittoral epiphytic diatom assemblage from a Long Island salt marsh. *Journal of Phycology*, **11**, 14–49.

Lee, J. J., Tietjen, J. H., Mastropaolo, C. and Rubin, H. (1977). Food quality and the heterogeneous spatial distribution of meiofauna. *Helgoländer wissenschaftliche Meeresuntersuchungen*, **30**, 280–282.

Lewin, J. C. (1962). Silicification. *In* "Physiology and Biochemistry of Algae" (Ed. R. A. Lewin), pp. 445–466. Academic Press, New York and London.

Lloyd, L. S. and Wilson, I. M. (1962). Development of the perithecium in *Lulworthia medusa* (Ell. and Ev.) Cribb and Cribb, a saprophyte on *Spartina townsendii*. *Transactions of the British Mycological Society*, **45**, 359–373.

Lofty, J. R. (1974). Oligochaetes. *In* "Biology of Plant Litter Decomposition" (Eds C. H. Dickinson and G. J. F. Pugh) vol. 2, pp. 467–488. Academic Press, New York and London.

Lopez, G. R., Levinton and Slobodkin, L. B. (1977). The effect of grazing by the detritivore *Orchestia grillus* on *Spartina* litter and its associated microbial community. *Oecologia*, **30**, 111–127.

Macer, C. T. (1967). The food web in Red Wharf Bay (North Wales), with particular reference to young plaice (*Pleuronectes platessa*). *Helgoländer wissenschaftliche Meeresuntersuchungen*, **15**, 560–573.

Mandels, M. (1975). Microbial sources of cellulases. *Biotechnology and Bioengineering Symposium*, **5**, 81–105.

Mann, K. H. (1972). Macroscopic production and detritus food chains in coastal areas. *In* "IBP-UNESCO Symposium on Detritus and its Ecological Role in Aquatic Ecosystems" (Supplement 29), pp. 353–382. Pallanza, Italy, Mem. Inst. Ital. Hydrobiol.

Margalef, R. (1968). "Perspectives in Ecological Theory". University of Chicago Press, London.

Marzolf, G. R. (1966). The trophic position of bacteria and their relation to the distribution of invertebrates. *Special Publication Pymatunig Laboratory of Ecology, University of Pittsburgh*, **4**, 131–135.

Marshall, N. (1970). Food transfer through the lower trophic levels of the marine benthic environment. *In* "Marine Food Chains" (Ed. J. H. Steele), pp. 52–56. Oliver and Boyd, New York.

Mason, C. F. (1976). Relative importance of fungi and bacteria in the decomposition of *Phragmites* leaves. *Hydrobiologia*, **51**, 65–69.

Matera, N. J. and Lee, J. J. (1972). Environmental factors affecting the standing crop of foraminifera in sublittoral and psammolittoral communities of a Long Island salt marsh. *Limnology and Oceanography*, **17**, 89–103.

May, M. S. (1974). Probable agents for the formation of detritus from the halophyte, *Spartina alterniflora*. *In* "Ecology of Halophytes" (Eds R. J. Reimold and W. H. Queen), pp. 429–447. Academic Press, New York and London.

McHugh, J. L. (1976). Estuarine fisheries: are they doomed? *In* "Estuarine Processes" (Ed. M. Wiley), vol. 1, pp. 15–27. Academic Press, New York and London.

McIntyre, A. D. (1964). Meiobenthos of sublittoral muds. *Journal of the Marine Biological Association of the United Kindgom*, **44**, 605–674.

McIntyre, A. D. (1968). The meiofauna and macrofauna of some tropical beaches. *Journal of Zoology, London*, **156**, 377–392.

McIntyre, A. D. (1969). Ecology of marine meiobenthos. *Biological Reviews*, **44**, 245–290.

McIntyre, A. D. (1971). Control factors on meiofaunal populations. *Thallasia Jugoslavica*, **7**, 209–215.

McIntyre, A. D. and Murison, D. J. (1973). The meiofauna of a flatfish nursery ground. *Journal of the Marine Biological Association of the United Kingdom*, **53**, 93–118.

Meyers, S. P. and Hopper, B. E. (1967). Studies on marine fungal-nematode associations and plant degradation. *Helgoländer wissenschaftliche Meeresuntersuchungen*, **15**, 270–281.

Meyes, S. P., Kamp, K. M., Johnson, R. F. and Shaffer, D. L. (1964). Thalassiomycetes. IV. Analysis of variance of ascospores of the genus *Lulworthia*. *Canadian Journal of Botany*, **42**, 519–526.

Millar, C. S. (1974). Decomposition of coniferous leaf litter. *In* "Biology of Plant Litter Decomposition" (Eds C. H. Dickinson and G. J. F. Pugh), vol. 1, pp. 105–128. Academic Press, New York and London.

Moriarty, D. J. W. (1975). A method of estimating the biomass of bacteria in aquatic sediments and its application to trophic studies. *Oecologia*, **20**, 219–229.

Moriarty, D. J. W. (1977). Improved method using muramic acid to estimate biomass of bacteria in sediments. *Oecologia*, **26**, 317–323.

Muller, W. A. (1975). Competition for food and other niche-related studies of three species of salt marsh foraminifera. *Marine Biology*, **31**, 339–351.

Muller, W. A. and Lee, J. J. (1969). Apparent indispensability of bacteria in foraminiferan nutrition. *Journal of Protozoology*, **16**, 471–478.

Muller, W. A. and Lee, J. J. (1977). Biological interactions and the realized niche of *Euplotes vannus* from the salt marsh aufwuchs. *Journal of Protozoology*, **24**, 523–527.

Mullin, M. M. (1963). Some factors affecting the feeding of marine copepods of the genus *Calanus*. *Limnology and Oceanography*, **8**, 239–250.

Muus, B. J. (1966). Notes on the biology of *Protohydra leukartia* Greef (Hydroidea, Protohydridae). *Ophelia*, **3**, 141–150.

Nassogne, A. (1970). Influence of food organisms on the development and culture of pelagic copepods. *Helgoländer wissenschaftliche Meeresuntersuchungen*, **20**, 333–345.

Newell, S. Y. (1976). Mangrove fungi: the succession in the mycoflora of red mangrove (*Rhizophora mangle* L) seedlings. *In* "Recent Advances in Aquatic Mycology" (Ed. E. B. G. Jones). pp. 51–91. Halsted Press, New York.

Nival, P. and Nival, S. (1976). Particle retention efficiencies of an herbivorous copepod, *Acartia clausi* (adult and copepodite stages): effects on grazing. *Limnology and Oceanography*, **21**, 24–38.

Odum, E. P. and de la Cruz, A. (1967). Particulate organic detritus in a Georgia salt marsh—estuarine ecosystem. *In* "Estuaries" (Ed. G. H. Lauff), pp. 383–388. American Association for the Advancement of Science. Publication No. 83. Washington, D.C.

Odum, W. E. and Heald, E. J. (1975). The detritus-based food web of an estuarine mangrove community. *In* "Estuarine Research" (Ed. L. E. Cronin), pp. 265–286. Academic Press, New York and London.

Owens, F. N. and Isaacson, H. R. (1977). Ruminal microbial yields: factors influencing synthesis and bypass. *Federation Proceedings*, **36**, 198–202.

Paffenhofer, G. A. (1976). Feeding, growth and food conversion of the marine planktonic copepod *Calanus helgolandicus*. *Limnology and Oceanography*, **21**, 39–50.

Parkinson, D. (1973). Techniques for the study of soil fungi. *In* "Modern Methods in the Study of Microbial Ecology" (Ed. K. Rosswall), pp. 2–36. Uppsala, Sweden.

Perkins, E. J. (1958). The food relationships of the microbenthos, with particular reference to that found at Whitstable, Kent. *Annals and Magazine of Natural History*, Ser. B, **13**, 64–77.

Perkins, E. J. (1974). The marine environment. *In* "Biology of Plant Litter Decomposition" (Eds C. H. Dickinson and G. J. F. Pugh), vol. 2, pp. 683–721. Academic Press, New York and London.

Pomeroy, L. R. (1970). The strategy of mineral cycling. *Annual Reviews of Ecological Systems*, **1**, 171–190.

Pomeroy, L. R., Shenton, R., Jones, H. and Reimoid, R. (1972). Nutrient flux in estuaries. *In* "Nutrients and Eutrophication" (Eds G. E. Likens and J. E. Hobbie). A.S.L.O. Special Publication, Lawrence, Kansas.

Provasoli, L. and Carlucci, A. F. (1974). Vitamins and growth regulators. *In* "Algal Physiology and Biochemistry" (Ed. W. D. P. Stewart), pp. 741–787. Blackwell, Oxford.

Provasoli, L., Shiraishi, K. and Lance, J. R. (1959). Nutritional idiosyncracies of *Artemia* and *Tigriopus* in monoxenic culture. *Annals of the New York Academy of Sciences,* **77**, 256–281.

Pugh, G. J. F. (1974). Terrestrial fungi. *In* "Biology of Plant Litter Decompositon" (Eds. C. H. Dickinson and G. J. F. Pugh), vol. 2, pp. 303–336. Academic Press, New York and London.

Rees, C. B. (1940). A preliminary study of the ecology of a mud flat. *Journal of the Marine Biological Association of the United Kingdom,* **24**, 158–199.

Rich, P. and Wetzel, R. (1978). Detritus in the lake ecosystem. *American Naturalist,* **112**, 57–70.

Richman, S. and Rogers, J. N. (1969). The feeding of *Calanus helgolandicus* on synchronously growing populations of the marine diatom *Ditylum brightwellii. Limnology and Oceanography,* **14**, 701–709.

Rubin, H. A. and Lee, J. J. (1976). Informational energy flow as an aspect of the ecological efficiency of marine ciliates. *Journal of Theoretical Biology,* **62**, 69–91.

Russell-Hunter, W. D. (1970). "Aquatic Productivity". Macmillan, New York.

Ryther, J. H. (1969). Relationship of phytosynthesis to fish production in the sea. *Science, New York,* **166**, 72–76.

Saks, N. M., Stone, R. J. and Lee, J. E. (1976). Autotrophic and heterotrophic nutritional budgets of salt marsh epiphytic algae. *Journal of Phycology,* **12**, 443–448.

Satchell, J. E. (1974). Litter-interface of animate-inanimate matter. *In* "Biology of Plant Litter Decomposition" (Eds C. H. Dickinson and G. J. F. Pugh), vol. 1, pp. xii–xliv. Academic Press, New York and London.

Saunders, G. W. (1977). Carbon flow in the aquatic system. *In* "Aquatic Microbial Communities" (Ed. J. Cairns, Jr.), pp. 417–440. Garland Publishing, New York.

Schliefer, K. H. and Kandle, O. (1972). Peptidoglycan types of bacterial cell walls and their taxonomic implications. *Bacteriological Reviews,* **36**, 407–477.

Sieburth, J. and Thomas, C. (1973). Fouling on eelgrass (*Zostera marine* L.). *Journal of Phycology,* **9**, 46–50.

Smidt, E. L. B. (1951). Animal production in the Danish Waddensae. *Medd. Fra. Kommis for Danmarks Fisk-og Havunders. Ser: Fiskeri,* **11**, 151.

Sternberg, D. (1976). *β*-glucosidase of *Trichoderma*: its biosynthesis and role in saccharification of cellulose. *Applied Environmental Microbiology,* **31**, 648–654.

Stotsky, G. (1973). Techniques to study interactions between micro-organisms and day minerals *in vivo* and *in vitro*. *In* "Modern Methods in the Study of Microbial Ecology" (Ed. J. Rosswall), pp. 17–28. Uppsala, Sweden.

Stout, J. D. (1974). Protozoa. *In* "Biology of Plant Litter Decomposition" (Eds C. H. Dickinson and G. J. F. Pugh), vol. 2, pp. 385–420. Academic Press, New York and London.

Teal, J. M. (1962). Energy flow in the salt marsh ecosystem of Georgia. *Ecology,* **43**, 614–624.

Tenore, K. R. (1977). Food chain pathways in detrital feeding benthic communities: a review, with new observations on sediment resuspension and detrital recycling. *In* "Ecology of Marine Benthos" (Ed. B. C. Coull), pp. 37–53. University of South Carolina Press, Columbia.

Tenore, K. R., Tietjen, J. H. and Lee, J. J. (1977). Effect of meiofauna on incorporation of aged eelgrass, *Zostera marina* detritus by the polychaete *Nephthys incisa. Journal of the Fisheries Research Board of Canada*, **34**, 563–567.

Thayer, G. W., Adams, S. M. and La Croiz, M. W. (1975). Structural and functional aspects of a recently established *Zostera marina* community. *In* "Estuarine Research" (Ed. L. E. Cronin), vol. 1, pp. 518–540. Chemistry, Biology and the Estuarine System. Academic Press, New York and London.

Tietjen, J. H. (1969). The ecology of shallow water meiofauna in two New England estuaries. *Oecologia*, **2**, 251–291.

Tietjen, J. H. (1971). Ecology and distribution of deep sea meiobenthos off North Carolina. *Deep Sea Research*, **18**, 941–957.

Tietjen, J. H. and Lee, J. J. (1977). Feeding behaviour of marine nematodes. *In* "Ecology of Marine Benthos" (Ed. B. C. Coull), pp. 22–36. University of South Carolina Press, Columbia.

Twinn, D. C. (1974). Nematodes. *In* "Biology of Plant Litter Decomposition" (Eds C. H. Dickinson and G. J. F. Pugh), vol. 2, pp. 421–465. Academic Press, New York and London.

Waid, J. S., Preston, K. J. and Harris, P. J. (1973). Autoradiographic techniques to detect active microbial cells in natural habitats. *In* "Modern Methods in the Study of Microbial Ecology" (Ed. T. Rosswall), pp. 317–322. Uppsala, Sweden.

Warner, G. F. and Woodley, J. D. (1975). Suspension feeding in the brittle-star, *Ophiothrix fragilis. Journal of the Marine Biological Association of the United Kingdom*, **55**, 199–210.

Watling, L. (1975). Analysis of structural variations in a shallow estuarine deposit feeding community. *Journal of the Experimental Marine Biology and Ecology*, **19**, 275–313.

Webster, J. and Lucas, M. (1961). Observations on British species of *Pleospora. Transactions of the British Mycological Society*, **44**, 417–436.

Wiebe, W. J. and Bancroft, K. (1975). Use of the adenylate energy charge ratio to measure growth state of natural microbial communities. *Proceedings of the National Academy of Sciences, U.S.A.*, **72**, 2112–2115.

Wiebe, W. J. and Pomeroy, L. R. (1972). Microorganisms and their assocation with aggregates and detritus in the sea: a microscopic study. *Memorie dell'Instituto italiano di Idrobiologia*, **29** (Suppl.), 325–352.

Wieser, W. (1953). Die Beziehung zwischen Mundhohlengestalt, Ernahrungsweise und Vorkommen bei freilebenden marininen Nematoden. *Arkiv für Zoologie*, **4**, 493–484.

Willoughby, L. G. (1974). Decomposition of litter in fresh water. *In* "Biology of Plant Litter Decomposition" (Eds C. H. Dickinson and G. J. F. Pugh), vol. 2, pp. 659–681. Academic Press, New York and London.

Wittcamp, M. and Crossley, D. A. (1966). The role of arthropods and microflora in breakdown of white oak litter. *Pedobiologia*, **6**, 293–303.

Wright, R. T. and Hobbie, J. E. (1966). The use of glucose and acetate by bacteria and algae in aquatic ecosystem. *Ecology*, **47**, 447–464.

Wright, R. T. and Shaw, N. M. (1975). The trophic role of glycolic acid in coastal seawater. I. Heterotrophic metabolism in seawater and bacterial cultures. *Marine Biology*, **33**, 175–186.

Yamane, K., Suzuki, H., Hirotani, M., Ozawa, H. and Nixizama, K. (1970). Effect of nature and supply of carbon sources of cellulase formation in *Pseudomonas fluorescens cellulosa. Journal of Biochemistry, Tokyo*, **67**, 9–18.

Yates, G. W. (1971). Feeding types and feeding groups in plant and soil nematodes. *Pedobiology*, **11**, 173–179.

Zimmerman, R. and Meyer-Reil, L. A. (1974). A new method for fluorescence of bacteria populations on membrane filters. *Kieler Meeresforschungen*, **30**, 24–27.

Furunculosis of fish—the present state of our knowledge

D. H. McCARTHY* and R. J. ROBERTS†

*MAFF Fish Disease Laboratory, Weymouth, Dorset, England
†Unit of Aquatic Pathobiology, University of Stirling, Scotland

*Present address: Tavolek Inc., 2779 152nd Avenue, N.E., Redmond, Washington, USA.

1 Introduction

To the pathologist engaged in the study of diseases of the higher vertebrates the term furunculosis is a misnomer for the condition of salmonid fish to which it was first applied. The necrotic ulcerated swellings from which the term derives are not typical of the furuncle or "boil" of the mammal and they are not a consistent or even a common feature of the variety of clinical diseases with which *Aeromonas salmonicida* is associated. This problem of nomenclature is further complicated by the subsequent finding that the microorganism is also a frequent pathogen of non-salmonid species, where the clinical picture may be very different. The term furunculosis is, however, now very firmly entrenched in the literature so that it is retained in the present review specifically for those *Aeromonas salmonicida* infections of salmonids in which the furuncle is extant, although the purview of its remit is considered to cover all of the diseases of teleost fish where *Aeromonas salmonicida* is implicated in the pathogenesis. The bibliography of furunculosis has already been well reviewed by McGraw (1952) with updating by Herman (1968) so the present study will not attempt to restate these but to present the most recent information on the bacteriology and pathology of *Aeromonas salmonicida* in the context of earlier studies, in an effort to provide a unified base for future research.

2 History

It is not known whether *Aeromonas salmonicida* infection is a relatively recent disease or has evolved with its host. The first definitive isolation of the microorganism was from brown trout (*Salmo trutta* L.) by Emmerich and Weibel in 1894 but there are a number of other, less reliable, reports from which the probability of furunculosis may be adduced but for which inadequate definition of the bacterial isolates exist (e.g. Forel, 1868; Fabré-Domergue, 1890; Fischel and Enoch, 1892; Charnin, 1893; Bataillon and Dubard, 1895).

The progressive record of diagnoses of the condition subsequent to the initial isolation and description by Emmerich and Weibel has suggested the possibility of its spreading from a central locus in Europe but this may be a reflection on the progressive spread of competent bacteriological diagnostic facilities as much as an actual progressive dissemination of the agent. Initially the disease as described by Emmerich and Weibel was considered to be solely a hatchery disease but Marianne Plehn, in a series of investigations between 1901 and 1911, showed that the disease was widely prevalent in Bavarian trout streams. Thereafter there was a series of references to the discovery of the condition in France, Austria, Belgium, Switzerland, Great Britain and Ireland (Fuhrman, 1909; Pittel, 1910; Surback, 1911; Masterman and Arkwright, 1911; Mettam, 1914). It is now also known to occur in salmonids in Bulgaria, Czechoslovakia, Denmark, Finland, Hungary, Italy, Poland, Sweden, USSR and Yugoslavia (Herman, 1968) and it is likely that it is even more widespread.

Its occurrence in the Americas was first shown by Marsh (1902) who considered it to be the cause of extensive losses in Michigan hatcheries. It was not reported from wild salmonids, however, until Duff and Stewart (1933) described it from wild Dolly Varden (*Salvelinus malma:* Walbaum) and cutthroat (*Salmo clarki:* Richardson) trout in British Columbia. It was first isolated from wild fish in the USA by Fish (1937) who demonstrated its existence in a wild self-propagating stock of brown trout in Wyoming. Subsequently it has been found from almost all parts of the United States and Canada.

There has not as yet been any indication that *Aeromonas salmonicida* infection occurs in salmonids in Australasia. Efforts to isolate it have been made by several workers but it appears to be absent from that continent (S. F. Snieszko, personal communication). Nevertheless, the disease is a problem in salmonid fishes in Japan and extensive disease epizootics in amago (*Oncorhynchus rhodurus:* Jordan and McGregor) have been described (Miyazaki and Kubota, 1975).

The possibilities of *A. salmonicida* infections in marine and estuarine conditions was studied by Scott (1968) who used hosts experimentally killed by *Aeromonas salmonicida* infection to disseminate the infection to her experimental fish. She showed that *Aeromonas salmonicida* could readily be transmitted from fish to fish even in full sea water. These experimental observations were confirmed by the clinical studies on haemorrhagic septicaemic mortalities in cultured Atlantic salmon (*Salmo salar* L.) by Håstein (1975) who described a condition distinct from the more commonly encountered vibriosis, caused by a Gram-negative bacterium, which was first considered to be a member of the genus *Pasteurella* but is now known to be an atypical strain of *A. salmonicida* (Bullock *et al.*, in preparation). Evelyn (1971) has also described the condition in the marine species *Anoplopoma fimbria* Pallas (the sable fish).

The apparent predisposition of *A. salmonicida* for salmonids may be simply an indication that this is the group on which the greatest volume of research has been carried out and because they are usually the most valuable species in developed countries which have good diagnostic bacteriological facilities. Certainly the range of non-salmonid fish species from which the bacterium has been isolated is wide and is spread across the class Osteichthyes. Indeed, there is one reference (Hall, 1963) to the bacterium causing mortality in captive petromyzonts.

The generally recognized but infrequently diagnosed post-traumatic septicaemia condition of *Aeromonas salmonicida* infection in non-salmonids was previously thought to be the only clinical entity associated with this bacterium in non-salmonids, but recently work by Fijan (1972) and Bootsma *et al.* (1977) has suggested that the clinically distinctive, and economically highly significant, condition of farmed cyprinids known as Carp Erythrodermatitis (CE) is also usually induced by the pathogenic activities of an atypical *Aeromonas salmonicida*. The precise relationship will become clearer as further investigations are carried out on a wider basis.

The Scandinavian workers Håstein *et al.* (1978) and Ljungberg (personal communication) have now extended the range of species for which detailed studies of naturally occurring *Aeromonas salmonicida* infections are available with their reports on the minnow (*Phoxinus phoxinus* L.) and the Northern pike (*Esox lucius* L.) respectively. Again atypical varieties of the microorganism were isolated.

3 Clinical and gross pathological features of *Aeromonas salmonicida* infection

Although the classical features of furunculosis are well recognized as darkening and lethargy of the affected fish, inappetence and the development of large swellings on the body surface, which ulcerate to release their sero-sanguinous fluid content into the water, there are a number of other syndromes in salmonids associated with *Aeromonas salmonicida* which are less frequently diagnosed because of their less specific clinical picture. The different clinical types fall well into the standard categories of peracute, acute, subacute and chronic.

3.1 PERACUTE FURUNCULOSIS

This condition is usually restricted to fingerling fish, especially young

Salmo salar in hatcheries fed with water-containing carrier wild fish. Affected fish are darker in colour, often tachybranchic, and may die rapidly with little more than slight exophthalmos or, if they survive slightly longer, may show small semilunar haemorrhages at the pectoral fin base, or a dark blotch between the dorsal and pelvic fins. Losses in farmed fish may be very high (Plehn, 1911; Davis, 1946; Roberts, unpublished).

The gross pathology of such infections is typical of any peracute septicaemia—injected vessels, and punctate haemorrhages in the parietal and visceral peritoneum, and over the myocardium. There may also be focal haemorrhages over the gills, which are congested perimortem.

3.2 ACUTE FURUNCULOSIS

Acute furunculosis is commonly found in growing fish and in adults. The clinical signs may include the development of furuncles or merely the standard features of an acute bacterial septicaemia, i.e. darkening in colour, inappetence, lethargy, tachybranchia and small haemorrhages at the base of the fins (Fig. 1). Affected fish usually die within two or three days and although any furuncles may rupture before death to release highly infective material this does not always occur, and differential diagnosis on a purely clinical basis is unreliable, due to the inconstancy of furuncle development.

At postmortem in clinically acute outbreaks there are usually all the signs of haemorrhagic septicaemia. Hyperaemia or, terminally, congestion (Davis, 1946) of all serosal surfaces with punctate or larger

Fig. 1. Acute furunculosis in a rainbow trout.

haemorrhages scattered over the abdominal walls and viscera and the heart are found, and the kidney tissue may be soft and friable, or liquefied. The spleen is invariably enlarged, with rounded edges, and is usually cherry red in colour (Davis, 1946) while the liver is generally pale and may have subcapsular haemorrhages or a mottled appearance due to focal necrosis of the underlying parenchymatous tissue. Although there is usually a blood-stained inflammatory exudate within the abdomen, it is rarely fibrinous and adhesions are not normally a feature of the acute condition.

The stomach and intestine are usually empty of food and the lumen may contain a congery of sloughed epithelial lining cells, mucus, and blood, which may be expressed from the vent (Plehn, 1911). The wall of the swim bladder is hyperaemic and clouded, and it may also have part of its internal lining sloughed into the lumen. Where this sloughing takes place in the area of the pneumatic duct, blockage leading to gaseous distension of the swim bladder (bloat) may occur.

Skin lesions, where they occur, may be haemorrhagic patches, or blotches, along the side of the body, or, more typically, raised umbonate furuncles. The furuncle, however, usually develops within the dermis and not the hypodermis and, in the marine environment, acute furunculosis can generally be differentiated from the more common vibriosis of salmonids by the lesser degree of myotomal involvement and the slightly more superficial nature of the lesion.

The marine form of furunculosis, described by Håstein (1975) and now known to be caused by atypical strains of *Aeromonas salmonicida*, is usually acute and the presenting signs are again depression, inappetence and darkening colour although these are usually accompanied by the development of small shallow haemorrhagic ulcers along the side of the body and haemorrhages along the base of the pectoral fins. Furunculosis associated with typical strains of the microorganism can also occur in salt water, when fish carrying it are transferred and the stress induces clinical disease (Needham, personal communication).

Furunculosis in the amago in salt water, which has been well described by Miyazaki and Kubota (1975) has as its most common clinical feature in growing fish marked petechiation of the gill.

3.3 SUBACUTE AND CHRONIC FURUNCULOSIS

Subacute and chronic furunculosis are more common in older fish and clinically they may show a lesser degree of darkening in colour, and inappetence, than in the acute condition. However, they are generally lethargic and have one or more, obvious, furuncles on the flank or dorsum. They usually have congested blood vessels at the base of fins, injection of the sclera and slight exophthalmia. The gills may be pale or congested and sero-sanguinous fluid can often be expressed from the nares and the vent. The furuncles may be large and when ruptured the fluid is more viscous and contains more formed, necrotic elements than the furuncle found in acute cases. Affected fish may live for some considerable time, and indeed may recover, although usually they eventually succumb (Fig. 2).

At postmortem, the main feature of the chronic furunculosis case is general visceral congestion and peritonitis. There is usually a small amount of sero-sanguinous fluid within the pericardial and abdominal cavities, and adhesions between viscera and between viscus and abdominal peritoneum are common. There may be haemorrhages, often of some size, over the pyloric area or the liver, and the spleen is swollen and rounded in outline. The kidney is soft and friable and there is often a large necrotic focus around the pneumatic duct and/or around the vent. The liver is usually clay coloured, and its surface may be mottled with foci of subcapsular parenchymatous necrosis. The digestive tract is usually empty at death, or else contains a small amount of necrotic epithelial debris and mucus, but it is not normally so severely affected as in the acute condition. Foci of myonecrosis may

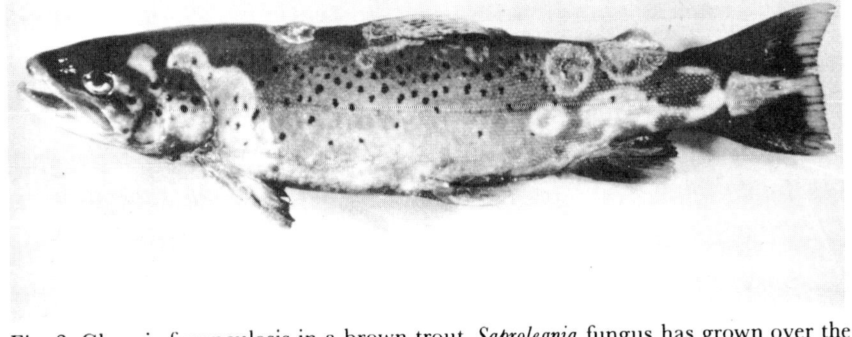

Fig. 2. Chronic furunculosis in a brown trout. *Saprolegnia* fungus has grown over the surface of some of the furuncles.

be found in the abdominal musculature or within the myotomal muscle but most commonly the furuncles in the skin will extend down to the muscle and the contents of the furuncle will include necrotic myofibrils as well as fibrin, fluid, inflammatory exudate and blood cell elements. The furuncle of the chronic disease is usually considered archetypal—large, umbonate, frequently leaking infectious exudate via a small ulcer at the apex, while beneath, the bulla containing largely formed, thick necrotic material is divided into several cavernous sinuses, sometimes extending deep into the muscle via the inter-myotomal fascial planes (Arkwright, 1912; Mackie *et al.*, 1935; Davis, 1946).

4 Histopathology

4.1 PERACUTE FURUNCULOSIS

Death of young fish from peracute furunculosis is often so rapid that suitable specimens for histopathological preparation may not be available. Where moribund specimens are available for fixation, the sections usually reveal the presence of small colonies of bacteria in the branchial mesenchyme, in the myocardium, in the anterior kidney and the spleen (Fig. 3). There is usually no inflammatory infiltrate of any sort and only a limited localized necrosis of tissue. However, in many cases there is considerable distension of the phagocytic cells lining the atrium (Ferguson, 1975) which may contain individual microorganisms or small colonies. Usually there is concommitant atrial myocardial necrosis and often severe pericardial inflammatory distension. It is probable that the cardiac damage is responsible for deaths in most cases, since the generalized toxic damage found in longer standing cases is usually not found.

4.2 ACUTE FURUNCULOSIS

Acute furunculosis can usually be distinguished from subacute and chronic furunculosis by the severity of the toxic septicaemic changes seen histopathologically and the rapid mortality which results. Klontz *et al.* (1966) have described the sequential pathology of experimental acute furunculosis in yearling rainbow trout (*Salmo gairdneri:* Richardson) injected intramuscularly with a virulent strain of the microorganism. Furuncles are an unusual feature of naturally occurring acute

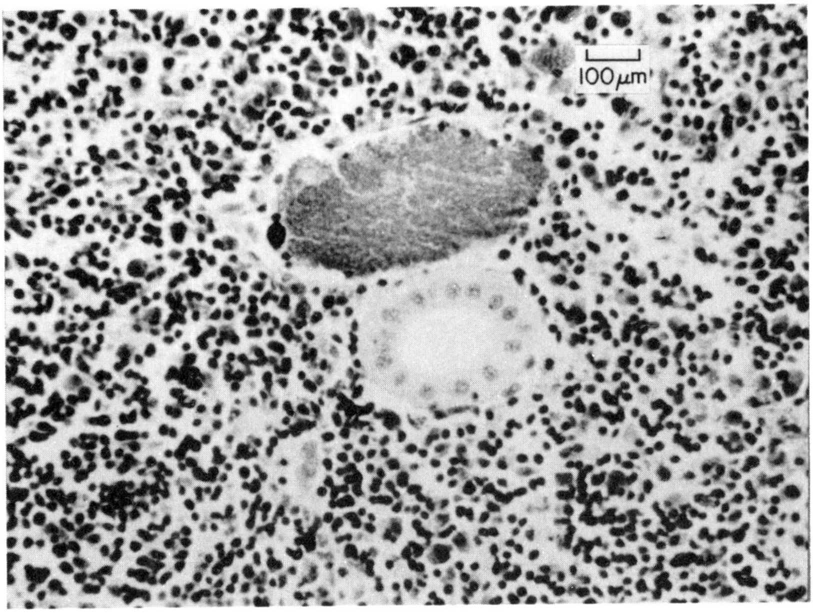

Fig. 3. Colony of *Aeromonas salmonicida* in haemopoietic tissue of the anterior kidney of a salmon parr. Humberstone stain.

furunculosis but, if injected experimentally, the organism does produce a focus of necrosis and so in the study by Klontz and his coworkers a focus of infection in the form of a furuncle was produced over a period of 72 hours. Histologically this lesion developed as a focus of myofibriller necrosis which progressed rapidly, to encompass vascular necrosis, with resultant haemorrhage. Initially there was a leucocytic response with macrophage infiltration but after about 50 hours this ceased, while the area of necrosis increased considerably in size and bacteria rapidly spread to other tissues. This dissemination of bacteria to other sites heralded overwhelming septicaemia, and death, but changes in haemopoietic tissues were found much earlier, with a marked leucocytosis in the early stages followed by considerable haemopoietic necrosis with congestion of sinusoids, and virtually complete depletion of the splenic and renal haemopoietic tissue.

In natural cases of acute furunculosis there may be a principal bacterial focus in any of a variety of organs, such as the anterior kidney, the spleen or the myocardium, or the infection may be multicentric, but toxic haemopoietic necrosis, myocardial and renal tubular degeneration and focal hepatic necrosis are consistently found irrespective of the site of primary localization. In branchial infection of amago, described by Miyazaki and Kubota (1975), initial lesions in the gill lamellar epithelium led to lamellar thrombosis. Embolic extension of infection from this site to the heart, liver, spleen and kidney was followed rapidly by generalized septicaemia and death of the fish.

4.3 SUBACUTE AND CHRONIC FURUNCULOSIS

In subacute and chronic furunculosis the lesions are similar but in the chronic case there is evidence of attempts at healing. It is in these stages that the classical furuncle (Fig. 4) is most frequently observed but it is by no means consistently found. The most detailed description of the pathology of subacute and chronic furunculosis is that of Ferguson and McCarthy (1978). They found the heart and the spleen to be the most conspicuously affected organs. In the heart, which was a focus of bacterial involvement in almost all of their cases, lesions varied but were predominantly epicardial and restricted almost entirely to the ventricle. Pronounced epicardial colonization was invariably accompanied

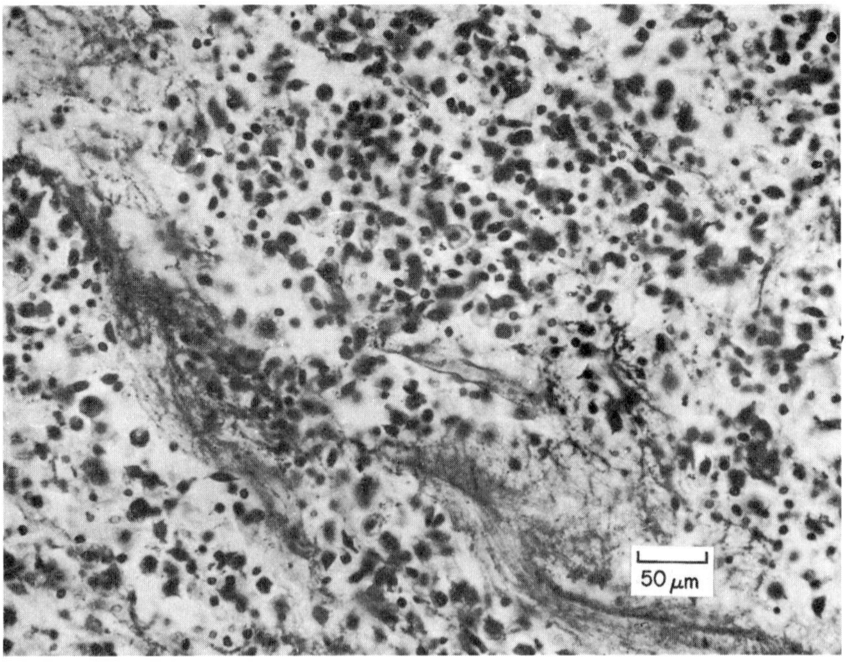

Fig. 4. Section through a typical furuncle showing cellular inflammatory debris with strands of fibrin along which bacteria are ranged.

by a fibrinous exudate and haemorrhage and congestion of the sub-epicardial space. Atrial involvement was restricted to occasional small bacterial colonies on the trabeculae and small endocardial mural thrombi. The degree of swelling and proliferation of lining cells of the atrium varied with the maturity of the infection, being marked in the early stages and very depleted later. Bacterial foci were prominent throughout the mycardial trabeculae of both compact and spongy layers. For the most part, and this is a consistent finding of other workers, there was no cellular inflammatory response other than occasional monocytes or neutrophils and the only effect on surrounding tissue was a limited liquefactive necrosis. Large mural thrombi were occasionally found on the endocardium of the ventricular trabeculae. These were usually red thrombi comprising an agglomerate of bacterial cells, thrombocytes and erythrocytes with a fibrinous stroma (Fig. 5).

Changes in the spleen are the most consistent features noted by all observers. Ferguson and McCarthy (1978) described them in the greatest detail and drew attention to the role of the ellipsoids in this process. In many cases these thick-walled vessels were completely destroyed by the bacteria seen suspended along their length (Fig 6). There was concomitant proliferation of the stromal cells, and lymphoid cuffing of the afferent splenic vessel as it passed through the

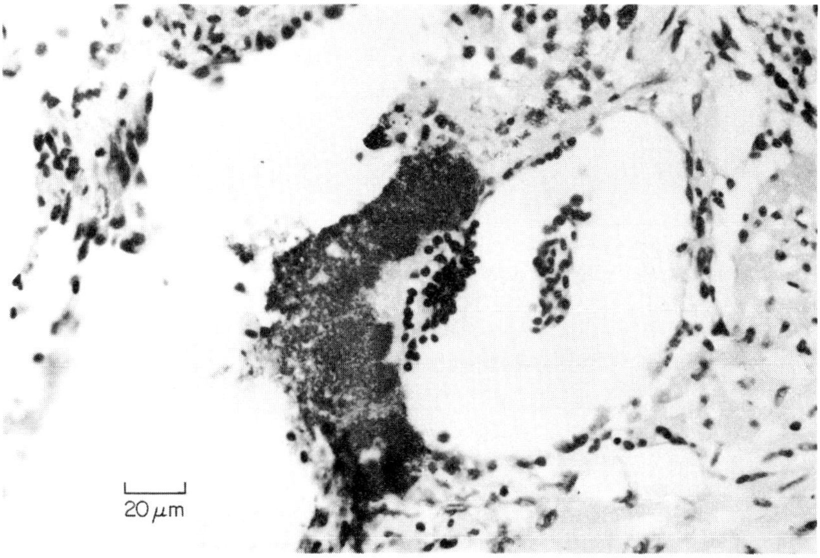

Fig. 5. Mural thrombus on wall of coronary vessel of brown trout.

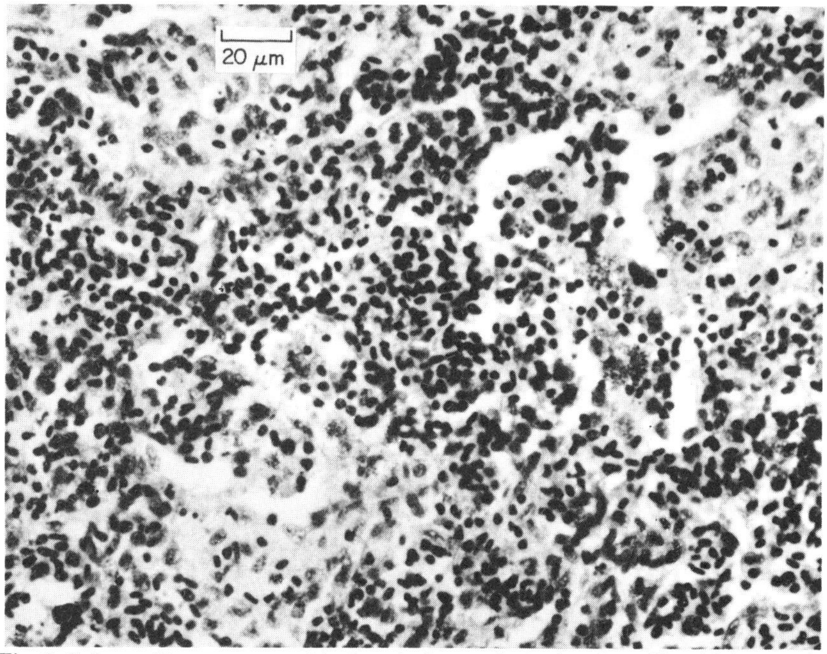

Fig. 6. Spleen of brown trout with chronic furunculosis showing ellipsoidal necrosis.

stroma. They considered renal involvement, in their series of subacute and chronic cases in brown trout, to comprise little more than focal colonies of bacteria with little host necrosis or inflammatory response (Fig. 7). This feature of the chronic disease has also been noted by others and is different from the acute condition where extensive haemopoietic and tubular necrosis are found; indeed Ferguson and McCarthy (1978) considered that in their chronic cases there was an enhancement of lymphoid cell numbers in the haemopoietic tissue suggesting the possibility that the persistent Gram-negative infection was acting in a mitogenic fashion. The liver, which despite Plehn's (1911) reference to its showing "enlargement of the phagocytic cells of von Kupffer" in furunculosis, does not in fact have phagocytic von Kupffer cells in any species of teleost in which it has been studied (Varitchak, 1938; Ferguson, 1975), and usually shows very little damage in chronic furunculosis apart from occasional foci of bacteria within periportal haemopoietic tissue, with associated focal hepatocyte necrosis (Fig. 8).

The histopathology of the furuncle varies depending on the state of its development but most authors who have studied it (Plehn, 1911;

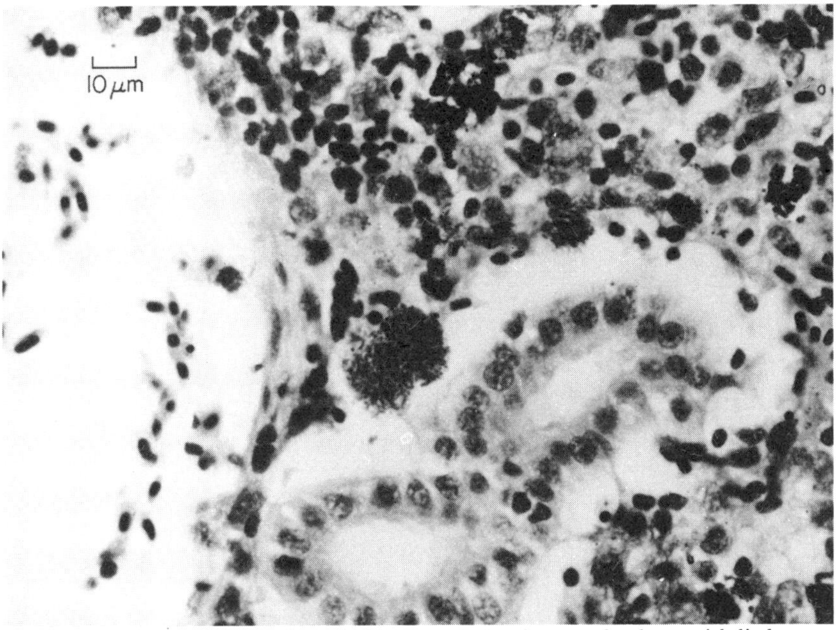

Fig. 7. Kidney of fish with chronic furunculosis showing renal colony with little or no host cellular reaction.

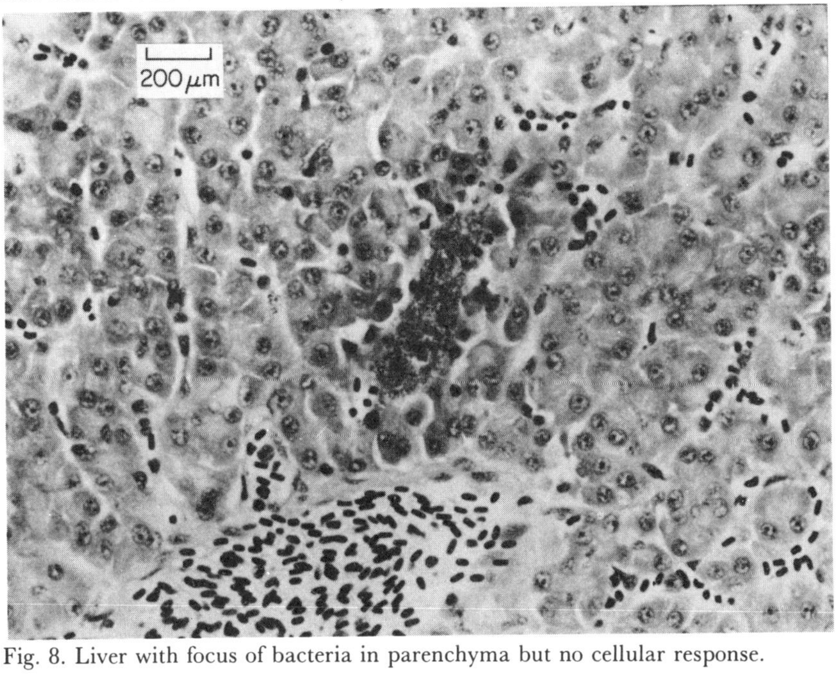

Fig. 8. Liver with focus of bacteria in parenchyma but no cellular response.

Williamson, 1929; Davies, 1946; Roberts, unpublished) emphasize its difference from the furuncle of higher vertebrates, which contains a necrotic mass of polymorphonuclear leucocytes as its main component. The furuncle of salmonids is usually comprised of tissue fluid exudate, a small component of macrophages, necrotic blood cells and necrotic elements of the host tissue within which the furuncle is formed. Bacteria can usually be seen spread along strands of fibrin and within the macrophages as well as being disseminated throughout the tissue fluid, which also contains considerable numbers of melanosomes from the ruptured dermal melanocytes. In large furuncles, extending down through the hypodermis to the skeletal muscle, there is extensive myofibrillar necrosis and liquefaction extending down myotomal fascial planes so that sinuses can be formed extending as far as the abdominal cavity or spinal cord.

No published reports are available on the neuropathological changes present in salmonids with chronic furunculosis but Roberts (unpublished) found congestion of meningeal vessels, and occasional bacterial thrombi in meningeal or chorioid vessels, associated with a slight inflammatory cellular infiltrate usually of lymphoid or monocytic elements.

Miyazaki and Kubota (1975), studying the pathology of the chronic condition in marine stages of the amago, found that where infection was percutaneous, a rather shallow dermal lesion was induced which extended subcutanously for some distance from the site of origin with extensive serofibrinous exudation. In more advanced cases this subcutaneous lesion extended down into the myotomes via sinuses. Ultimately there was development of metastatic lesions and fatal septicaemia. Chronic furunculosis in the amago more often originated in the hypodermis where there was marked serofibrinous exudation and bacterial localization with a limited infiltrate of mononuclear cells. Extension from this primary site involved bacterial invasion of the intermyotomal fascia with marked necrosis, haemorrhage and exudation. Once embolic dissemination had occurred the spleen and kidney were affected and the usual pattern of mortality ensued.

5 *Aeromonas salmonicida* infection in non-salmonids

The *Aeromonas* group is associated with occasional outbreaks of acute bacterial septicaemia in a wide variety of teleost fishes and although there are few records of specific involvement of *Aeromonas salmonicida* recent findings suggest that it may well be responsible for a wide range of bacterial septicaemic syndromes in non-salmonids (McCarthy, 1975b). The two syndromes with which it has most recently been particularly associated are post-traumatic septicaemia in centrarchids and the carp erythrodermatitis syndrome. Full clinical and pathological reports on these conditions are not yet available so that it is not possible to describe these clinical or pathological features in the same detail as is available for salmonid infections.

6 Post-traumatic septicaemia of centrarchids

Bulkley (1969) was the first to describe this syndrome in his study of an epizootic causing high mortality in wild yellow bass (*Morone mississipiensis:* Jordan and Eigenmann). The main clinical feature was depression and high mortality and many fish had heavy fungal infection of the skin. Internal organs of such fish appeared normal except for occasionally observed enteric inflammation. The microorganism was isolated readily from internal organs and Bulkley considered that the high mortality, coupled with the lack of internal pathology, indicated an acute or peracute infection. Drawing on evidence from his extensive studies of the population over the previous year, Bulkley postulated that the inductive trauma was the rise of temperature in spring to optimal growth temperatures for the growth of the microorganism coinciding with the onset of sexual maturation in a population of fish which were at a peak of density and on a very low plane of nutrition. Le Tendre *et al.* (1972) described an outbreak in wild small-mouth bass (*Micropterus dolomieui* Lacépède) where there was a consistent relationship between net damage and infection. The main presenting signs were skin lesions and high mortality. The skin lesions varied from whitish discolouration to shallow haemorrhagic ulcers. The bacterium was readily isolated from the ulcers but was only present in kidney and spleen when the fish were moribund. The high incidence of the disease was correlated with skin trauma in vulnerable fish due to changes that had taken place in netting levels following an extension of the commercial netting season in the area of Lake Ontario where the outbreak occurred.

7 Carp erythrodermatitis

Carp erythrodermatitis (CE) is part of the "carp dropsy syndrome", the aetiology of which was a source of great confusion until the seminal work of Fijan (1972) who showed that the syndrome comprised at least two separate diseases, carp viral haemorrhagic septicaemia, caused by a virus he called *Rhabdovirus carpei*, and carp erythrodermatitis, a predominantly skin infection, caused by a transmissible, antibiotic sensitive, non-filtrable agent. Efforts at isolating the agent were ultimately successful when Bootsma and Fijan (1977) claimed to have isolated a small Gram-negative bacterium, which they subsequently classified as an atypical strain of *Aeromonas salmonicida* from such lesions from typical outbreaks of the disease in Yugoslavia.

The principal clinical feature of CE is the development of a large circular, very shallow, haemorrhagic ulcer on the fish. The ulcers may be anywhere on the surface of the body but are generally found on the sides of the trunk. Affected fish are depressed, inappetent, and darker in colour. If the infection becomes generalized, there may also be extensive abdominal and retrobulbar oedema. The initial ulcer is a small dark area which gradually enlarges and develops a necrotic centre. It rarely extends below the *stratum spongiosum* of the dermis but may extend laterally. The edge is haemorrhagic and slightly raised with a contiguous external ring of darker skin. The centre becomes extremely inflamed and haemorrhagic, and secondary infection by fungus or other bacteria is common. The outcome is either resolution of the lesion and healing, or generalization of infection, septicaemia and death. When resolution takes place, the large healed ulcer which is formed is recognizable as an extensive grey–black scar and often contraction of the collagen of the scar tissue can result in serious deformity which reduces the commercial value of the fish considerably (Fijan, 1972). Mawdesley-Thomas (1969), in a very extensive clinico-pathological investigation of an outbreak of an ulcerative condition of goldfish (*Carassius auratus* L.) showed that, in this outbreak at least, typical pigment-producing strains of the pathogen were well able to induce a chronic ulcerative septicaemic condition. Affected fish developed ulcers of varying size and depth, and some fish died shortly after infection. Others showed lethargy, loss of orientation and abnormal swimming movements and their ulcers frequently became secondarily infected with *Saprolegnia* fungus. Healing was manifested by an increase in pigmentation around the ulcers and treated fish recovered completely. The condition showed a number of differences from CE, apart from the biotype of the pathogen. The ulcers were deeper and more extensive than those found in CE and renal and splenic changes consonant with those found in subacute furunculosis of salmonids were found at a much earlier stage than in CE where internal lesions are confined to the later stages of the disease if it generalizes.

8 Immunology

Although there are a number of reports on aspects of the immune response of salmonids and other teleosts to *Aeromonas salmonicida* there is no sequential study of the development of immunity during a natural infection *per se*. All studies have been oriented towards one of two applied concepts, namely the possibility of producing an acceptable vaccine and the use of antibody titres as indicators of the carrier state, or previous infection.

A wide variety of vaccine preparations have been used to stimulate an agglutinin response in salmonids, but this has rarely been related to actual protection against the disease and there is no study of what would almost certainly be at least as significant a source of protection: the cell-mediated immune response (Roberts, 1975). Among the vaccines used have been oral chloroform-killed whole cells (Duff, 1942), parenteral inoculation of formalinized whole cells, with or without an oily adjuvant (Reddecliffe, 1960; Krantz *et al.*, 1963), and parenteral inoculation of alum-based vaccines such as alum-precipitated culture sonicate (Anderson and Klontz, 1970) or alum-absorbed endotoxin (Paterson and Fryer, 1974). Udey and Fryer (1978) showed that highly virulent strains of *Aeromonas salmonicida* have an additional layer external to their cell wall which is not present in avirulent strains. Such virulent strains were also shown to aggregate, and to adhere to cultured fish tissue cells. They therefore considered that choice of vaccinal strain was crucial in production of a protective vaccine. They also found that oral vaccination even with virulent strains failed to confer protection whereas parenteral vaccination in conjunction with Freunds complete adjuvant conferred a high degree of protection.

Most recent studies have involved the use of vaccine cells in suspension in which fish are bathed following hyper-osmotic shock (Amend and Fender, 1976; Croy and Amend, 1977). Such techniques which offer great ease of antigen exposure would, if proved successful, have great potential for large-scale vaccination programmes but critical long-term evaluation of the technique is not yet available.

The best reported results both for agglutinins and precipitins using traditional vaccination methods have usually been obtained using adjuvanted formalinized vaccines (Paterson and Fryer, 1974) but Krantz and Heist (1970) using rainbow trout, induced agglutinin titres ranging from maxima of 1:20 to 1:320 which persisted at detectable levels for at least a year with formalin-killed bacteria. These compared to maxima of 1:160 in fish from natural outbreaks and 1:2560 from fish given a dose of live bacteria.

Experimental investigations of antibody production have been accompanied by serological surveys, in a number of species, to detect the prevalence of naturally occurring agglutinins in both wild and farmed populations. Krantz and Heist (1970) found great variations in the proportion of a population that possessed an antibody titre. This variation ranged from 0·9 per cent to 87 per cent within 24 populations of brown and brook trout (*Salvelinus fontinalis* Mitchill). They did not detect agglutinins in individuals less than one year old but the prevalence of antibodies, which increased with age, was twice as great in farms where furunculosis was enzootic. A post-epizootic search for antibodies in the sera of yellow bass was carried out by Schwartz (1973). He had failed to detect bacteria from any of the fish sampled but succeeded in demonstrating specific agglutinins in the serum of approximately 10 per cent of those fish old enough to have survived the epizootic.

Temperature has long been known to moderate the rate of the antibody response of teleosts (Bissett, 1948; Avtalion *et al.*, 1973) and obviously in vaccination studies earliest possible protection is desirable. In their study Paterson and Fryer found that specific antibody could be measured within the serum of trout challenged by endotoxin within one week at 17·8°C but that a month was required to obtain a demonstrable response in even a proportion of test fish when temperature was held at 6·7°C. The endotoxic preparation used was highly immunogenic at levels as low as 1 μg, so over the range of doses used no significant variation of response correlatable with dose could be obtained.

The use of antibody levels to determine previous infection or the immune status of a population has been used by a number of workers.

Often, however, such studies have been bedevilled by low levels of anti-*A. salmonicida* antibodies in most normal stocks of fish, which show up readily because of the high sensitivity of the antigen–antibody system used (Paterson and Fryer, 1974). Krantz and Heist (1970) in their detailed serological survey of 24 brown or brook trout populations in hatcheries in Pennsylvania, USA, found very great variation between populations, with 8 per cent to 87 per cent reactive individuals depending on the population of origin.

Antibodies were not detected in fish less than one year old and reactivity increased with age. In populations where furunculosis was considered to be enzootic the prevalence of antibody detected was approximately half the frequency found in those populations where regular epizootics occurred. By contrast a post-epizootic search for *Aeromonas salmonicida* antibody in yellow bass by Schwartz (1973) showed that in this species bacteria could not be cultured from any of 777 fish examined two or three years after the epizootic and agglutinins were only found in sera of 5 fish. These were all specimens considered to have been old enough to have survived the very severe epizootic. Schwartz offered a number of reasons as to why his experience with this non-salmonid species differed from that of those reporting on salmonid epizootics but considered that it was still likely that the microorganism persisted somewhere within the lacustrine environment.

9 Haematology

Haematological studies of fish are rendered extremely difficult by the lack of consistency that environmental, seasonal and nutritional variations impose on the blood picture and particularly on the difficulties associated with interpreting and comparing results from samples removed from fish in varying degrees of stress by diverse routes. However there are two reports on blood parameters in salmonids infected with *Aeromonas salmonicida*. Foda (1973) studied changes in haematocrit and haemoglobin concentration in Atlantic salmon in a Canadian hatchery. He detected a lowering of haematocrit and haemoglobin levels by approximately 50 per cent. He also gave values for blood levels in recently dead fish but it is unlikely that these can be reliable because of the very rapid degenerative changes that occur in fish tissues, including blood, *post mortem*. Foda considered his findings to be attributable to a haemorrhagic rather than a haemolytic anaemia but this is probably an oversimplification as Roberts (unpublished) has consistently found extensive haemosiderin deposits in the melanomacrophage centres of haemopoietic tissue in chronic furunculosis, indicating extensive haemolysis even where significant haemorrhage was not manifest. A decreased haemotocrit in brook trout with furunculosis was also reported by Schumacher (1956).

10 Virulence mechanisms of *Aeromonas salmonicida*

The Furunculosis Committee (Mackie *et al.*, 1930, 1933, 1935) failed to demonstrate any toxin production by *Aeromonas salmonicida* by injection of ultra-filtrates of broth cultures and diseased fish tissue into healthy fish. However they emphasized the organism's marked proteolytic activity and also noted the paradoxical lack of leukocytic infiltration in necrotic muscle lesions. In addition, they observed a variation in colony morphology among isolates (presumably rough and smooth varients) but stated that this phenomenon was not accompanied by a difference in virulence. In conclusion, they proposed, as a hypothesis for the mechanism of virulence of the microorganism, based partly on their failure to detect toxin production and partly on their extensive and detailed clinical observations, that the bacterium's pathogenic activity would be attributed to its abundant growth in the host's blood and tissues which interfered with blood supply and thus resulted in infarctions, anoxic cell necrosis and eventually death.

Another early worker Duff (1939) also noted the existence of rough and smooth variants of *A. salmonicida* and compared their virulence by injecting goldfish with either rough *A. salmonicida* isolates or with the corresponding smooth dissociant. He found that all fish injected with the smooth strain died within 9 days but those receiving the rough variant were apparently unaffected. Unfortunately, the number of viable cells in each injection was not assessed, but ". . . 0·1 cc of a No. 4 McFarland suspension made from a 48-hour broth culture . . ." was given.

Field *at al.* (1944) carried out an interesting study of the sequential chemical pathology of experimental *Aeromonas salmonicida* infection in the carp (*Cyprinus carpio* L.). The most significant feature of their study was a rapid fall in blood sugar levels resulting in hypoglycaemic shock, which they considered could be fatal *per se* on occasion. They suggested that the hypoglycaemia resulted from a very rapid utilization of blood glucose by the multiplying microorganism. They also detected an increase in creatine levels as a result of muscle breakdown and an increased blood urea which they interpreted as a result of progressive renal failure. This latter suggestion is not tenable, however, as all teleosts, including the carp, excrete most of their nitrogeneous waste via the gills. A much more likely cause is reduction of branchial circulation due to endothelial and myopathic activity by the micro-organism or its toxins.

Griffin (1954) made several theoretical observations regarding the virulence mechanisms of *A. salmonicida*: first, he suggested that the hypoglycaemia observed by Field *et al.* (1944) might be caused by a "pharmacologically-active compound"; second, he speculated that the paradoxical observation by the Furunculosis Committee, that marked cytolytic tissue necrosis did not appear to be accompanied by leukocytic infiltration, might be the result of leukocidin production *in vivo* by the bacterium; third, he was of the opinion that, although *A. salmonicida* was known to produce haemolysin *in vitro*, it was unlikely to play a significant role in the disease process because of the haematological findings of Field *et al.* (1944). It is worthy of note, however, that Karlsson (1962) found fish erythrocytes to be more sensitive *in vitro* to *A. salmonicida* haemolysin than human, horse or guinea-pig erythrocytes.

Klontz *et al.* (1966) followed the sequential pathological changes in rainbow trout following intramuscular injection of a lethal dose of virulent *A. salmonicida* cells. Although initially a cellular infiltrate response was initiated, after 56 hours the number of inflammatory cells present had diminished and this was accompanied by marked degenerative changes in the reticuloendothelial tissue. Bacterial phagocytosis was not observed at any time. Tissues examined 72 h post-injection were entirely devoid of inflammatory cells and undergoing marked necrosis. Finally, at 120 h, when some fish had already died, a massive bacteraemia occurred in the remainder. They also reported that subcutaneous injection of an ultra filtrate of a saline-soluble extract of *A. salmonicida* cells resulted in haemotological changes in the kidney indistinguishable from those exhibited by furunculosis-infected fish. Although a marked and progressive leukopenia was observed after 56 h in experimentally infected fish, values for erythrocyte count, haematocrit and haemoglobin concentration were not significantly different from those for healthy control fish.

The marked lack of response of fish to injections of Gram-negative endotoxin has been noted and emphasized by several investigators. Wedemeyer *et al.* (1969) reported no clinical effect in salmonids injected with 80 mg kg^{-1} of crude *A. salmonicida* lipopolysaccharids (LPS). Furthermore, Paterson and Fryer (1974) failed to cause mortalities in salmonids with an intraperitoneal injection containing in excess of 700 mg of *A. salmonicida* LPS per kg of fish. Lipopolysaccharide distribution and concentration can be conveniently detected in fish tissues by titrating extracts against sensitized latex particles (McCarthy, 1975a). In a study utilizing this technique, LPS levels were measured following intramuscular injection of concentrations of crude *A. salmonicida* LPS similar to those used by Paterson and Dryer (1974). Levels of LPS in liver, kidney and spleen were undiminished over 48 hours; and there was no evidence of clinical disturbance (McCarthy unpublished). Clearly, then, the role played by LPS in the pathogenesis of furunculosis is not a major one.

Further work on the significance of smooth and rough variants of *A. salmonicida* was published by Anderson (1972a, b), who compared the virulence of smooth and rough variants of several *A. salmonicida* isolates by measuring LD_{50}s in coho salmon (*Oncorhynchus kisutch* Walbaum) and found that, with the exception of one of his smooth strains, LD_{50}s, for smooth variants were much higher and many passages were required to enhance virulence. Most workers have used a maximum of four passages and failed to make any comparative assessment of whether virulence had, in fact, been increased. Clearly, the stability of virulence of laboratory-stored strains of *A. salmonicida* is of importance when contemplating experimental pathogenesis or vaccinal studies. For instance, it has been found (McCarthy, unpublished) that serial subculture of a virulent strain of *A. salmonicida* on tryptone soya agar (Oxoid Ltd.) resulted in a progressive loss of virulence, the minimum lethal dose to rainbow trout determined by intramuscular injection rising from $3{\cdot}4 \times 10^3$ at the start of the experiment to $5{\cdot}1 \times 10^5$ after the eleventh passage. Nevertheless, at pass 15, when the experiment was discontinued, the MLD was $5{\cdot}3 \times 10^6$, still a significant measure below the MLD of its avirulent precursor ($9{\cdot}3 \times 10^9$) from which it had been derived by serial fish passage. Storage of this strain on agar slopes at 8°C in the dark also resulted in a progressive loss of virulence as follows: MLD after one month—$4{\cdot}8 \times 10^3$; two months—$4{\cdot}8 \times 10^4$; three months—$4{\cdot}7 \times 10^6$; four months—$5{\cdot}0 \times 10^6$; eight months—$8{\cdot}1 \times 10^7$. Freshly isolated strains from acute furunculosis may possess a minimum lethal dose as low as $10^2 - 10^3$ viable cells, whereas an isolate obtained from a carrier fish artificially stressed by injection with prednisolone acetate requires the injection of about 10^5 viable cells to kill non-carrier brown trout.

Fuller *et al.* (1977) isolated a glycoprotein from broth culture supernatants of a virulent strain of *A. salmonicida*. In addition to demonstrating that it was cytotoxic for fish leukocytes and that intravenous injection into rainbow trout produced a pronounced leukopenia, they claimed that when injected along with a live culture of *A. salmonicida* it resulted in a reduction of the number of such bacteria required to kill test fish. They also found that a virulent strain of *A. salmonicida* produced much more of this factor than an avirulent strain and so concluded that the material was one of the virulence factors of *A. salmonicida*.

Similar work (McCarthy and Salsbury, unpublished) using ultra-filtered crude extracts of fresh furuncle material and agar culture extracellular material has confirmed the above report. Suspension of bacterial cells of an avirulent strain of *A. salmonicida* in either material reduced its MLD from $9{\cdot}5 \times 10^9$ viable cells to $5{\cdot}7 \times 10^3$ viable cells when injected into rainbow trout held at 13°C. Furthermore, this phenomenon was observed when either virulent or avirulent isolates were used to produce extracts and the active factor appeared to be both heat stable and non-dialysable.

11 Serology

The first worker to contemplate a serological study of *A. salmonicida* was Williamson (1929); however, when faced with persistent auto-agglutination of her strains in saline solutions she abandoned the attempt. Blake (née Williamson) and Anderson (1930) successfully overcame the problem by abandoning agglutination and using a complement fixation technique instead. They studied 82 strains of *A. salmonicida*, 60 from the UK and the remainder from other countries. They found that all *A. salmonicida* strains produced a positive reaction in their test, while 21 strains of various other bacteria isolated by them from soil and water failed to react. Duff (1939) by suspending his strains in 0·05 per cent saline, reported that he was able to compare by agglutination the serological composition of eight rough strains with their respective smooth variants, previously obtained by dissociation in lithium chloride broth. Duff reported the presence of an additional antigen in the rough variant of five of his eight strains represented by him as R = S + n, and in the smooth variant of the remaining three strains (S = R + n).

In 1961 Ewing *et al.* examined 21 *A. salmonicida* strains and various other aeromonads by a variety of serological methods and demonstrated strong reactions of homology between the *A. salmonicida* strains. They also showed them to cross-react with an *A. hydrophila* antiserum. Liu (1961), in a gel-diffusion precipitin study of the serological specificity of extracellular antigens of a variety of bacterial species demonstrated shared antigens among the two *A. salmonicida* strains included. Moreover, he also found three of the four *A. hydrophila* isolates to cross-react with the single *A. salmonicida* antiserum, although both *A. salmonicida* strains failed to react with the single *A. hydrophila* antiserum tested.

In similar studies, albeit employing different enzymes as antigen, Karlsson (1962) using haemolysin and Gonzales (1963) using lipase, demonstrated iso-enzyme specificity within *A. salmonicida* and cross-reaction with *A. hydrophila.* In a more detailed investigation employing both precipitin and agglutinin tests, Karlsson (1964) was unable to detect any differences in the thermolabile and theromstable antigens of 12 *A. salmonicida* isolates. Although he was unable to detect cross-reaction with *A. hydrophila* using simple precipitin and agglutinin tests, by using a double-gel diffusion technique he demonstrated the presence of several common thermolabile components. Karlsson, like Williamson earlier, encountered serious difficulties on producing satisfactory suspensions of rough strains in saline and was forced to resort to ultrasonic dispersion of clumped cells. Further evidence of cross-reaction between soluble antigens of *A. salmonicida* and *A. hydrophila* was furnished by Bullock (1966) and in another enzymo-serological study, Sandvick and Hagan (1968) demonstrated serological cross-reaction between the casein-precipitating enzymes of these two species.

A report of antigenic variation among strains of *A. salmonicida* was made by Klontz and Anderson (1968), who examined smears of 24 isolates with three different antisera using an indirect fluorescent antibody test. On the basis of non-reactivity of certain strains with one or more of their antisera, they postulated the existence of at least seven different serotypes among the 24 isolates studied. However, the fluorescent antibody technique, while a valuable diagnostic method for detecting and localization of bacteria and other antigens in tissues, is not generally considered a suitable method for serological analysis of laboratory cultures, being particularly prone to technical difficulties. Such antigenic variation, therefore, remains to be confirmed by the more reliable yet technically simpler whole-cell double cross-absorption technique. Kimura (1969b) compared the thermolabile and thermostable somatic antigens of his atypical "*masoucida*" isolate with a typical *A. salmonicida* isolate by double cross-absorption and reported the presence of an additional thermolabile component in his sub-species. Also in 1969, Popoff compared the serological composition of 39 strains of *A. salmonicida* in a detailed study but failed to detect antigenic differences. Whang (1972) unsuccessfully tested nine strains of *A. salmonicida* for production of the common enterobacterial antigen.

Serological work undertaken to establish a basis for development of rapid serological techniques (McCarthy and Rawle, 1975), provided support for Popoff's (1969) findings. Whole-cell agglutination and double cross-absorption of smooth strains and passive haemagglutination and double cross-absorption of rough strains were used to study both thermolabile and thermostable somatic antigens of *A. salmonicida* and their relationship to certain other bacteria. Cross-reaction titres for both thermolabile and thermostable antigens were generally high and very weak cross-reaction (titres *c.* 1:4) occurred between the single *A. hydrophila* and three of the six thermostable *A. salmonicida* antisera incorporated in the study. The single strains of *Vibrio anguillarum* and *Pseudomonas fluorescens* studied, on the other hand, both failed to react with any *A. salmonicida* test antisera. The passive haemagglutination reactions were more sensitive, titres obtained for positive reactions being tenfold higher than those recorded for the same reaction when tested by whole cell agglutination. A double diffusion method was also used to study cell-free extracts prepared from the bacteria used for the somatic antigen study. Strong cross-reaction among *A. salmonicida* strains was found with this technique and cross-reaction between *A. salmonicida* and *A. hydrophila* and *V. anguillarum* strains and, to a lesser extent, *P. fluorescens* was observed. Essentially, the study failed to demonstrate qualitative differences in serological composition among *A. salmonicida* strains, but it did indicate that prolonged laboratory storage of certain *A. salmonicida* strains resulted in progressive loss of serological reactivity and negative reactions were obtained with particular antisera. For example, an old laboratory strain of ATCC 14174 was used in the study and failed to react in the passive haemagglutination test, whereas a fresh culture of the same strain obtained from the American Type Culture Collection reacted fully. When contemplating serological or vaccination studies it is thus important to include only recent isolates since fish passage of the non-reacting isolate fails to restore reactivity. Examination of the surface antigens of the currently emerging atypical strains of *A. salmonicida* could well repay itself in that they may indicate homologies differentiating them as a group from the typical strain.

The first isolation of bacteriophage specific for *A. salmonicida* was made by Todd (1933), and a subsequent study by Popoff (1971) clearly demonstrated the feasibility of phage typing *A. salmonicida* isolates.

The slide agglutination test (Rabb *et al.*, 1964) is regularly used for preliminary identification of *A. salmonicida* but it cannot be used with rough (auto-agglutinating in 0·9 per cent NaCl) isolates. Although the method of reducing the electrolyte level of the suspending fluid used by Duff (1939) is adequate to permit examination of some rough strains, it is unsatisfactory for use with those that auto-agglutinate strongly. Rough strains that produce friable colonies are, without doubt, much more frequently isolated, at least in England and Wales, than smooth strains with butyrous colonies. A search of the early literature on colony descriptions of *A. salmonicida* indicates that the workers describing their isolates were describing the colonial morphotype of the rough organisms. Serological identification by the passive haemagglutination method, reported by McCarthy and Rawle (1975), allows rapid identification and also serological examination of both rough and smooth strains of *A. salmonicida.* In a search for rapid specific diagnosis McCarthy (1975a) introduced the latex agglutination test, which allows detection of *A. salmonicida* antigen in diseased fish tissue, allowing laboratory diagnosis of furunculosis even more rapidly (approximately 2 h). Infected tissue extracts give strongly positive reactions of up to 1:1 000 000 especially when furuncle extracts are used. The technique is also valuable in allowing specific diagnosis of furunculosis in material submitted in a state rendering it unfit for bacterial isolation, even diagnosis from formal-fixed material being feasible. An even more rapid serological method, which permits diagnosis to be made in less that 15 min from receipt of the specimen by direct microscopic examination of diseased tissue smears for the presence of *A. salmonicida*, was devised by McCarthy and Whitehead (1977). The technique utilizes the specific India ink immuno-staining reaction originally used by Geck (1971) for detection of enteric pathogens in human faeces.

A modification of the latex test (McCarthy, 1977c) now allows direct serological examination of mixed cultures of *A. salmonicida* from primary isolation plates. This is an important advance since isolates recovered from skin lesions of non-salmonid fish are frequently overgrown by other aquatic bacteria. The technique has proved particularly useful for rapid identification of rough, non-pigmented, and slow-growing strains of *A. salmonicida* and their preliminary differentiation from apparently similar bacteria that may cause disease in fish, e.g. *Pasteurellae* since the antiserum used, although raised against a single typical isolate of *A. salmonicida*, has reacted with all 54 atypical isolates so far examined as well as the typical ones.

12 Ecology of *Aeromonas salmonicida*

The precise route of transmission of furunculosis is not known. However, contaminated water, infected fish or fish farm materials are all possible sources of infection. Carrier fish, showing no overt sign of disease but harbouring the bacterium in their tissues, also appear to be involved in horizontal or vertical transmission. Such carrier fish are presumed to act as a reservoir responsible for its maintenance in fish populations, but the possibility of a free-living life-cycle for *A. salmonicida* has also been considered, since its existence would greatly complicate disease prevention and control in the fish farm environment.

There are a number of possible sources of *A. salmonicida* in the fish's environment: diseased or carrier fish; other animals; water; river or pond mud, and fish farm implements such as nets and tanks. Although the possibility that animals other than fish harbour and transmit the bacterium obviously exists, it seems unlikely in view of the report of Cornick *et al.* (1969) who examined 2954 vertebrates and invertebrates (including leeches) taken from fish ponds during an active outbreak of furunculosis for the presence of *A. salmonicida* and did not find any evidence for their infection.

Although there have been many published reports describing studies on the existence and survival of *A. salmonicida* in water, most were concerned with viability of the organism in distilled water or sterilized substrates. Since such highly artificial environments are far removed from natural water or fish pond situations where the organism must compete in a heterogeneous microbial community, they are of limited practical value. However, since a satisfactory selective medium for the isolation of *A. salmonicida* is not available, it is not surprising that most survival studies have been on sterilized water samples since *A. salmonicida* is notoriously difficult to isolate from mixed microbial populations.

This problem was avoided by McCarthy (1977b) who used a streptomycin-resistant mutant strain of *A. salmonicida,* providing the facility of isolation on streptomycin media, to study survival of the organism in unsterilized, fresh, brackish and full sea water in the presence of live salmonid fish. In these experiments survival was most prolonged in brackish water (24 days) but the organism also survived for 17 days in fresh water and 8 days in sea water. The results for survival in brackish and fresh water were closer to those reported by Smith (1962): river water 7–19 days; brackish water 16–25 days longer than those published (*c.* 1–2 days) by Williamson (1929) and Lund (1967). The result for survival in seawater (8 days), however, was closer to that reported by Lund (1967) (4–6 days) than the result found by I. W. Smith (personal communication) (21–26 days). A recent publication by Dubois-Darnaudpeys (1977) showed that survival of *A. salmonicida* in river water was related to pH and temperature. In river water (pH 7·9) she found the bacterium to survive longest at 4°C (50 days), survival at 14°C and 22°C being reduced to 21 days and 15 days respectively. Although these data demonstrate the organism's ability to survive for a significant length of time under natural conditions, both Cornick *et al.* (1969) and Kimura (personal communication) failed to isolate *A. salmonicida* from fish pond water during epizootics of furunculosis, although, of course, it is possible that the organism may have been present, perhaps in low numbers, and for the technical reasons stated above not detected. An effective selective medium for all strains of *A. salmonicida* would greatly facilitate such studies.

It is likely that the mud and detritus present in earth-bottomed fish ponds or even in rivers will become contaminated with *A. salmonicida* as a result of an epizootic of furunculosis and certain members of the family Enterobacteriacae are capable of metabolizing sediments; consequently, it is possible that *A. salmonicida* may survive for a considerable period of time in such substrates. By utilizing the streptomycin-resistant mutant technique (McCarthy, unpublished) the viability of the organism in pond mud has been determined. On termination of the experiment after 29 days, 10^5 viable cells of *A. salmonicida* were still present. This is highly significant since the survival time for this number of cells when released into fresh water is at least 14 days. Similar results were obtained by Dubois-Darnaudpeys (1977) who, in addition, demonstrated actual multiplication of the organism. Further experiments with the streptomycin-resistant mutant emulsified in fish slime and faeces and then dried on fish nets demonstrated the organism's capability of survival at least beyond the six days' duration of the experiment.

Carcasses of fish that have succumbed to furunculosis are heavily contaminated with *A. salmonicida*, muscle lesions, for example, containing as many as 10^8 viable cells per ml of necrotic tissue. Consequently, the presence of such fish, which cannot be conveniently removed, constitutes a significant source of contamination. They are readily taken by piscivorous birds, and may be dropped into uninfected waters. McCarthy (1977b), again using a streptomycin-resistant mutant, studied survival of *A. salmonicida* in rainbow trout experimentally infected by intramuscular injection by assaying every four days both infected fish cadavers and tank water at 10°C for the presence of the organism. The results of this experiment demonstrated that the number of viable bacteria in diseased fish tissue gradually decreased until by the 24th day they were fewer in number than those present in water, which now began to increase due to disintegration of the test fish. Viable cells were recovered from fish tissue for up to 30 days. They were still present in the tank water for a further eight days despite an open circulation of water through the tank. It was presumed, therefore, that the test organism must have contaminated the tank sides and was being gradually released from that site. Prolonged survival of *A. salmonicida* in dead fish, in addition to presenting a possible source of infection for healthy fish either by contamination of the tank water or by direct ingestion, adequately demonstrates the danger of feeding scrap fish or offals which may be carrying the organism. This is of particular importance in view of the report by Cornick *et al.* (1969) who showed that the organism can survive for up to 49 days in infected trout tissues stored at −10°C.

It was mentioned earlier that apparently healthy carrier fish have long been suspected of being involved in transmission of furunculosis, but a technical problem which has hindered research in this area has been the difficulty involved in detecting such fish with any certainty; for example, culture isolation methods are at present far too insensitive. Bullock and Stuckey (1975), however, found that injection of the cortico-steroid triamcinolone acetonide activated latent infections in carrier fish and facilitated their detection. McCarthy (1977b) used a similar cortico-steroid (prednisolone acetate) and carried out a series of experiments designed to assess the prevalence and significance of carrier fish in the fish farm environment. Essentially, the method consisted of injecting 20 mg kg^{-1} of prednisolone acetate intramuscularly and then subjecting the fish to thermal stress at 18°C. Carrier fish then died of clinical infection in 4–10 (mean 5) days. In the first experiment one-year-old brown trout from four different commercial fish farms were found to have a carrier rate in the range 40–80 per cent while similar populations of rainbow trout had either a very low incidence (less than 5 per cent) or were apparently free. Two further experiments were carried out with brown trout from one of the farms with a high carrier rate. In the first, a sample of fish was taken from all age groups on the farm and tested with prednisolone acetate. The following percentage population carrier rates obtained; 1+ (fish > 1 year old), 96 per cent; 2+ (fish > 2 years old), 100 per cent; 3+ (fish > 3 years old), 50 per cent. The second experiment consisted of testing samples of the 1+ population at different times throughout the year, when it was found that the carrier rate did not differ markedly. It is ironic that the particular farm on which this study was performed had never reported clinical furunculosis.

The focus of infection by *A. salmonicida* in carrier fish is obviously important from the point of view of the ecology of the organism and specifically its transmission, but is not precisely known. Most workers believe that the kidney is involved, although some have suggested the intestinal tract is a significant site of harbourage. When the kidney, spleen, liver, intestinal tract and heart blood of carrier fish, injected with prednisolone acetate, are sampled bacteriologically at regular intervals up to 96 h intervals after injection, the kidney appears to be the primary site of carriage although the organism can also be isolated from the intestinal tract. It is recognized that fish surviving a furunculosis epizootic (and therefore possibly carrier fish) have demonstrable agglutinins (Kimura, 1970). The possible usefulness of this assay of specific circulating anti-*A. salmonicida* agglutinins as a detection method for carrier fish has been investigated by sampling 30 brown trout from a known carrier population. The fish were bled, individually tagged and then injected with prednisolone acetate to determine the proportion of carriers. When serum antibody titres for each fish was assessed, however, no consistent difference was detected between fish shown to be carriers and non-carrier fish.

It is worth emphasizing here, because of the existence of a high carrier rate in brown trout without a history of furunculosis, the necessity of research workers taking into account the possible pitfalls of using such fish in experiments. For example, when carrying out virulence studies it is important to bear in mind the possibility that mortalities in test fish might be due at least in part to activation of a latent infection. Furthermore, during animal passage procedures the very real possibility of contaminating or replacing a vaccine strain with the organism being carried should also be borne in mind.

Aeromonas salmonicida may thus survive for a considerable time in various substrates, either in rivers or on fish farms.

Transmission of furunculosis by contact with contaminated water and infected fish was demonstrated by Blake and Clark (1931) when they placed seven brown and eight rainbow trout in cohabitation with an infected brown trout and found that all seven browns but none of the rainbows succumbed, demonstrating the now generally accepted relative resistance of rainbow trout to furunculosis. McCarthy (1977b) also infected brown trout but failed to infect rainbow trout by simple addition of viable bacteria to tank water. However, if the lateral integument of the rainbow trout was abraded, infection and death from furunculosis followed. In this particular study, a single brown trout failed to succumb but, when later challenged with prednisolone acetate, it also died of furunculosis. Further experiments with larger numbers of fish (50) from a non-carrier population of brown trout, transferred to a tank receiving water from infected and dying fish during a natural summer epizootic, further demonstrated this phenomenon. Mortalities among test fish commenced on the 8th day and by the time they had ceased 28 days later 41 fish had died. The nine remaining fish were then brought back to the laboratory and challenged with prednisolone acetate; all succumbed. This experiment demonstrated both the apparent ease with which the disease spreads through water and also indicates that non-carrier brown trout are extremely susceptible to furunculosis and that any survivors will, in all probability, become carriers. A similar experiment was carried out to compare the relative susceptibility of non-carrier and carrier brown trout. Mortalities among the two test populations ceased 15 days after their introduction to contaminated water, at which time 13 of the 15 non-carriers, but only 7 of the 16 carrier fish, had died. Fourteen days later all fish were injected with prednisolone acetate, when both surviving non-carriers, but only 3 of the 9 surviving carrier fish, succumbed. It is difficult to interpret these results precisely, but they indicate that non-carrier fish are apparently more liable to succumb to clinical furunculosis than carrier fish. The experiment also indicated that, although non-carrier fish may be preferable, the extra cost involved in raising such fish for angling re-stocking may not be justified in so far as a virtually enzootic disease such as furunculosis is concerned: indeed, under certain circumstances, such fish may be at a distinct disadvantage. However, MacDermott and Berst (1968) reported no evidence of transmission of furunculosis in a Canadian stream during a two-year period after stocking with wild populations of both infected and non-infected brook trout.

The creation of carrier fish following an active epizootic of furunculosis is a common occurrence and generally well accepted. But the introduction of furunculosis to non-carrier populations without the occurrence of active disease does not appear to be generally recognized. McCarthy (1977b) reported on such a case of the introduction of furunculosis to a specific pathogen-free population of farmed brown trout without the occurrence of active furunculosis. The population was used as a source of experimental non-carrier fish and was routinely tested with prednisolone acetate in order to ensure that the population was still free from carriers. Unexpectedly, a positive result was obtained and a carrier rate of about 70 per cent subsequently established at the farm among the previously non-carrier population. Although the pond was normally supplied solely with spring water, the infection was traced to the use of water from a nearby stream used as a supplement during severe drought conditions. Subsequent sampling of this very small stream yielded 40 wild brown trout, two of which manifested furunculosis following injection of prednisolone acetate. Clinical furunculosis did not occur during the use of the stream water or subsequently. It is intriguing that a population of apparently healthy wild fish with a very low carrier rate could be responsible for infection of a very large fish farm population without clinical disease occurrring at any time in either. A history of freedom from clinical furunculosis must not therefore be taken as evidence of absence of the disease. Certainly large populations of non-carrier fish may be readily infected, without clinical evidence of disease, by very small numbers of carrier fish.

Infection via contaminated food has been postulated as a route of infection. McCarthy (1977b) failed to infect brown trout by feeding food soaked with a saline suspension containing about 10^8 viable cells of *A. salmonicida* per ml; in addition, test fish did not succumb to subsequent prednisolone acetate challenge. This result conflicts with reports by Plehn (1911) and Blake and Clark (1931) who both succeeded in infecting fish by feeding contaminated food. However, Krantz *et al.* (1964b), who also failed to infect brown trout by feeding food containing 10^8 viable cells per g, suggested that discrepencies in such results might be due to differences in experimental and environmental conditions. McCarthy (1977b) carried out a further experiment by oral intubation of 10^7 viable cells of a streptomycin-resistant mutant in rainbow trout. Although the organism was detected in kidney tissue from 5 h onwards, no mortalities occurred during the experiment and the bacterium was completely eliminated from the fish within 48 hours of feeding. Although this procedure again failed to provoke mortalities, infection via the gastrointestinal tract must remain a significant possibility in view of the report of Klontz and Wood (1972) who observed clinical furunculosis in the sable fish, apparently resulting from ingestion of carrier coho salmon.

Snieszko (1974) regarded vertical or transovarian transmission of *A. salmonicida* as a possible route of infection. However, published evidence is inconclusive; Smith (1939; cited by McFadden, 1969) claimed that the organism was carried on the egg surface, while Plehn (1911) and Mackie *et al.* (1930) considered that the bacterium was unable to infect fish eggs. Moreover, although Mackie *et al.* (1930) and Lund (1967) both failed to isolate *A. salmonicida* from fertilized eggs obtained from experimentally infected fish, Lund (1967) described a positive isolation from the interior of fertile eggs from naturally infected mature fish. The organism has also been isolated from the ovary and testes of mature fish (Lund, 1967; McDermott and Berst, 1968).

Since the carrier state is present in fish throughout the year (*vide supra*) there is a possibility that such fish may transmit the bacterium with their sex products. McCarthy (1977b) carried out two experiments to investigate this possibility: in the first, male and female brood stock taken from a known carrier population were stripped and the eggs fertilized in the usual way. Although, when injected with prednisolone acetate, five of the eight brood stock fish (3 male and 2 female) proved to be carriers, *A. salmonicida* was not isolated from the fertilized egg sample following culture of an egg macerate. A further group of fish from the same population were treated similarly except that fish were first injected with prednisolone acetate and four days allowed to elapse to allow for activiation of the latent disease to occur. Fish were then stripped and the eggs tested for the presence of *A. salmonicida*. Again, although *A. salmonicida* was isolated from the brood stock fish, the organism was not recovered from the egg sample.

Although active epizootics of furunculosis have been reported during winter (Klontz and Wood, 1972) they appear to be infrequent. The significance of overt infection on ovarian infection in ripe brood stock was studied by McCarthy (1977b) by injecting brood fish with 10 times the minimum lethal dose of a streptomycin-resistant strain and then stripping on the first appearance of the clinical signs of furunculosis. Both fish organs and fertilized eggs were positive for *A. salmonicida*. Although *c*. 10^6 viable cells of *A. salmonicida* per ml of egg macerate were present when first tested, numbers rapidly declined thereafter and were not detected five days after commencement of incubation. It appeared that incubation in hatchery water resulted in *A. salmonicida* being quantitatively displaced by organisms (myxobacteria and pseudomonads) similar to those described by Trust (1972). These experiments indicated that vertical transmission of *A. salmonicida* was not a significant route of transmission and, moreover, in the unlikely event that frankly infected fish were used for stripping, the organism was unlikely to survive to the eyed-egg stage at which they are sold.

Epizootics among salmonids held in sea water may also be initiated by carrier fish. Furthermore, the natural susceptibility of the strictly marine sable fish (Evelyn, 1971) and the successful experimental infection of the plaice (*Pleuronectes platessa* L.) by I. W. Smith (personal communication) and also possible transmission of the disease in seawater (Scott, 1968) are of considerable epizootiological importance to those workers contemplating mixed culture of salmonids and marine fish. In addition, the possibility that scavenger fish such as saithe *Pollachius virens* (L) which are attracted to sea cages by excess food, acting as carriers of the disease must be considered, since they will eat moribund or dead salmonid carrier fish and Klontz and Wood (1972) have shown that the sable fish can be infected in this way. Therefore, although non-carrier salmonids may be at a disadvantage when used to stock fresh waters because these may contain or be subsequently stocked with carrier fish, they have considerable advantage for use in marine fish farming in uninfected areas.

13 Classification and identification of the causative organism

McCraw's review (1952) of the earlier literature (1894–1952) on *Bacterium* or *Bacillus salmonicida*, as the organism was known at that time, revealed a significant measure of disagreement among the results published and these discrepancies led Griffin *et al.* (1953) to undertake the first comprehensive bacteriological study of the organism. They demonstrated a marked homogeneity of taxonomic features among their strains. They were also of the opinion that *B. salmonicida* should be reclassified in the genus *Aeromonas*, a suggestion subsequently implemented by Snieszko in the 7th edition of *Bergey's Manual of Determinative Bacteriology* (Breed, 1957). Snieszko's classification of the genus *Aeromonas* appears to have stimulated a great deal of interest because, following a period of relative inactivity, a series of detailed publication on the genus (including *A. salmonicida*) began to appear during the 1960s.

The first of these studies was reported by Eddy (1960) who, although principally concerned with the classification of motile aeromonads, also included *A. salmonicida* strains, and, like Griffin *et al.* (1953), found them to be bacteriologically homogeneous. However, although he supported their contention that the organism was correctly classified in the genus *Aeromonas*, he pointed out that, being non-motile and unable to produce 2, 3-butanediol from glucose *A. salmonicida,* they failed to conform in two of the three principal characteristics which had originally led Kluyver and van Neil (1936) to create the genus *Aeromonas.* In 1961 Ewing *et al.* carried out a similar study and reported that all 21 *A. salmonicida* strains examined reacted consistently and also differed sufficiently from the motile aeromonads to warrant separate species status with the genus. They, too, were cautious and suggested that retention of these organisms within *Aeromonas* should be temporary.

An important contribution by Smith (1963) described a study in which 42 *A. salmonicida* strains were compared with motile aeromonads by Adansonian analysis. On the basis of certain biochemical differences which emerged, she concluded that *A. salmonicida* should be removed from the genus *Aeromonas* to a newly created genus *Necromonas* as *N. salmonicida.* It is worth noting here that this suggestion has now been implemented by Cowan and Steel (1974) who list the organism under this name in the 2nd edition of their "Manual for the Identification of Medical Bacteria". In addition to the 42 typical brown-pigmented *A. salmonicida* isolates, Smith also included 6 non-pigmented strains in the study and distinguished them from the pigmented isolates, suggesting the creation of a new species, *Necromonas achromogenes.* Although this appears to be the first report of isolation of non-pigmented *A. salmonicida* strains from diseased fish, it must be remembered that variations among pigmented isolates, both in the amount of pigment produced and also the length of time taken for its appearance, were reported in the literature many years ago (Horne, 1928; Mackie and Menzies, 1938). Moreover, non-pigmented varieties of typically pigmented strains may arise on subculture (Duff and Stewart, 1933; Evelyn, 1971). Furthermore, both of Smith's "achromogenes" strains, lodged with the National Collection of Marine Bacteria, produce pigment, albeit slowly, when grown on suitable culture medium.

Another atypical isolate was recovered by Ojala (1966) from diseased fish from Finland. This isolate possessed, among other unusual biochemical characteristics, the ability to produce acetylmethylcarbinol (Ojala, 1968), thus confirming the earlier report of Liu (1962) who first demonstrated the presence of this biochemical pathway among strains of *A. salmonicida.*

In 1967 Schubert compared one of Smith's achromogenic isolates with a single pigmented *A. salmonicida* strain and, because of insufficient variance between them, disagreed with her proposal to create a new species and instead reclassified her organism as *A. salmonicida* subspecies *achromogenes.* The study of Popoff (1969) should be mentioned in passing since this worker examined 93 strains of *A. salmonicida* with an extended series of bacteriological tests, but apart from noting their close similarity made no comment on their classification: presumable no atypical strains were available for inclusion. Also in 1969, Kimura (1969a) isolated an atypical strain of *A. salmonicida* from diseased Pacific salmon (*Oncorhynchus masou*) Walbaum in Japan. The isolate, which was genuinely non-pigmented and also produced acetoin, was considered by Kimura to differ sufficiently from other *A. salmonicida* strains to merit subspecies status, and, in a subsequent publication (Kimura, 1969b), he proposed the name *A. salmonicida* subspecies *masoucida* for it.

In his most recent publication on *A. salmonicida* on which he bases the classification for the 8th edition of "Bergey's Manual", Schubert (1969) compares the bacteriological characteristics of single strains of the three aberrant organisms described above with a typical *A. salmonicida* isolate. As a result, he reaffirmed his earlier views that Smith's strains (1963) merited subspecies status as *A. salmonicida* subspecies *achromogenes* and in addition supported Kimura's (1969a, b) assertion that his isolate should be treated similarly, Ojala's (1966) strain being classified as *A. salmonicida* subspecies *masoucida*.

In a subsequent publication Popoff (1970), however, questioned Schubert's classification and suggested that the *achromogenes* and *masoucida* strains may, in fact, lie closer to the non-motile *A. hydrophila* isolates. Many further aberrant strains of *A. salmonicida* have been isolated since the classification of *A. salmonicida* in "Bergey's Manual" was made. Evelyn (1971) isolated such a strain from a strictly marine host, the sable fish and also from certain Pacific salmon. He considered that his isolates shared certain atypical characteristics with Kimura's *A. salmonicida* subspecies *masoucida*, notably instability of pigment production (one strain became non-pigmented following agar-subculture over two years) and acetylmethylcarbinol production, but he did not classify this strain. McCarthy (1975b) isolated a non-pigmented acetoin-producing *A. salmonicida* strain from diseased silver bream (*Blicca bjoerkna*) and certain other non-salmonid fish, which also shared certain characteristics with Kimura's strain. Since these publications, McCarthy (unpublished) has isolated, or received for identification, a significant number (54) of these strains, suggesting that they are far more common than can be assumed from the published literature.

Clearly then, since the classification for atypical strains of *A. salmonicada* published in the 8th edition of "Bergey's Manual" (Buchanan and Gibbons, 1974) was based on the examination of the few strains extant at that time, detailed re-examination and possible revision was necessary. Accordingly, a comparative study of 29 such atypical *A. salmonicida* strains isolated from a wide variety of fish hosts and geographic sources was carried out and provisinally reported on by McCarthy (1977a). A total of 145 other bacteria consisting mainly of typical strains of *A. salmonicida* and culture collection strains of *A. hydrophila* were also included in this detailed phenotypic and genotypic study which involved the application of 188 characterization tests and also DNA-base composition estimations and DNA-DNA hybridization studies. Briefly, the results obtained by the phenetic study, when subjected to various numerical taxonomic computer analyses, indicated the existence of four main phenons: (a) typical *A. salmonicida* strains; (b) atypical *A. salmonicida* strain from salmonid fish; (c) atypical *A. salmonicida* strains from non-salmonid fish; and (d) *A. hydrophila*. As expected, typical strains of *A. salmonicida* formed a very compact group indeed and the member strains could be distinguished from isolates comprising the two atypical *A. salmonicida* groups on the basis of certain bacteriological characterisitcs. Care must, however, be observed when making use of the important characteristic pigment production for separation of strains, since some atypical strains are pigmented and occasional typical isolates show instability of pigment production.

Although the phenetic analysis revealed the existence of three distinct and recognizable groups among the strains of *A. salmonicida* studied, these groupings were not unequivocally defined by either DNA-base composition analysis or the DNA-DNA homology studies. It is, however, important for epizootiological reasons that they be established as distinct taxa.

Group 1 consists of typical *A. salmonicida* strains; Group 2 of atypical strains isolated from salmonid fish and includes both Schubert's (Buchanan and Gibbons, 1974) subspecies *achromogenes* and *masoucida*; while Group 3 is composed of a new group of atypical strains which are associated with disease syndromes of non-salmonoid fish. The taxonomic status and nomenclature which McCarthy proposed for these organisms was:

Group 1 strains—*A. salmonicida* subspecies *salmonicida*
Group 2 strains—*A. salmonicida* subspecies *achromogenes*
Group 3 strains—*A. salmonicida* subspecies *nova*

The minimal divergence in DNA homology observed among the *A. salmonicida* strains studied, most of which were of a widely separated geographic origin, may have indicated either minimal evolutionary divergence or relatively recent wide dispersion of the organism with commercial fish movements.

It was noted above that *A. hydrophila* could be distinguished from *A. salmonicida* phenotypically and this difference was confirmed by the genotypic analysis. DNA-base ratio compositions although similar were significantly different, but the degree of DNA binding between the two species (56–65 per cent) indicated a relationship at the generic level, thus settling the long-standing argument as to whether *A. salmonicida* should be removed from *Aeromonas* to a new genus *Necromonas* as proposed by Smith (1963). This similarity is emphasized by the results for percentage DNA binding between both *A. salmonicida* and *A. hydrophila* and members of other genera, for example *Pseudomonas*, *Vibrio*, *Escherichia*, etc. which was uniformly low (20–37 per cent).

Isolation and subsequent identification of *A. salmonicida* from uncomplicated cases of furunculosis is readily achieved using routine culture media such as tryptone soya agar and the serological techniques described earlier. However, when skin lesions are advanced and super-infection with other bacteria occurs, isolation of both typical and atypical strains is complicated. This is primarily because the presence of numerous contaminating bacteria on culture plates markedly suppresses the growth of *A. salmonicida* and the close proximity of colonies of other bacteria is also well recognized as a cause of inhibition of pigment production by typical strains. Furthermore, although with typical strains the appearance of pigment produced by the more isolated colonies seen on mixed culture plates usually indicates the presence of *A. salmonicida*, this is not of course apparent with non-pigmented atypical strains. It is recommended, therefore, that when chronic lesions are present and involvement of *A. salmonicida* suspected, at least six fish should be examined from each outbreak and samples for bacterial examination should be taken from skin lesions in all stages of development. In the absence of a selective medium for *A. salmonicida*, heavily contaminated culture plates should be examined by the modified latex test (McCarthy, 1977c).

Classical bacteriological tests such as the oxidase test, etc. may be used for differentiation of *A. salmonicida* isolates from the majority of other Gram-negative fish pathogens. Although technically sophisticated and time consuming it is recommended that DNA-base composition is estimated for any strains on which research work is to be based, especially the aberrant strains that are being isolated with increased frequency.

14 Therapy and control

The successful application of chemotherapy to diseases such as furunculosis has been, and continues to be, of great benefit to fish culture. Nevertheless, there are a number of limiting factors which are beginning to express themselves. For reasons such as legislative requirements and practical cost-effectiveness the range of efficacious antibacterial drugs of low toxicity suitable for use in fish destined for human consumption is very narrow. Furthermore the extensive use and abuse of these compounds has resulted in increasing bacterial drug resistance, which is approaching the level where a rotation policy with suitable antibiotics will become necessary.

As with most other diseases of intensively cultured livestock, the contribution of husbandry stresses to the pathogenesis of furunculosis is significant (Roberts, 1978). Consequently, especially in view of the problems associated with antibiotic usage detailed above, investment in husbandry improvements will generally be more productive, in the absence of a successful furunculosis vaccine, than continuous expenditure on chemotherapy, which is usually at best palliative.

Most of the husbandry improvements that aid in the control of furunculosis are aimed at improvement of the water quality or fish handling systems. Ideally a farm should be sited on a spring water supply and be self-sufficient in egg supply but such ideal conditions are difficult to achieve and usually a compromise between ideal environmental conditions and economic viability has to be accepted.

Where eggs must be purchased from elsewhere, although vertical transmission has not been unequivocally demonstrated. Herman (1972) has quite properly advised routine disinfection of eggs. Acriflavine was recommended for this purpose by Blake (1930) but MacFadden (personal communication) asserted that an iodophor compound, povidone iodine, was more suitable for disinfection of eggs contaminated with *Aeromonas hydrophila*. In 1972, Ross and Smith demonstrated that two commercial iodophore compounds, Wescodyne® and Betadyne®, rapidly killed saline suspensions of *A. salmonicida*. Since then, routine disinfection of eggs with iodophores has been recommended (Snieszko, 1974). However McCarthy (1977b) compared Wescodyne® and Acriflavine® for their ability to disinfect both green eggs produced by artificially infected brood stock and eyed eggs artificially infected *in vitro* with *A. salmonicida*. Although standard methods were used, the iodophore failed to disinfect both types of eggs at the usually recommended levels (50 and 100 ppm). Furthermore, although treatment with 200 ppm successfully disinfected eggs, this concentration is above the safe non-toxic limit (100 ppm) recommended for iodophor compounds by Amend (1974). Acriflavine solutions are more effective at alkaline pHs, but although the pH of the disinfecting solution was maintained at 7·2, a stronger solution (1:1000) than recommended in earlier literature (1:2000) was found to be necessary for successful disinfection of both types of eggs. Acriflavine is markedly non-toxic to fish eggs and can safely be used at the stronger concentration. It could be argued that, since vertical transmission of furunculosis may not occur and *A. salmonicida* is rapidly eliminated from viable fish eggs, routine disinfection of incoming eyed eggs is unnecessary. On balance, however, it is wise to disinfect eyed eggs since it is conceivable that they could nevertheless have been infected by exposure to contaminated hatchery water at or around the time of transportation.

Although attempts at selective breeding (increase in genetic resistance) of salmonids for resistance to furunculosis (Erhlinger, 1977) have been generally discouraging, this approach requires further research. Although use can be made of the relative resistance of rainbow trout when compared to brown trout, in the short term, successful mass vaccination (non-genetic increase in resistance) appears to hold the only real promise of natural protection.

Husbandry skills notwithstanding, furunculosis outbreaks occur and then therapy must be instituted immediately. Once infection has become established diseased fish cease to feed and since medication is usually incorporated in the feed such fish will be lost if treatment is delayed. In addition, once clinical signs are well established, pathological damage appears to be so severe that subsequent injection of large amounts of antibiotic into such fish fails to elicit a therapeutic effect (Roberts unpublished observations). It is also very important to correct (if possible) any environmental or physiological stress factor that may have been responsible for precipitating the epizootic, since both sulphamerazine and oxytetracycline, the drugs commonly used for treatment of furunculosis, are merely bacteriostatic, inhibiting multiplication of the invading organism and relying on defences of the host to clear the infection. But if the fish's capacity to clear the infection is impaired by stress, the well-recognized phenomenon of recrudescence of furunculosis may occur once tissue levels of antibiotic are reduced below therapeutic levels by natural excretion. Treatment of furunculosis is, of course, largerly impracticable in open waters such as rivers and reservoirs and where feasible affected fish should be removed and incinerated.

In the United States there are strict regulations concerning the use of drugs for treatment of fish destined for human consumption and in the United Kingdom such drugs are available only on the prescription of a veterinary surgeon. The usual treatments recommended to achieve control of furunculosis are as follows: sulphamerazine (200 mg day^{-1} per kg of fish) and tetracycline (75 mg day^{-1} per kg of fish), both drugs being incorporated in food and applied for 10–14 days. If drug resistance (see below) is encountered other antibacterial drugs may be required, but it is important to remember that in the United States, at least, they may not be cleared for use. Post (1962) recommended furizolidone for treatment of furunculosis and McCarthy *et al.* (1974a, b, c) demonstrated that a potentiated sulphonamide (sulphamethylphenazole and trimethoprim) was highly effective. Combination of trimethoprim with a sulphonamide, in addition to resulting in a striking mutual enhancement of their individual antibacterial activities, results in a bacteriocidal action and, because a double blockage is formed in the same biochemical pathway, resistance is consequently less likelty to develop (Table 1).

Following treatments of fish destined for human consumption it is important that a sufficient period is allowed to elapse for natural

TABLE 1
Sensitivity of fish-pathogenic bacteria to a potentiated sulphonamide

Organism	Number of strains	m.i.c. (Hg ml^{-1}) sulphonamides Sulphadiazine alone	Sulphadiazine with trimethoprim (20:1)
Aeromonas punctata	8	4–10	0·2–0·8
Aeromonas punctata	14	11–>50	0·2–1·5
Aeromonas salmonicida	6	10–23	0·6–1·7
Aeromonas salmonicida	12	24–>50	1·1–1·6
Vibrio anguillarum	4	8–17	0·6–0·8
Pseudomonas sp.	2	>50, 20	3·1, 25
Haemophilus piscium	2	30, 18	1·0, 1·0

From McCarthy *et al.* (1974a).

elimination of drug residue from fish tissue. United States regulations call for a three-week period but four weeks is probably a safer period, particularly at low water temperatures when drug elimination is more protracted.

If chemotherapy is to be effective it is important that full therapeutic dosage is given for the correct period of time. On the rare occasions when prophylactic antibiotic therapy may be justified it must again be administered at the full therapeutic level. Transferable drug resistance has been known to occur in *A. salmonicida* since 1959 when Aoki *et al.* (1971) demonstrated the presence of R-factors. The frequency of occurrence of transferable drug resistance reflects the extent of antibiotic usage so that given the limited number of antibiotics available for fish therapy and in the absence of a commercial vaccine, the only reliable long-term control for the disease would appear to be high standards of husbandry.

References

Amend, D. F. (1974). Comparative toxicity of two iodophors to rainbow trout eggs. *Transactions of the American Fisheries Society,* **103**, 73–78.

Amend, D. F. and Fender, F. C. (1976). Uptake of bovine serum albumin by rainbow trout from hyperosmotic solutions: a model for vaccinating fish. *Science,* **192** (4241): 793–794.

Anderson, D. P. (1972a). Virulence and persistance of rough and smooth forms of *Aeromonas salmonicida* inoculated into Coho salmon (*Oncorhynchus kisutch*). *Journal of the Fisheries Research Board of Canada,* **29**, 204–206.

Anderson, D. P. (1972b). Investigations of the lipopolysaccharide fractions from *Aeromonas salmonicida* smooth and rough forms. *FI:EIFAC 72/SC II—Symposium 20.*

Anderson, D. P. and Klontz, G. W. (1970). Precipitating antibody against *Aeromonas salmonicida* in serums of inbred albino rainbow trout (*Salmo gairdneri*). *Journal of the Fisheries Research Board of Canada,* **27**, 1389–1393.

Aoki, T., Egusa S., Kimura, T. and Watanabe, T. (1971). Detection of R factors in naturally occurring *Aeromonas salmonicida* strains. *Applied Microbiology,* **22**, 716–717.

Arkwright, J. A. (1912). An epidemic disease affecting salmon and trout in England during the summer of 1911. *Journal of Hygiene,* **12**, 391–413.

Avtalion, R. R., Wojdani, A., Malik, Z., Shahrabani, R. and Duczyminer, M. (1973). Influence of environmental temperature on the immune response in fish. *In* "Current Topics in Microbiology and Immunology", vol. 61, pp. 1–35. Springer, Heidelberg.

Batallion, M. E. and Dubard, M. F. (1895). *Compte rendu des seances de la Société de Biologie, Paris,* **14**, 353–355. (After Williamson, 1929.)

Bissett, K. A. (1948). The effect of temperature on antibody production in cold blooded vertebrates. *Journal of Pathological Bacteriology,* **60**, 87–92.

Blake, I. (1930). The external disinfection of fish ova with reference to the prophylaxis of furunculosis. *Fishery Board for Scotland. Salmon Fisheries,* **2**, HMSO, Edinburgh.

Blake, I. J. F. and Anderson, E. J. M. (1930). The identification of *Bacillus salmonicida* by the complement fixation test—a further contribution to the study of furunculosis of the salmonidae. *Fishery Board for Scotland, Salmon Fisheries,* **1**, HMSO, Edinburgh.

Blake, I. and Clarke, J. C. (1931). Observations on experimental infection of trout by *B. salmonicida,* with particular reference to "carriers" of furunculosis and to certain factors influencing susceptability. *Fishery Board for Scotland. Salmon Fisheries,* **7**, 1–13.

Bootsma, R., Fijan, N. and Blommaer, T. T. (1977). Isolation and preliminary identification of the causative agent of carp erythrodermatitis. *Veterinarski Arhiv. Zagreb,* **6**, 291–302.

Breed, R. S. (1957), (Ed.). "Bergeys Manual of Determinative Bacteriology". Williams and Wilkins, Baltimore, Maryland.

Buchanan, R. E. and Gibbons, N. E. (1974), (Ed.). "Bergey's Manual of Determinative Bacteriology", 8th edition. Williams and Wilkins, Baltimore, Maryland.

Bulkley, R. V. (1969). A furunculosis epizootic in clear lake Yellow Bass. *Bulletin of the Wildlife Diseases Association,* **5**, 322–327.

Bullock, G. L. (1966). Precipitins and agglutinin reaction of aeromonads isolated from fish and other sources. *Bulletin of the International Office of Epizootics,* **65**, 805–824.

Bullock, G. L. and Stuckey, H. M. (1975). *Aeromonas salmonicida:* detection of asymptomatically infected trout. *Progressive Fish Culturist,* **37**, 237–239.

Charnin, M. A. (1893). The bacterium of fish disease. *Compte rendu memoranda Du Société Biologique; Paris,* **45**, 331–333.

Cornick, J. W., Chudyk, R. V. and MacDermott, L. A. (1969). Habitat and viability studies on *Aeromonas salmonicida,* causative agent of furunculosis. *Progressive Fish Culturist,* **31**, 90–93.

Croy, T. R. and Amend, D. F. (1977). Immunisation of sockeye salmon (*Oncorhynchus nerka*) against vibriosis using hyperosmotic infiltration. *Aquaculture,* **12**, 317–325.

Cowan, S. T. and Steel, K. J. (1974). "Manual for the Identification of Medical Bacteria". Cambridge University Press, London.

Davis, H. S. (1946). Care and diseases of trout. *Research Report U.S. Fisheries and Wildlife Service,* **12**, 98.

Dubois-Darnaudpeys, A. (1977). The epidemiology of furunculosis in salmonids. Parts I, II and III. (This work was submitted as a doctoral thesis to the Pierre and Marie Curie University, Paris, in 1976.)

Duff, D. C. B. (1939). Some serological relationships of the S, R and G phases of *Bacillus salmonicida. Journal of Bacteriology,* **38**, 91–101.

Duff, D. C. B. (1942). The oral immunisation of trout against *Bacterium salmonicida. Journal of Immunology,* **44**, 87–94.

Duff, D. C. B. and Stewart, B. J. (1933). Studies on furunculosis of fish in British Columbia. *Contributions to Canadian Biology and Fisheries,* **8**, 103–122.

Eddy, B. P. (1960). Cephalotrichous, fermentative Gram negative bacteria: the genus *Aeromonas. Journal of Applied Bacteriology,* **23**, 216–248.

Emmerich, R. and Weibal, E. (1894). Uber eine durch Bakterien erzeugte Seuche unter den Forellen. *Archiv für Hygiene und Bakteriologie,* **21**, 1–21.

Erhlinger, N. F. (1977). Selective breeding of trout for resistance to furunculosis. *New York Fish and Game Journal,* **24**, 25–36.

Evelyn, T. P. T. (1971). An aberrant strain of the bacterial fish pathogen *Aeromonas salmonicida* isolated from a marine host, the Sablefish (*Anoplopoma fimbria*) and from two species of Cultured Pacific salmon. *Journal of the Fisheries Research Board of Canada,* **28**, 1629–1634.

Ewing, W. H., Hugh, R. and Johnson, J. G. (1961). Studies on the *Aeromonas* group. U.S. Department of Health, Education and Welfare. Public Health Service Communicable Disease Centre, Atlanta, Georgia.

Fabre-Domergue, J. P. (1890). On a disease of fish. *Compte rendu memoranda du Société Biologique, Paris*, **42**, 127–129.

Ferguson, H. (1975). Phagocytosis by the endocardial lining cells of the atrium of plaice. *Journal of Comparative Pathology*, **85**, 561–569.

Ferguson, H. and McCarthy, D. H. (1978). Histopathology of furunculosis in brown trout *Salmo trutta* L. *Journal of Fish Diseases*, **1**, 165–174.

Field, J. B., Gee, L. L., Elvehjem, C. A. and Juday, C. (1944). The blood picture in furunculosis induced by *Bacterium salmonicida* in fish. *Archives of Biochemistry*, **3**, 277–284.

Fijan, N. N. (1972). Infectious dropsy of carp—a disease complex. *Proceedings of the Symposia of the Zoological Society of London*, **30**, 39–57.

Fischel, D. and Enoch, H. P. (1892). Quoted by Williamson (1929).

Fish, F. F. (1937). Furunculosis in wild trout. *Copeia*, **1**, 37–40.

Foda, A. (1973). Changes in haematocrit and haemoglobin in Atlantic Salmon (*Salmo salar*) as a result of furunculosis disease. *Journal of the Fisheries Research Board of Canada*, **30**, 467–468.

Forel, F. (1868). Red disease of trout. *Bulletin de la Société Vandoise des Sciences Naturelles, Lausanne*, **9**, 599–608.

Fuhrman, O. (1909). La Furunculose. *Bulletin Suisse de Peche et Pisciculture*, **12**, 193–195.

Fuller, D. W., Pilcher, K. S. and Fryer, J. L. (1977). A leukocytolytic factor isolated from cultures of *Aeromonas salmonicida*. *Journal of the Fisheries Research Board of Canada*, **34**, 1118–1125.

Geck, P. (1971). India ink immuno-reaction for the rapid detection of enteric pathogens. *Acta Microbiologica Academiae Scientiarum Hungaricae*, **18**, 191–196.

Gonzales, C. (1963). Neutralisation croisee de la lipase d'*Aeromonas hydrophilia* et *salmonicida* var l'antilipase: difference antigenique entre la lipase de quelques bactéries des Genras *vibrio* et *Aeromonas*. *Zentralblatt für Bakteriologie, Parasitenkunde, Infectionskrankheiten und Hygiene, Abt 2*, **191**, 565–571.

Griffin, P. J. (1954). Nature of bacteria pathogenic to fish. *Transactions of the American Fisheries Society*, **83**, 241–253.

Griffin, P. J., Snieszko, S. F. and Friddle, S. B. (1953). A more comprehensive description of *Bacterium salmonicida*. *Transactions of the American Fisheries Society*, **82**, 129–138.

Hall, J. D. (1963). An ecological study of the chestnut lamprey, *Ichthyomyzon castaneus*. Girard in the Manistee river, Michigan. Ph.D. Thesis, University Michigan, Ann Arbor. 106.

Håstein, T. (1975). Vibriosis in fish. Ph.D. Thesis, University of Stirling, Scotland.

Håstein, T., Saltveit, S. J. and Roberts, R. J. (1978). Mass mortality among minnows. *Phoxinus phoxinus* (L) in Lake Tveivatn, Norway, due to an aberrant strain of *Aeromonas salmonicida*. *Journal of Fish Diseases*, **1**, 241–250.

Herman, R. L. (1968). Fish furunculosis. *Transactions of the American Fisheries Society*, **97**, 221–230.

Herman, R. L. (1972). A review of the prevention and treatment of furunculosis. *FI:EIFAC 72/SC II-symposium*, **19**, 1–6.

Horne, J. H. (1928). Furunculosis in trout and the importance of carriers in the spread of the disease. *Journal of Hygiene, Cambridge*, **28**, 67–78.

Karlsson, K. A. (1962). An investigation of the *Aeromonas salmonicida* haemolysin. *9th Nordic Veterinary Congress.* Kobenhavn.

Karlsson, K. A. (1964). Serological studies of *Aeromonas salmonicida. Zentralblatt für Bacteriologie, Parasitenkunde, Infectionsrankheiten und Hygiene,* **194**, 73–80.

Kimura T. (1969a). A new subspecies of *Aeromonas salmonicida* as an etiological agent of furunculosis on "Sakuramasu" (*Oncorhynchus masou*) and pink salmon (*O. gorbuscha*) rearing for maturity, Part 1. On the morphological and physiological properties. *Fish Pathology,* **3**, 34–44.

Kimura, T. (1969b). A new subspecies of *Aeromonas salmonicida* as an etiological agent of furunculosis on "Sakuramasu" (*Oncorhynchus masou*) and pink salmon (*O. gorbuscha*) rearing for maturity, Part 2. On the serological properties. *Fish Pathology,* **3**, 45–52.

Kimura, T. (1970). Studies on a bacterial disease of adult "Sakuramasu" (*Oncorhynchus masou*) and pink salmon (*O. gorbuscha*) reared for maturity. *Scientific report of the Hokkaido Salmon Hatchery,* **24**, 9–100.

Klontz, G. W. and Anderson, D. P. (1968). Fluorescent antibody studies of isolates of *Aeromonas salmonicida. Bulletin of the International Office for Epizootics,* **69**, 1149–1157.

Klontz, G. W. and Wood, J. W. (1972). Observations on the epidemiology of furunculosis disease in juvenile Coho salmon (*Oncorhynchus kisutch*). *FI:EIFCA 72/SC II–Symposium,* **27**, 1–8.

Klontz, G. W., Yasutake, W. T. and Ross, A. J. (1966). Bacterial diseases of the salmonidae in the Western United States: Pathogenesis of furunculosis in rainbow trout. *American Journal of Veterinary Research,* **27**, 1455–1460.

Kluyver, A. J. and Van Neil, C. B. (1936). Prospects for a natural classification of bacteria. *Zentralblatt für Bacteriologie, Parasitenkunde, infections-krankheiten und Hygiene, Abt. 2,* **94**, 369.

Krantz, G. E. and Heist, C. E. (1970). Prevalence of naturally acquired agglutinating antibodies against *Aeromonas salmonicida* in hatchery trout in Central Pennsylvania. *Journal of the Fisheries Research Board of Canada,* **27**, 969–973.

Krantz, G. E., Reddecliff, J. M. and Heist, C. E. (1963). Development of antibodies against *A. salmonicida* in trout. *Journal of Immunology,* **91**, 757–760.

Krantz, G. E., Reddecliff, J. M. and Heist, C. E. (1964b). Immune response of trout to *Aeromonas salmonicida.* Part 2. Evaluation of feeding techniques. *Progressive Fish Culturist,* **26**, 65–69.

Le Tendre, G. C., Schneider, C. P. and Ehlinger, N. F. (1972). Net damage and subsequent mortality from furunculosis in small mouth bass. *New York Fish and Game Journal,* **19**, 73–82.

Liu, P. V. (1961). Observations on the specificities of extra-cellular antigens of the genera *Aeromonas* and *Serratia. Journal of General Microbiology,* **24**, 145–153.

Liu, P. V. (1962). Fermentation reaction of *Pseudomonas caviae* and its serological relationships to aeromonads. *Journal of Bacteriology,* **83**, 750–753.

Lund, M. (1967). A study of the biology of *Aeromonas salmonicida.* MSc Thesis. Department of Agriculture, University of Newcastle-upon-Tyne.

MacDermott, L. A. and Berst, A. H. (1968). Experimental plantings of brook trout *Salvelinus fontinalis* from furunculosis infected stock. *Journal of the Fisheries Research Board of Canada,* **25**, 2643–2649.

Mackie, T. J. and Menzies, W. J. M. (1938). Investigations in Great Britain of furunculosis of the salmonidae. *Journal of Comparative Pathology and Therapeutics,* **51**, 225–234.

Mackie, T. J., Arkwright, J. A., Pryce-Tannatt, T. E., Mottram, J. C., Johnston, W. P.

and Menzies, W. J. M. (1930, 1933, 1935). Interim, second and final reports of the furunculosis committee. HMSO, Edinburgh.
Marsh, M. C. (1902). *Bacterium truttae,* a new bacterium pathogenic to trout. *Science,* **16**, 706.
Masterman, R. B. and Arkwright, J. A. (1911). After Fish (1938).
Mawdesley-Thomas, L. E. (1969). Furunculosis in the goldfish, *Carassius auratus* (L). *Journal of Fish Biology,* **1**, 19–23.
McCarthy, D. H. (1975a). Detection of *Aeromonas salmonicida* antigen in diseased fish tissue. *Journal of General Microbiology,* **88**, 384–386.
McCarthy, D. H. (1975b). Fish furunculosis caused by *Aeromonas salmonicida var. achromogenes. Journal of Wildlife Diseases,* **2**, 489–493.
McCarthy, D. H. (1977a). The identification and significance of atypical strains of *Aeromonas salmonicida. Bulletin of the International Office for Epizootics,* **87**, (5–6), 459–463.
McCarthy, D. H. (1977b). Some ecological aspects of the bacterial fish pathogen, *Aeromonas salmonicida. Aquatic Microbiology, SAB Symposium,* **6**, 299–324.
McCarthy, D. H. (1977c). A latex test for rapid identification of both fine and mixed agar cultures of *Aeromonas salmonicida. Fish Health News,* **6**, 146–147.
McCarthy, D. H. and Rawle, C. T. (1975). The rapid serological diagnosis of fish furunculosis caused by smooth and rough strains of *Aeromonas salmonicida. Journal of General Microbiology,* **86**, 185–187.
McCarthy, D. H. and Whitehead, P. (1977). An immuno-India ink technique for rapid laboratory diagnosis of fish furunculosis. *Journal of Applied Bacteriology,* **42**, 429–431.
McCarthy, D. H., Stevenson, J. P. and Roberts, M. J. (1974a). Combined *in-vitro* activity of trimethoprim and sulphonamides on fish-pathogenic bacteria. *Aquaculture,* **3**, 87–91.
McCarthy, D. H., Stevenson, J. P. and Roberts, M. J. (1974b). A comparative pharmacokinetic study of a new sulphonamide potentiator, trimethoprim, in rainbow trout (*Salmo gairdneri,* Richardson). *Aquaculture,* **4**, 299–303.
McCarthy, D. H., Stevenson, J. P. and Roberts, M. J. (1974c). Therapeutic efficacy of a potentiated sulphonamide in experimental furunculosis. *Aquaculture,* **4**, 407–410.
McCraw, B. M. (1952). Furunculosis of fish. *United States Fish and Wildlife Services, Special Scientific Report,* **84**, p. 87.
McFadden, T. W. (1969). Effective disinfection of trout eggs to prevent egg transmission of *Aeromonas liquefaciens. Journal of the Fisheries Research Board of Canada,* **26**, 2311–2318.
Mettam, A. E. (1914). Report on the outbreak of furunculosis in the river Liffey in 1913. Scientific Investigations of the Fisheries Branch Department of Agriculture, etc. for Ireland, 1914, p. 19.
Miyazaki, T. and Kubota, S. S. (1975). Histopathological studies on furunculosis in Amago. *Fish Pathology,* **9**, 213–218.
Ojala, O. (1966). Isolation of an anaerogenic bacterium resembling *Aeromonas salmonicida* in spawning lake trout. *Bulletin of the International Office for Epizootics,* **65**, 793–804.
Ojala, O. (1968). Observations on the occurrence of *Aeromonas salmonicida* and *A. punctata* in fish. *Bulletin of the International Office for Epizootics,* **69**, 1107–1123.
Paterson, W. D. and Fryer, J. L. (1974). Immune response of juvenile coho salmon (*Oncorhynchus kisutch*) to *Aeromonas salmonicida* cells administered intraperitoneally in Freud's complete adjuvant. *Journal of the Fisheries Research Board of Canada,* **31**, 1751–1755.

Pittel, L. (1910). La furonculose observée sur la Bordeliere, la Nase et le Barbeau. *Bulletin Suisse de Peche et Pisculture*, **2**, 171–173.

Plehn, M. (1911). Die furunkulose der Salmoniden. *Zentrablatt für Bakteriologie, Parasitenkunde, infectionskrankheiten und Hygiene, Abt. 1*, **60**, 609–624.

Popoff, M. (1969). *Aeromonas salmonicida:* biochemical and antigenic properties. *Annales Recherche Veterinaire*, **3**, 49–57.

Popoff, M. (1970). Bacterioses, generalities—*Aeromonas* et Aeromonose. Mimeograph, p. 38. Institut National de la Recherche Agronsomique.

Popoff, M. (1971). Studies on *Aeromonas salmonicida* II. Characterisation and phage typing of bacteriophages active on *A. salmonicida. Annales de Recherche Veterinaire*, **2**, 33–45.

Post, G. (1962). Furazolidone for control of furunculosis in trout. *Progressive Fish Culturist*, **24**, 182–184.

Rabb, L., Cornick, J. W. and MacDermott, L. A. (1964). A microscopic slide agglutination test for the presumptive diagnosis of furunculosis in fish. *Progressive Fish Culturist*, **26**, 118–120.

Reddecliffe, J. M. (1960). Immune response of trout to *Aeromonas salmonicida*. MSc. Thesis, Pennsylvania State University, Philadelphia.

Roberts, R. J. (1975). Experimental pathogenesis of lymphocystis in the plaice. (*Pleuronectes platessa*). *In* "Wildlife Diseases" (Ed. L. A. Page), pp. 431–441. Plenum Press, New York.

Roberts, R. J. (1978). "Fish Pathology", Baillière, Tindall and Cassell, London.

Ross, A. J. and Smith, C. A. (1972). Effect of two iodophors on bacterial and fungal fish pathogens. *Journal of the Fisheries Research Board of Canada*, **29**, 1359–1361.

Sandvick, O. and Hagan, O. (1968). Serological studies on proteinases produced by *Aeromonas salmonicida* and other aeromonads. *Acta veterinaria Scandinarica*, **9**, 1–9.

Schubert, R. H. W. (1967). The taxonomy and nomenclature of the genus *Aeromonas* Kluyver and Von Niel (1936). Part II Suggestions of the taxonomy and nomenclature of the anaerogenic aeromonads. *International Journal of Systematic Bacteriology*, **17**, 273–279.

Schubert, R. H. W. (1969). On the taxonomy of *Aeromonas salmonicida. Subsp. achromogenes* (Smith, 1963) Schubert 1967 and *Aeromonas salmonicida* Subsp. *masoucida*, Kimura, 1969. *Zentralblatt für Bacteriologie Parasitenkunde, Infectionskrankheiten und Hygiene*, **211**, 413–417.

Schumacher, R. E. (1956). Sedimentation rates of brook trout affected by furunculosis. *Progressive Fish Culturist*, **18**, 147–149.

Schwartz, J. J. (1973). A post-epizootic search for furunculosis in a warm water fish population. *Journal of Wild Life Diseases*, **9**, 106–110.

Scott, M. (1968). The pathogenicity of *Aeromonas salmonicida* (Griffin) in sea and brackish waters. *Journal of General Microbiology*, **50**, 321–327.

Smith, I. W. (1962). Furunculosis in kelts. Department of Agriculture and Fisheries for Scotland. *Freshwater and Salmon Fisheries Research*, **27**, 1–5.

Smith, I. W. (1963). The classification of *Bacterium salmonicida. Journal of General Microbiology*, **33**, 263–274.

Snieszko, S. F. (1974). Fish furunculosis. *EIFAC Symposium FID: CFD/74/inf.*, 197–199.

Surback, G. (1911). La furunculose des poissons dans les eaux libres. *Bulletin Suisse de Peche et Pisciculture*, **10**, 162–165; **11**, 173–175.

Todd, C. (1933). The presence of a bacteriophage for *B. salmonicida* in river waters. *Nature, London*, **131**, 360.

Trust, T. J. (1972). The bacterial population in vertical flour tray hatcheries during incubation of salmonid eggs. *Journal of the Fisheries Research Board of Canada,* **29**, 567–571.

Udey, L. R. and Fryer, J. L. (1978). Immunization of fish with bacterins of *Aeromonas salmonicida. Marine Fisheries Review,* **40**, 12–17.

Varitchak, T. (1938). Studies on endothelial cells in the liver of fishes. *Zeitschrift für Zellforschung und Microskopishe Anatomie,* **27**, 46–51.

Wedemeyer, G., Ross, A. J. and Smith, L. (1969). Some metabolic effects of bacterial endotoxins in fish. *Journal of the Fisheries Research Board of Canada,* **26**, 115–122.

Whang, H. Y. (1972). Production by *Aeromonas* of common enterobacterial antigen and its possible taxonomic significance. *Journal of Bacteriology,* **110**, 161–164.

Williamson, I. J. F. (1929). Furunculosis of the salmonidae. *Fishery Board for Scotland, Salmon Fisheries,* **5**, HMSO, Edinburgh.

Subject Index

Index of Authors

Index of Titles